ARITHMETIC PROGRESSION

$$a_1, \quad a_2 = a_1 + d, \quad a_3 = a_1 + 2d, \cdots, \quad a_n = a_1 + (n-1)d$$

$$a_1 + a_2 + \cdots + a_n = \frac{1}{2}n(a_1 + a_n) = \frac{1}{2}n[2a_1 + (n-1)d]$$

GEOMETRIC PROGRESSION

$$a_1, \quad a_2 = a_1 r, \quad a_3 = a_1 r^2, \cdots, \quad a_n = a_1 r^{n-1}$$

$$a_1 + a_1 r + a_1 r^2 + \cdots + a_1 r^{n-1} = a_1 \frac{1-r^n}{1-r} \qquad r \neq 1$$

$$a_1 + a_1 r + a_1 r^2 + \cdots = \frac{a_1}{1-r} \qquad |r| < 1$$

PERMUTATIONS AND COMBINATIONS

$$n! = 1 \cdot 2 \cdot 3 \cdots n \qquad 0! = 1$$

The number of permutations of n objects k at a time is

$$P(n, k) = n(n-1)(n-2)\cdots(n-k+1) = \frac{n!}{(n-k)!}$$

The number of combinations of n objects k at a time is

$$\binom{n}{k} = \frac{P(n,k)}{k!} = \frac{n(n-1)(n-2)\cdots(n-k+1)}{k!} = \frac{n!}{k!(n-k)!}$$

BINOMIAL THEOREM

$$(x+y)^n = x^n + \binom{n}{1}x^{n-1}y + \cdots + \binom{n}{k}x^{n-k}y^k + \cdots + \binom{n}{n-1}xy^{n-1} + y^n$$

$$= \sum_{k=0}^{n} \binom{n}{k} x^{n-k} y^k$$

COMPLEX NUMBERS

$$i^2 = -1$$

$$(a + bi) + (c + di) = (a + c) + (b + d)i$$

$$(a + bi)(c + di) = (ac - bd) + (ad + bc)i$$

$$\overline{a + bi} = a - bi$$

$$\overline{a + b} = \overline{a} + \overline{b}$$

$$\overline{ab} = \overline{a}\overline{b}$$

$$|a + bi|^2 = a^2 + b^2$$

$$|\alpha\beta| = |\alpha||\beta|$$

$$\left|\frac{\alpha}{\beta}\right| = \frac{|\alpha|}{|\beta|}$$

Preface

AIMS OF THIS BOOK

Our aims are to provide students with a solid working knowledge of algebra and a feeling for applications of the subject. We hope to achieve these aims by

1. presenting the material in a clear, down-to-earth manner.
2. emphasizing practical skills, problem solving, and good computational technique.
3. providing a large number of worked examples together with ample and varied exercise sets.
4. motivating the student by including interesting examples and up-to-date applications.
5. avoiding excessive formalism without sacrificing mathematical soundness.

FEATURES

Style

Too often students treat their math texts as portable collections of exercises and formulas, not as books to be read. We believe that a math book can be readable and have tried hard to write this book that way.

We aim for a clear and lively style. We use a conversational tone and direct, unfussy language. Often we throw in asides to the reader in the form of remarks, warnings, and suggestions. We change the pace; we try to keep the audience interested and awake.

Our hope is that the straightforward style will motivate students actually to read the book—and learn from it.

Examples

The worked examples are the backbone of the text, for it is through them that the reader sees how mathematics is done. This book contains a large number of formal worked examples for its size; there are 296, many of which have several parts. In addition, there are many informal examples that are not numbered.

Having a large number of examples allows us to illustrate both important manipulative techniques and interesting applications.

Exercises

The exercise sets are ample; the book contains about 2840 exercises. There are plenty of drill problems and a few challenging ones. But in addition, there is a variety of interesting *medium level* problems. We believe that such problems are important. Because of the moderate level, students will be able to do most of them; and because of the real-life flavor, students will feel they are accomplishing something worthwhile.

Most exercises have a worked prototype in the text. Each chapter ends with a set of review exercises.

The answer section in the back of the book contains answers to *all* the odd-numbered exercises. We emphasize *all* because we think it unfair to omit answers just because they are not short or not convenient to set in type. For example, many texts omit answers to problems in mathematical induction because such answers require explanation. How, then, are students to check their induction assignments?

Word Problems

The dread of most students. Why? Because they are taught to read and manipulate a foreign language called "algebra" but not to *write* it. Students dread story problems because they cannot translate the problems into algebra.

In Chapter 3, we try to meet this difficulty head-on. We devote two full sections to setting up equations. Section 1 is "Translating Words into Math," in other words, learning the vocabulary of the language. Section 2 is "Translating Problems into Algebra," that is, writing sentences (equations) using this vocabulary.

Once students gain some confidence from these sections, their anxiety over word problems should diminish. They may find that many word problems are not hard at all. In order to maintain problem-solving skills we have tried to sprinkle word problems throughout the book.

Rigor

We are convinced that a formal definition–theorem–proof style at this level is deadly. That is why we use an informal approach, leading up to general principles naturally, through specific examples. We do include some proofs and derivations when they are instructive in themselves and not abstract. But, in general, our experience shows that nothing turns off students faster than proofs for their own sake.

Use of Calculators

The hand-held calculator is a fact of modern life. We feel that an up-to-date college algebra course should use the calculator as a support tool. This benefits students in two ways. First, it heightens motivation by giving the subject a modern flavor and by reducing computational tedium. Second, it develops calculator skills that will be of value in future courses and beyond.

We introduce various uses of the calculator as the need arises. For example, while discussing exponentials and radicals, it is natural to mention the keys $\boxed{y^x}$ and $\boxed{\sqrt{}}$. And what better way to illustrate scientific notation than to square a 6-digit integer on a scientific calculator!

Actually, calculators play a minor role except in Chapter 7 on exponentials and logarithms, where we encourage their use in computations formerly done by logarithms. We do not insist on calculators, however; the reader has the option of using logarithms. Two sections of instructions on computing with logarithms are contained in the appendix, and there is a 4-place log table at the back of the book.

Let us emphasize that this text is no more a "calculator college algebra" than earlier books were "log table college algebras." The calculator replaces logarithmic *computations*, but nothing can replace the logarithm *function*; that is here to stay.

One further advantage of the calculator: it is a valuable teaching aid. Even a hypothetical calculator can often be useful. There is something very concrete about pressing buttons, or even thinking about pressing buttons. Therefore, we include a number of calculator exercises designed to reinforce such concepts as priority of algebraic operations, functions, inverse functions, and so on. (See Exercises 25–32, page 16; Exercises 21–22, page 150; and Exercises 43–46, page 203).

CONTENTS

Chapter 1 is a review of the real numbers—brief, but not too brief. It points out how basic properties such as the associative and distributive laws are used in algebra. It pays special attention to priority of algebraic operations, calculator priorities, and use of parentheses.

Chapter 2 covers exponents, radicals, polynomials, rational functions, and factoring, stressing sound techniques. The final section, on avoiding and correcting common mistakes, should be instructive.

Chapter 3 deals with setting up and solving equations: linear, quadratic, and other types. There is a large selection of interesting applications and exercises. We emphasize the technique of completing the square, for its applications both to quadratic equations here, and to inequalities and maximization of quadratic functions in Chapter 4. Section 8, we believe, is unique; it discusses problem solving and illustrates typical errors by giving five "solutions" to the same simple problem. The chapter ends with our famous Wyatt Earp principle.

Chapter 4 discusses inequalities, an important subject that deserves better than it usually gets. We apply inequalities to the lost art of estimation and to some realistic problems.

Chapter 5 introduces the coordinate plane, the distance and mid-point formulas, and graphs of equations. It also introduces the important concept of function. We try to give a sound understanding of this concept, e.g. via many examples and the black box interpretation. We stress functional notation and include exercises like this: if $f(t)$ is the temperature t hours after midnight, express the temperature at 1:45 AM. This chapter also covers graphs of functions and composite functions and makes a special effort to convey the idea of inverse functions. Example: after driving t hours, I have traveled $s = f(t)$ miles. Express the inverse function in words.

Chapter 6 takes up further topics in functions and graphs, in particular graphs of factored polynomials and rational functions, and an introduction to conic sections in standard position. It emphasizes intelligent graphing, via symmetry, zeros, asymptotes, and so on.

Chapter 7 deals with exponential and logarithm functions, their graphs, logs to various bases, and exponential equations. We stress the precise meaning of exponential growth. (Recently, the general public has "discovered" exponential growth and now uses the term incorrectly to describe any rapid increase.) This chapter contains an unusually rich collection of real-life problems including business applications to compound interest, annuities, etc.

Chapter 8 covers systems of equations and inequalities with emphasis on solving linear systems by Gaussian elimination using matrix notation. Basic properties of determinants and Cramer's rule follow. Finally, there is a discussion of plane regions defined by inequalities and an introduction to linear programming.

Chapter 9 treats polynomials and their zeros, starting with synthetic division and the Factor and Remainder theorems. Section 2 examines the interplay between synthetic division, nested factored form of a polynomial, and numerical evaluation of polynomials. This is a very appealing subject which does not seem to appear in other texts. Sections 3–5 discuss rational zeros, complex numbers, and complex zeros of polynomials. Section 6 treats existence of zeros of polynomials and numerical approximations of their zeros (also numerical solutions of other equations). This section incorporates a number of important mathematical ideas: zeros of a continuous function, inequalities and estimation, linear interpolation, and calculator techniques.

Chapter 10 studies arithmetic and geometric sequences, mathematical induction, permutations and combinations, and the binomial theorem. It contains a number of unusual and interesting applications. In particular, the section on induction does more than the standard summation formulas, and provides an especially varied set of exercises.

The Appendix is devoted to computing technique. For readers who wish to improve their calculator skills, Sections 1 and 2 discuss general calculator know-how plus efficient use of memory-related keys, parentheses, constant factors, and so on. For readers who prefer logarithms, Sections 3 and 4 discuss use of log tables and computations by logarithms.

ACKNOWLEDGMENTS

We are grateful to Paul Blanchard (University of Wisconsin), Thomas Butts (Ohio State University), Douglas Hall (Michigan State University), Stanley Lukawecki (Clemson University) and Donald Sherbert (University of Illinois), who reviewed the manuscript and supplied valuable criticism and encouragement.

The answer section was prepared by Harold Schwalm (Drexel University) and checked by Marc Glucksman (El Camino College). The Solutions Manual was written by Judith Silverman (student at Harvard University) and Harold Schwalm.

Marc Glucksman has written an excellent study guide to accompany this text, including lists of concepts and procedures, additional examples and exercises, and chapter review tests. In the process he has also contributed to the accuracy of this book.

The staff at Saunders College Publishing has been most helpful and professional, particularly editors Jim Porterfield, Jay Freedman, and Carol Field; Art Director Richard Moore; and artist Sharon Iwanczuk. Finally, we thank our typist, Judy Snider, for her speedy and accurate work.

Justin J. Price
Harley Flanders

To Bob, Marian, Sue
Dave, and Judy

Table of Contents

Chapter 1   REAL NUMBERS

1	Introduction	1
2	Addition and Multiplication	3
3	Subtraction and Division	7
4	Priority of Operations	13
5	Order and Absolute Values	16
6	About Calculators	22
	Review Exercises	22

Chapter 2   FUNDAMENTALS OF ALGEBRA

1	Integral Exponents	24
2	Polynomials	30
3	Polynomials in Several Variables	34
4	Factoring	37
5	Rational Expressions	43
6	Roots and Radicals	47
7	Rational Exponents	53
8	Avoiding Common Errors	58
	Review Exercises	60

Chapter 3   EQUATIONS

1	Translating Words into Math	61
2	Translating Problems into Algebra	65
3	Equations; Linear Equations	70
4	Applications of Linear Equations	75
5	Quadratic Equations	80

6	The Quadratic Formula; Applications	84
7	Other Types of Equations	91
8	Suggestions for Solving Problems	96
	Review Exercises	97

Chapter 4   INEQUALITIES

1	Basic Properties	99
2	Solving Inequalities	106
3	Inequalities Involving Absolute Values; Quadratic Inequalities	111
	Review Exercises	115

Chapter 5   COORDINATES, LINES, FUNCTIONS

1	Coordinates in the Plane	117
2	Equations of Lines	124
3	Functions	132
4	Graphs of Functions	137
5	Operations on Functions; Inverse Functions	144
	Review Exercises	151

Chapter 6   FUNCTIONS AND GRAPHS

1	Linear and Quadratic Functions	153
2	Polynomial Functions of Degree Greater than Two	159
3	Rational Functions	165
4	Variation	171
5	Conic Sections	177
	Review Exercises	185

Chapter 7   EXPONENTIAL AND LOGARITHM FUNCTIONS

1	Exponential Functions	187
2	Logarithms and Logarithm Functions	193

3	Common Logarithms; Exponential Equations	199
4	Logarithms to Other Bases	204
5	Applications	208
	Review Exercises	215

Chapter 8 SYSTEMS OF EQUATIONS AND INEQUALITIES

1	Systems of Equations in Two Variables	217
2	Systems of Equations in Three Variables	223
3	Solution of Linear Systems in Matrix Notation	228
4	Determinants	232
5	General Determinants and Their Properties	237
6	Systems of Linear Inequalities and Linear Programming	242
	Review Exercises	249

Chapter 9 ZEROS OF POLYNOMIALS AND COMPLEX NUMBERS

1	Division of Polynomials; Synthetic Division	251
2	Polynomial Evaluation	257
3	Rational Zeros of Polynomials	260
4	Complex Numbers	264
5	Complex Zeros of Polynomials	268
6	Real Zeros of Polynomials and Approximation of Zeros	274
	Review Exercises	279

Chapter 10 DISCRETE ALGEBRA

1	Sequences and Series	281
2	Arithmetic Sequences	285
3	Geometric Sequences	290
4	Mathematical Induction	296
5	Permutations	302
6	Combinations	306
7	The Binomial Theorem	310
	Review Exercises	314

APPENDIX ON COMPUTATION

1	Use of Calculators	316
2	Further Calculator Techniques	321
3	Tables of Logarithms	324
4	Computations with Logarithms	327

Answers to Odd-Numbered Exercises   331

Table of Logarithms   T.1

Index   i

1

Real Numbers

1 INTRODUCTION

The subject of algebra grew out of arithmetic. It deals with the operations of arithmetic and relations between numbers expressed by these operations. To make numerical statements as clear and as brief as possible, algebra uses a notation that is the shorthand of mathematics. For example, consider the statement

"The area of a rectangle is equal to its length multiplied by its width."

Certainly not all of these 14 words and 57 letters are needed to convey the meaning. The same information is contained in these shorter versions:

$$\text{Area equals length times width}$$
$$\text{Area} = \text{length times width}$$
$$\text{Area} = \text{length} \cdot \text{width}$$
$$A = L \cdot W$$
$$A = LW$$

The final version, $A = LW$, expresses the given statement boiled down to its bare bones.

As another example, consider the statement

"To convert Celsius to Fahrenheit, multiply by $\frac{9}{5}$ then add 32."

In algebraic notation, these 11 words are shortened to

$$F = \tfrac{9}{5} C + 32$$

Algebraic notation is brief, but even more important, it is general. The formula $A = LW$ is a general statement about every rectangle, not just one. Similarly, the formula $A = \pi r^2$ gives the area of *every* circle. In particular:

The area of a circle of radius 4 is $\pi \cdot 4^2$
The area of a circle of radius 17 is $\pi \cdot 17^2$
The area of a circle of radius 951.6 is $\pi \cdot 951.6^2$

and so on. We could spend a lifetime writing out the information contained in $A = \pi r^2$ and never finish.

2   Real Numbers

Algebraic statements use letters. Look at a page of any algebra book and you will see lots of letters and also symbols such as

$$+ \quad \times \quad \sqrt{} \quad ()^2$$

The letters represent numbers, and the symbols represent various operations on numbers. So to prepare for algebra, we will devote this chapter to a review of the number system and the operations of arithmetic.

Real Numbers

The familiar numbers of everyday life and of science are called **real numbers**. The set of all real numbers together with their properties of arithmetic is the **real number system**.

The real number system contains the **positive integers** (or **natural numbers**)

$$1, 2, 3, 4, 5, \cdots$$

the **negative integers**

$$-1, -2, -3, -4, -5, \cdots$$

and 0. The set of positive and negative integers together with 0 is called the **integers**.

The real number system also contains the **rational numbers**, those real numbers representable as fractions. For example,

$$\frac{1}{5}, \quad \frac{3}{8}, \quad \frac{-70}{19}, \quad \sqrt{\frac{1}{4}} = \frac{1}{2}, \quad 0.333333 \cdots = \frac{1}{3}$$

are rational numbers. Note that integers are rational numbers because each integer n can be represented as the fraction $n/1$.

Not all real numbers are rational. For example,

$$\sqrt{2}, \quad \sqrt[3]{6}, \quad \pi$$

are **irrational**; they cannot be expressed as fractions.

Each real number can be represented by a decimal. For a rational number, the decimal expression is finite, such as

$$\frac{1}{2} = 0.5 \qquad \frac{9}{4} = 2.25 \qquad \frac{482}{125} = 3.856$$

or repeating, such as

$$\frac{1}{3} = 0.333333 \cdots \qquad \frac{4}{7} = 0.571428\ 571428\ 571428 \cdots$$

For an irrational number, it is non-repeating, such as

$$\pi = 3.1415926539 \cdots$$

(See Example 8, page 77.)

The real number system comes equipped with the four operations of arithmetic: addition, subtraction, multiplication, and division. We review their properties in Sections 2, 3, and 4, emphasizing those that will be helpful in algebra.

Besides these four operations, the real number system has an important feature called **order**, discussed in Section 5: given two real numbers, one is smaller and one is larger. This allows comparison of numbers and of quantities represented by real numbers—temperatures, distances, prices, etc.

2 ADDITION AND MULTIPLICATION

For each pair of real numbers a, b, there is a real number $a + b$, their **sum**, and a real number $a \times b$, their **product**. The product is also written $a \cdot b$, or simply ab. **Addition** is the process of forming a sum; **multiplication** is the process of forming a product. In the sum $a + b$, the numbers a and b are called **summands** or **terms**. In the product ab, they are called **factors**.

Let us examine the algebraic properties of addition and multiplication. Throughout this chapter all letters will represent real numbers.

Commutative Laws	$a + b = b + a$	$ab = ba$
Associative Laws	$(a + b) + c = a + (b + c)$	$(ab)c = a(bc)$

The commutative laws say that addition and multiplication of two real numbers can be done in either order. For example,
$$3 + 5 = 5 + 3 \qquad 7 \cdot 6 = 6 \cdot 7$$
The associative laws give unique meanings to sums and products of three real numbers such as $a + b + c$ and abc. For example, $2 + 3 + 4 = 9$ whether we compute the sum as
$$(2 + 3) + 4 = 5 + 4 = 9 \qquad \text{or as} \qquad 2 + (3 + 4) = 2 + 7 = 9$$
and $2 \cdot 3 \cdot 4 = 24$ whether we compute the product as
$$(2 \cdot 3) \cdot 4 = 6 \cdot 4 = 24 \qquad \text{or as} \qquad 2 \cdot (3 \cdot 4) = 2 \cdot 12 = 24$$

Sums and products of more than three real numbers also have unique meanings, and can be computed in any arrangement and grouping. For example,
$$\begin{aligned} 1 + 2 + 3 + 4 &= (1 + 2 + 3) + 4 = 6 + 4 = 10 \\ &= (1 + 4) + (2 + 3) = 5 + 5 = 10 \\ &= 3 + (1 + 4 + 2) = 3 + 7 = 10, \text{ etc.} \end{aligned}$$

Similarly,
$$\begin{aligned} 2 \cdot 3 \cdot 4 \cdot 2 \cdot 5 &= (2 \cdot 3 \cdot 4)(2 \cdot 5) = 24 \cdot 10 = 240 \\ &= (2 \cdot 3)(4 \cdot 2 \cdot 5) = 6 \cdot 40 = 240 \\ &= (2 \cdot 2) \cdot (3 \cdot 5) \cdot 4 = 4 \cdot 15 \cdot 4 = 240, \text{ etc.} \end{aligned}$$

> A sum or a product of real numbers can be arranged and grouped in any way.

EXAMPLE 1 Compute

(a) $2 \cdot 4 \cdot \frac{1}{3} \cdot 25 \cdot 5 \cdot 3$ \qquad (b) $1 + 2 + 3 + \cdots + 98 + 99 + 100$

4 Real Numbers

SOLUTION (a) The products $\frac{1}{3} \cdot 3$, $2 \cdot 5$, and $4 \cdot 25$ are especially easy, so rearrange and regroup this way:

$$2 \cdot 4 \cdot \tfrac{1}{3} \cdot 25 \cdot 5 \cdot 3 = (\tfrac{1}{3} \cdot 3) \cdot (2 \cdot 5) \cdot (4 \cdot 25)$$
$$= 1 \cdot 10 \cdot 100 = \underline{1000}$$

(b) Rearrange and regroup this way:

$$1 + 2 + 3 + \cdots + 98 + 99 + 100 = (1 + 100) + (2 + 99) + (3 + 98)$$
$$+ \cdots + (49 + 52) + (50 + 51)$$

Each term on the right side is 101 and there are exactly 50 terms. Hence the sum is $50 \cdot 101 = \underline{5050}$.

An important property connects the operations of addition and multiplication.

Distributive Law $\qquad a(b + c) = ab + ac \qquad (b + c)a = ba + ca$

We use the distributive law all the time in ordinary arithmetic. For example, to compute $5 \cdot 31$ mentally, we think this way:

$$5 \cdot 31 = 5(30 + 1) = 5 \cdot 30 + 5 \cdot 1 = 150 + 5 = 155$$

That's the distributive law in action.

EXAMPLE 2 Verify using the distributive law

$$(a + b) \cdot (c + d) = ac + ad + bc + bd$$

SOLUTION Think of $(a + b)$ as a single number multiplying the sum $c + d$. By the distributive law,

$$(a + b) \cdot (c + d) = (a + b)c + (a + b)d$$

Now apply the distributive law to each term on the right:

$$(a + b) \cdot (c + d) = (ac + bc) + (ad + bd)$$
$$= \underline{ac + ad + bc + bd} \qquad \text{[rearrangement]}$$

Identity and Inverse Elements

The real number system contains numbers 0 and 1 whose algebraic properties deserve special mention.

Identity Elements

Addition	Multiplication
There is a real number 0, called the **additive identity,** such that	There is a real number 1, called the **multiplicative identity,** such that
$0 + a = a + 0 = a$	$1 \cdot a = a \cdot 1 = a$
for every real number a.	for every real number a.

Next we observe that real numbers come in pairs that "neutralize" each other. For example, a 4° rise in temperature "neutralizes" or "undoes" a 4° drop in temperature, and a 4° drop neutralizes a 4° rise. In other words,
$$4 + (-4) = 0$$
The numbers 4 and -4 neutralize each other by addition. Similarly, doubling a salary neutralizes a $\tfrac{1}{2}$ cut, and a $\tfrac{1}{2}$ cut undoes a doubling. In other words,
$$2 \cdot \tfrac{1}{2} = \tfrac{1}{2} \cdot 2 = 1$$
The numbers 2 and $\tfrac{1}{2}$ neutralize each other by multiplication.

Inverses	
Addition	*Multiplication*
For each real number a, there is a unique real number $-a$, its **additive inverse** or **negative**, such that $$a + (-a) = (-a) + a = 0$$	For each real number a different from 0, there is a unique real number $\dfrac{1}{a}$, its **multiplicative inverse** or **reciprocal**, such that $$a \cdot \dfrac{1}{a} = \dfrac{1}{a} \cdot a = 1$$

EXAMPLE 3 Justify the cancellation laws:

(1) If $a + b = a + c$, then $b = c$.   (2) If $ab = ac$ and $a \neq 0$, then $b = c$.

SOLUTION Given
$$a + b = a + c \qquad ab = ac \quad a \neq 0$$
we must show that $b = c$ in each case. On the left, we add $-a$ to both sides. On the right, we multiply both sides by $\dfrac{1}{a}$:

$$-a + (a + b) = -a + (a + c) \qquad \dfrac{1}{a} \cdot ab = \dfrac{1}{a} \cdot ac$$
$$(-a + a) + b = (-a + a) + c \qquad \left(\dfrac{1}{a} \cdot a\right) \cdot b = \left(\dfrac{1}{a} \cdot a\right) \cdot c$$
$$0 + b = 0 + c \qquad\qquad\qquad 1 \cdot b = 1 \cdot c$$
$$\underline{b = c} \qquad\qquad\qquad\qquad \underline{b = c}$$

Rules for Inverses

(1) $-(-a) = a$ (2) $-(a + b) = (-a) + (-b)$

(3) $\dfrac{1}{\frac{1}{a}} = a$ (4) $\dfrac{1}{ab} = \dfrac{1}{a} \cdot \dfrac{1}{b}$

(5) $(-a)b = a(-b) = -ab$ (6) $(-a)(-b) = ab$

[In (3) and (4), we assume $a \neq 0$ and $b \neq 0$.]

6 Real Numbers

Examples
(1) $-(-4) = 4$
(2) $-(8 + 3) = (-8) + (-3)$
(3) $\dfrac{1}{\frac{1}{2}} = 2$
(4) $\dfrac{1}{4 \cdot 5} = \dfrac{1}{4} \cdot \dfrac{1}{5}$
(5) $(-3) \cdot 7 = 3 \cdot (-7) = -21$
(6) $(-3) \cdot (-7) = 21$

Rules (2) and (4) apply also to three or more terms or factors. For example,

$$-(a + b + c) = (-a) + (-b) + (-c) \qquad \dfrac{1}{abc} = \dfrac{1}{a} \cdot \dfrac{1}{b} \cdot \dfrac{1}{c}$$

The number 0 has some important properties with respect to multiplication.

Properties of Zero
(1) $a \cdot 0 = 0 \cdot a = 0$ for every real number a.
(2) The number 0 has no reciprocal.
(3) If $ab = 0$, then either $a = 0$ or $b = 0$ (or both).

EXAMPLE 4 If $abc = 0$, show that at least one of the factors is 0.

SOLUTION Write
$$abc = a(bc) = 0$$
By Property (3), either $a = 0$ or $bc = 0$. If $a = 0$, there is nothing more to prove. If $a \neq 0$, then $bc = 0$. But then by Property (3), either $b = 0$ or $c = 0$. Thus in all cases, at least one factor is 0.

EXERCISES

Justify the statement
1. $6(5 + 4) = 6 \cdot 5 + 6 \cdot 4$
2. $(8 + 1) + 3 = 8 + (1 + 3)$
3. $2 \cdot (8 \cdot 7) = (2 \cdot 8) \cdot 7$
4. $(-5) + 5 = 0$
5. $(4 + 0) \cdot (5 + 0) = 4 \cdot 5$
6. $3 \cdot [2 + (-2)] = 0$
7. $(-1) \cdot 3 = -3$
8. $(-2) \cdot (-4) = 8$
9. $\tfrac{1}{2} \cdot \tfrac{1}{5} = \tfrac{1}{10}$
10. $(-3)(-\tfrac{1}{3}) = 1$

Compute
11. $51 + 52 + 53 + \cdots + 99 + 100$
12. $2 + 4 + 6 + \cdots + 98 + 100$
13. $1 + 3 + 5 + \cdots + 197 + 199$
14. $1 + 4 + 7 + \cdots + 55 + 58$
15. $2 \cdot 3 \cdot 4 \cdot 5 \cdot \tfrac{1}{2} \cdot \tfrac{1}{3} \cdot \tfrac{1}{4}$
16. $2 \cdot 4 \cdot 8 \cdot 5 \cdot 25 \cdot 125$
17. $42 \cdot 23$ as $(40 + 2)(20 + 3)$
18. $36 \cdot 111$ as $(30 + 6)(100 + 10 + 1)$

Show that
19. $3(5a) = 15a$
20. $a(5b) = 5(ab)$

21. $2a \cdot 2b = 4ab$
22. $a \cdot 2b \cdot 3c = 6abc$
23. $a + b + (-a) = b$
24. $-(a + b) + b = (-a)$
25. $a(b + c + d) = ab + ac + ad$
26. $a\left(b + \dfrac{1}{a}\right) = ab + 1$
27. $a(bc + bd) = ab(c + d)$
28. $a(b + c)d = abd + acd$
29. $(-a)(-b)(-c) = -abc$
30. $(-a)(-b)(-c)(-d) = abcd$

Justify the statement

31. If $c + 4 = d + 4$, then $c = d$.
32. If $9r = 9s$, then $r = s$.
33. If $x + y = x + z + 1$, then $y = z + 1$.
34. If $a + b + c = b + c + d$, then $a = d$.
35. If $a \neq 0$ and $ab + 1 = 3a + 1$, then $b = 3$.
36. If $a \neq 0$ and $a(b + c) = a(b + 2)$, then $c = 2$.
37. If $abcd = 0$, then at least one of the factors is 0.
38. If $abcd = 1$, then $\dfrac{1}{c} = abd$.
39. If $(a + 1)(a + 2) = 0$, then either $a = -1$ or $a = -2$.
40. If a is an integer, then $(a - \tfrac{1}{2})(a + \tfrac{4}{3})$ cannot be 0.
41. If $a \neq 0$, then none of the numbers $a, 2a, 3a, 4a, \cdots$ is 0.
42. If any of the numbers $a, 2a, 3a, 4a, \cdots$ is 0, then they are all 0.
43. Can a real number be its own negative?
44. Can a real number be its own reciprocal?

Given that $a \cdot 0 = 0 \cdot a = 0$ for every real number a, explain why

45. 0 has no reciprocal.
46. If $ab = 0$, then either $a = 0$ or $b = 0$. [Hint If $a \neq 0$, then a has a reciprocal.]

3 SUBTRACTION AND DIVISION

In this section, we define subtraction in terms of addition and division in terms of multiplication.

> **Definition of Subtraction**
>
> $a - b$ is defined as $a + (-b)$

The number $a - b$ is the **difference** of a and b. It is that number x such that $b + x = a$. (See Exercise 35.)

8 Real Numbers

Examples
$$5 - 2 = 5 + (-2) = 3 \qquad 4 - 9 = 4 + (-9) = -5$$
$$1 - (-7) = 1 + [-(-7)] = 1 + 7 = 8$$

EXAMPLE 1 Justify this form of the distributive law:
$$a(b - c) = ab - ac$$

SOLUTION
$$\begin{aligned}
a(b - c) &= a[b + (-c)] & \text{[definition of subtraction]} \\
&= ab + a(-c) & \text{[distributive law]} \\
&= ab + (-ac) & \text{[rule (5) for inverses]} \\
&= \underline{ab - ac} & \text{[definition of subtraction]}
\end{aligned}$$

As an application of Example 1, try to compute $45 \cdot 98$ mentally. Since $98 = 100 - 2$, you can do it this way:
$$\begin{aligned}
45 \cdot 98 &= 45(100 - 2) = 45 \cdot 100 - 45 \cdot 2 \\
&= 4500 - 90 = 4410
\end{aligned}$$

Longer expressions such as $a - b - c$ and $a - b + c - d - e$ are also defined in terms of addition. For instance,

$a - b - c$ is defined as $a + (-b) + (-c)$
$a - b + c - d - e$ is defined as $a + (-b) + c + (-d) + (-e)$

Being sums, expressions of this type can be rearranged and regrouped in any way.

EXAMPLE 2 Show that $\quad a - b - c + d - e = (a + d) - (b + c + e)$

SOLUTION
$$\begin{aligned}
a - b - c + d - e &= a + (-b) + (-c) + d + (-e) & \text{[by definition]} \\
&= (a + d) + [(-b) + (-c) + (-e)] & \text{[regrouping]} \\
&= (a + d) + [-(b + c + e)] & \text{[property of inverses]} \\
&= \underline{(a + d) - (b + c + e)} & \text{[def. of subtraction]}
\end{aligned}$$

Example 2 illustrates a familiar idea. For example, suppose you make deposits in a bank account of amounts a and d, and make withdrawals of amounts b, c, and e. The bank computes your balance by subtracting the total of the withdrawals from the total of the deposits, $(a + d) - (b + c + e)$.

EXAMPLE 3 Compute
$$21 + 22 + 23 + \cdots + 29 + 30 - 1 - 2 - 3 - \cdots - 9 - 10$$

SOLUTION By definition, the sum means
$$21 + 22 + \cdots + 30 + (-1) + (-2) + \cdots + (-10)$$

Regroup, writing the sum as

$[21 + (-1)] + [22 + (-2)] + \cdots + [30 + (-10)]$
$= (21 - 1) + (22 - 2) + \cdots + (30 - 10)$
$= 20 + 20 + \cdots + 20$ [10 terms]
$= 10 \cdot 20 = \underline{200}$

Now let us state the definition of division:

Definition of Division

$\dfrac{a}{b}$ is defined as $a \cdot \dfrac{1}{b}$ if $b \neq 0$

$\dfrac{a}{0}$ is not defined

We call a/b the **quotient** of a by b. It is that number x such that $b \cdot x = a$. (See Exercise 36.) The expression a/b is a **fraction** with **numerator** a and **denominator** b. Here a and b are not necessarily integers.

Division by 0 is undefined. For $a/0$ should be a number x such that $x \cdot 0 = a$. But $x \cdot 0 = 0$, so if $a \neq 0$ there is no such number x. If $a = 0$, then any x would do. Hence $0/0$ would mean any real number, which is absurd.

Two quotients may be equal. For example, $\frac{2}{3} = \frac{4}{6}$. There is a simple test for equality of quotients. (See Exercises 47 and 48.)

Equality of Quotients

$\dfrac{a}{b} = \dfrac{c}{d}$ if and only if $ad = bc$ $b \neq 0, d \neq 0$

Examples

$\frac{7}{1008} = \frac{1}{144}$ because $7 \cdot 144 = 1008 \cdot 1$

$\frac{37}{200} \neq \frac{5}{27}$ because $37 \cdot 27 = 999$ but $200 \cdot 5 = 1000$

Another basic fact about equality of quotients is that

$\dfrac{a}{b} = \dfrac{ka}{kb}$ for every $k \neq 0$

---

which holds because $a \cdot kb = b \cdot ka$. This property allows us to multiply the numerator and denominator by any non-zero number without changing the quotient. Thus

$$\frac{2}{3} = \frac{4}{6} = \frac{-6}{-9} = \frac{20}{30} = \frac{2(1.759)}{3(1.759)} = \frac{2\pi}{3\pi} \quad \text{etc.}$$

It also allows us to cancel a common non-zero factor from the numerator and denominator. Thus,

$$\frac{21}{28} = \frac{7 \cdot 3}{7 \cdot 4} = \frac{3}{4} \qquad \frac{9\pi}{5\pi} = \frac{9}{5} \qquad \frac{81ab}{162a} = \frac{(81a) \cdot b}{(81a) \cdot 2} = \frac{b}{2}$$

EXAMPLE 4 Assuming $b \neq 0$ and $d \neq 0$, show that

$$\text{(a)} \quad \frac{a}{b} + \frac{c}{b} = \frac{a+c}{b} \qquad \text{(b)} \quad \frac{a}{b} \cdot \frac{c}{d} = \frac{ac}{bd}$$

SOLUTION

(a) $\dfrac{a}{b} + \dfrac{c}{b} = a \cdot \dfrac{1}{b} + c \cdot \dfrac{1}{b}$

$\qquad = (a + c) \cdot \dfrac{1}{b}$ [distributive law]

$\qquad = \dfrac{a + c}{b}$ [definition of quotient]

(b) $\dfrac{a}{b} \cdot \dfrac{c}{d} = \left(a \cdot \dfrac{1}{b}\right)\left(c \cdot \dfrac{1}{d}\right)$

$\qquad = ac\left(\dfrac{1}{b} \cdot \dfrac{1}{d}\right)$ [regrouping]

$\qquad = ac \cdot \dfrac{1}{bd}$ [property of inverses]

$\qquad = \dfrac{ac}{bd}$ [definition of quotient]

Example 4 gives two of the basic rules for computing with quotients. Here is a more complete list:

Rules for Quotients

(1) $\dfrac{a}{b} \pm \dfrac{c}{b} = \dfrac{a \pm c}{b}$ \qquad (2) $\dfrac{a}{b} \pm \dfrac{c}{d} = \dfrac{ad \pm bc}{bd}$

(3) $\dfrac{a}{b} \cdot \dfrac{c}{d} = \dfrac{ac}{bd}$ \qquad (4) $\dfrac{\frac{a}{b}}{\frac{c}{d}} = \dfrac{a}{b} \cdot \dfrac{d}{c}$

(5) $a \cdot \dfrac{b}{c} = \dfrac{ab}{c}$ \qquad (6) $\dfrac{-a}{b} = \dfrac{a}{-b} = -\dfrac{a}{b}$

(all denominators non-zero)

These rules include the rules for ordinary fractions but they apply with any real values of *a*, *b*, *c*, *d*, not just integers.

EXAMPLE 5 Compute and simplify $\dfrac{\frac{a}{4}}{\frac{a}{2} + \frac{b}{3}}$

SOLUTION The denominator is
$$\frac{a}{2} + \frac{b}{3} = \frac{3a + 2b}{6} \qquad \text{[Rule (2)]}$$

Therefore,
$$\frac{\frac{a}{4}}{\frac{3a + 2b}{6}} = \frac{a}{4} \cdot \frac{6}{3a + 2b} \qquad \text{[Rule (4)]}$$

$$= \frac{6a}{4(3a + 2b)} \qquad \text{[Rule (3)]}$$

$$= \frac{3a}{2(3a + 2b)} \qquad \text{[canceling common factor]}$$

Warning Rule (2) says that
$$\frac{1}{a} + \frac{1}{b} = \frac{b + a}{ab} \qquad \text{NOT} \qquad \frac{1}{a + b}$$

When you are not sure about a rule, try substituting numbers. For example,
$$\frac{1}{2} + \frac{1}{4} = \frac{3}{4} \qquad \text{NOT} \qquad \frac{1}{2 + 4} = \frac{1}{6}$$

EXERCISES

Compute. Simplify when possible

1. $5 - (-3)$
2. $-4 - (-10)$
3. $10 - 1 + 3 - 2$
4. $12 - 5 - 4 + 2$
5. $1 - 2 + 3 - 4 + - \cdots + 29 - 30$
6. $100 + 99 + 98 + \cdots + 51 - 50 - 49 - 48 - \cdots - 2 - 1$
7. $(\frac{4}{9} - \frac{1}{3})(\frac{8}{5} - \frac{5}{4})$
8. $(\frac{1}{5} - \frac{1}{6})(\frac{1}{4} - \frac{1}{5})$
9. $\frac{1}{4}(\frac{11}{6} - 5 \cdot \frac{2}{9})$
10. $\frac{2}{3}(\frac{3}{2} + 2 \cdot \frac{5}{8})$
11. $\frac{5}{12} \div \frac{2}{9}$
12. $\frac{4}{7} \div \frac{16}{21}$

13. $\dfrac{\dfrac{3}{4}}{\dfrac{1}{3}-\dfrac{1}{2}}$

14. $\dfrac{1}{2\left(\dfrac{1}{4}+\dfrac{3}{5}\right)}$

15. $\dfrac{\dfrac{5}{3}-\dfrac{3}{2}}{\dfrac{7}{3}+\dfrac{1}{6}}$

16. $\dfrac{\dfrac{1}{2}-\dfrac{1}{3}}{\dfrac{1}{2}+\dfrac{1}{3}}$

17. $\dfrac{ab}{\dfrac{1}{a}+\dfrac{1}{b}}$

18. $ab\left(\dfrac{1}{a}-\dfrac{1}{b}\right)$

Mental arithmetic

19. $32 \cdot 99$

20. $19 \cdot 75$

21. $\tfrac{1}{3} \cdot 2999$

22. $\tfrac{1}{5} \cdot 4995$

Which of these statements is true?

23. $\dfrac{7}{2\pi} = \dfrac{49}{14\pi}$

24. $\dfrac{-4}{17} = \dfrac{12}{-51}$

25. $\dfrac{a}{0} = 0$

26. $\dfrac{0}{a} = 0$

27. $\dfrac{a+b}{3} = \dfrac{a}{3}+\dfrac{b}{3}$

28. $\dfrac{3}{a+b} = \dfrac{3}{a}+\dfrac{3}{b}$

Justify the statement

29. $\dfrac{a}{1} = a$

30. $\dfrac{a}{a} = 1$

31. $\dfrac{a}{d}+\dfrac{b}{d}+\dfrac{c}{d} = \dfrac{a+b+c}{d}$

32. $\dfrac{a}{b} \cdot \dfrac{c}{d} \cdot \dfrac{e}{f} = \dfrac{ace}{bdf}$

33. $a \cdot \dfrac{b}{c} = \dfrac{a}{c} \cdot b$

34. $\dfrac{ac+b}{c} = a+\dfrac{b}{c}$

35. $a - b$ is a number x such that $b + x = a$.

36. a/b is a number x such that $b \cdot x = a$.

Give a numerical example showing that

37. $\dfrac{a+b}{c+d}$ does not equal $\dfrac{a}{c}+\dfrac{b}{d}$

38. $\dfrac{a+b}{a+c}$ does not equal $1+\dfrac{b}{c}$

Prove the cancellation law

39. If $a - c = b - c$, then $a = b$.

40. If $a - b = a - c$, then $b = c$.

41. If $\dfrac{a}{c} = \dfrac{b}{c}$, then $a = b$.

42. If $\dfrac{c}{a} = \dfrac{c}{b}$ and $c \neq 0$, then $a = b$.

Justify the rule for quotients

43. Rule (2)
44. Rule (4)
45. Rule (5)
46. Rule (6)

47. If $a/b = c/d$, show that $ad = bc$.

48. If $ad = bc$, where $b \neq 0$ and $d \neq 0$, show that $a/b = c/d$.

4 PRIORITY OF OPERATIONS

How much is $3 + 4 \times 5$? Is it
$$(3 + 4) \times 5 = 7 \times 5 = 35 \quad \text{or} \quad 3 + (4 \times 5) = 3 + 20 = 23?$$
There is a convention that gives a precise meaning to such computations:

> In a sequence of operations, multiplications and divisions must be done before additions and subtractions.

No doubt about it: $3 + 4 \times 5 = 3 + 20 = 23$.

EXAMPLE 1 Compute
(a) $2 \times 3 + 4 \times 5 - 7$ (b) $5 - 18 \div 6 + 2 \times 4$

SOLUTION (a) Do the multiplications first:
$$2 \times 3 + 4 \times 5 - 7 = (2 \times 3) + (4 \times 5) - 7$$
$$= 6 + 20 - 7 = \underline{19}$$

(b) Do the division and multiplication first:
$$5 - 18 \div 6 + 2 \times 4 = 5 - (18 \div 6) + (2 \times 4)$$
$$= 5 - 3 + 8 = \underline{10}$$

Remark To remember the priority of operations, think of My Dear Aunt Sally. The initials **MDAS** suggest that Multiplication and Division come before Addition and Subtraction.

Parentheses

The product of $3 + 7$ by $5 - 2$ is $10 \times 3 = 30$. But if we write $3 + 7 \times 5 - 2$, that means $3 + (7 \times 5) - 2 = 3 + 35 - 2 = 36$. To express the given product correctly, write
$$(3 + 7) \times (5 - 2)$$
because whatever is inside a pair of parentheses is considered as a single number.

> In a computation, all calculations within parentheses must be done first.

With parentheses, you can disobey Aunt Sally. For example, writing $(3 + 7) \times (5 - 2)$ lets you do the addition and subtraction *before* the multiplication. As another example, the average of numbers a and b is half their sum, $\frac{1}{2}(a + b)$. The parentheses let you compute the sum first then multiply by $\frac{1}{2}$. Writing $\frac{1}{2}a + b$ is wrong because
$$\frac{1}{2}a + b = (\frac{1}{2}a) + b \quad \text{NOT} \quad \frac{1}{2}(a + b) = \frac{1}{2}a + \frac{1}{2}b$$

Often parentheses are essential for clarity. For instance, to subtract from a the number that is 1 larger than b, write
$$a - (b + 1) \quad \text{NOT} \quad a - b + 1$$
For example, write
$$10 - (5 + 1) = 10 - 6 = 4 \quad \text{NOT} \quad 10 - 5 + 1 = 6$$
In case of nested parentheses (parentheses within parentheses (like this)), compute the inner ones first. Then, if more nested parentheses remain, repeat the process.

In a computation with nested parentheses compute the innermost ones first.

EXAMPLE 2 Compute $14 - ((15 + 9) - (7 + 4(9 - 2 \cdot 3)))$

SOLUTION First compute $(9 - 2 \cdot 3) = 3$. The calculation becomes
$$14 - ((15 + 9) - (7 + 4 \cdot 3))$$
Now compute the two inner parentheses. The calculation reduces to
$$14 - (24 - 19) = 14 - 5 = \underline{9}$$

Remark Because expressions with many levels of parentheses are hard to read, we often replace some of the parentheses by brackets [ ] or braces { }. For example, instead of
$$14 - ((15 + 9) - (7 + 4(9 - 2 \cdot 3)))$$
we can write
$$14 - \{(15 + 9) - [7 + 4(9 - 2 \cdot 3)]\}$$

Suppose we want to subtract a sum in parentheses such as $(a - b - c + d)$. We can either compute $a - b - c + d$ and subtract, or we can remove the parentheses according to the following rule:

To subtract a sum in parentheses, change the sign of each term inside and remove the parentheses.

Example $20 - (15 - 7 - 4 + 2) = 20 - 15 + 7 + 4 - 2 = 14$

EXAMPLE 3 Compute $a + b + c - (b + d - 1) - (a - d - e + 2)$

SOLUTION Remove the parentheses and change the sign of each term inside; then regroup:
$$a + b + c - b - d + 1 - a + d + e - 2$$
$$= (a - a) + (b - b) + c + (-d + d) + e + (1 - 2)$$
$$= 0 + 0 + c + 0 + e + (1 - 2)$$
$$= \underline{c + e - 1}$$

Calculator Priorities

Some calculators use **algebraic logic;** that is, they do multiplications and divisions before additions and subtractions. These calculators compute the key sequence

$$3\boxed{+}\,2\boxed{\times}\,9\boxed{=}$$

as

$$3 + (2 \times 9) = 3 + 18 = 21$$

Other calculators use **left-to-right** logic doing each operation as it is keyed in. These calculators compute the given key sequence as

$$(3 + 2) \times 9 = 5 \times 9 = 45$$

EXAMPLE 4 Find the result of the computation

$$8\boxed{+}\,7\boxed{\times}\,4\boxed{+}\,6\boxed{\div}\,3\boxed{=}$$

with (a) algebraic logic (b) left-to-right logic

SOLUTION (a) The calculator does the multiplication and the division before the additions. It computes

$$8 + (7 \times 4) + (6 \div 3) = 8 + 28 + 2 = \underline{38}$$

(b) The calculator does the operations as they are keyed in: first $8 + 7 = 15$, then $15 \times 4 = 60$, then $60 + 6 = 66$, and finally $66 \div 3 = 22$. Thus, it computes

$$\{[(8 + 7) \times 4] + 6\} \div 3 = \underline{22}$$

EXAMPLE 5 Find a sequence of calculator steps to compute $45 + 3/16$ using (a) algebraic logic (b) left-to-right logic

SOLUTION (a) The first sequence that comes to mind is

$$4\,5\,\boxed{+}\,3\,\boxed{\div}\,1\,6\,\boxed{=}$$

This is right; with algebraic logic, the sequence produces

$$45 + (3/16) = 45.1875$$

(b) With left-to-right logic, this sequence produces

$$(45 + 3) \div 16 = 48 \div 16 = 3$$

which is wrong. But wait, don't throw away your left-to-right calculator! Since the division must come first, just do it first: compute $(3/16) + 45$. The correct sequence is

$$3\,\boxed{\div}\,16\,\boxed{+}\,45\,\boxed{=}$$

EXERCISES

Compute

1. $8 \div 2 + 6$
2. $8 + 6 \div 2$

16 Real Numbers

3. $5 \times 4 - 7 \times 6$
4. $9 \times 3 + 2 \times 5$
5. $2 \times 3 - 4 \times 5 + 6 \times 7$
6. $10 \div 2 + 2 \times 3 + 6 \div 3$
7. $6 \div 7 \times 14 - 4$
8. $9 \times 10 \div 3 + 2$
9. $(2 + 3 \cdot 5)(4 - 1)$
10. $(9 - 4)(6 \div 2 + 1)$
11. $1 + 2(3 + 4)$
12. $(7 - 5) \cdot 6 + 5$
13. $(15 + 3) \div (1 + 2) - (4 - 10)$
14. $(8 \div 4 + 1) - (5 - 2) \times (6 - 2)$
15. $[3(11 - 4) - 2(6 - 1)] [3(8 + 4) - 29]$
16. $[2(4 + 5) - 4(8 - 1)] [20 - 4(9 - 5)]$
17. $6 + 5(4 + 3(2 + 1))$
18. $1 - (2 - (3 - 4))$
19. $3 + [22 - 7\{8(1 + 2) - 4(1 + 2 \cdot 3)\}]$
20. $2(8[1 + 5] - \{3[2 \cdot 7 - 12] + \frac{1}{2}[16 - 2(10 - 6 + 1)]\})$
21. $(a + b + c) - (b - c + 2)$
22. $a + b - 1 - (b - c - 2) - (c - d + 3)$
23. $3 + 4 + 5 + \cdots + 20 - (1 + 2 + 3 + \cdots + 18 + 19)$
24. $(12{,}814 - 9629) + (9629 - 857) + (857 - 198) + (198 - 14)$

Find the result of the calculator sequence assuming
 (a) algebraic logic (b) left-to-right logic

25. $8\boxed{\div}4\boxed{-}6\boxed{\div}2\boxed{=}$
26. $8\boxed{+}4\boxed{\div}6\boxed{+}2\boxed{=}$
27. $10\boxed{\times}2\boxed{\div}4\boxed{+}2\boxed{=}$
28. $10\boxed{\times}2\boxed{+}4\boxed{\div}2\boxed{=}$
29. $2\boxed{\times}3\boxed{\times}4\boxed{-}5\boxed{\times}6\boxed{=}$
30. $7\boxed{-}6\boxed{+}2\boxed{\times}3\boxed{\times}4\boxed{=}$
31. $a\boxed{-}b\boxed{\times}c\boxed{\div}d\boxed{=}$
32. $a\boxed{\times}b\boxed{+}c\boxed{-}d\boxed{\div}e\boxed{=}$

Find a sequence of calculator steps using only the keys $\boxed{+},\boxed{-},\boxed{\times},\boxed{\div},\boxed{=}$ to compute the given quantity. Assume (a) algebraic logic (b) left-to-right logic.

33. $a + 2b$
34. $\frac{a}{3} + b$
35. $5(a - b)$
36. $a + 3(b - c)$
37. $a + \frac{bc}{d}$
38. $a + \frac{b}{c} + d$
39. $6\left(\frac{a}{b} + c\right)$
40. $\frac{a + b}{c}$

5 ORDER AND ABSOLUTE VALUE

Let us picture the real number system geometrically. Imagine a horizontal line extending indefinitely in both directions. Choose a point and label it 0. Then fix a unit of length and mark off 1, 2, 3, . . . units to the right and $-1, -2, -3, \ldots$

to the left (Fig. 1). (Taking the positive direction to the right is just a convention. Maybe on some planet somewhere in the universe, they take the positive direction to the left.)

FIG. 1

Figure 1 shows the integers corresponding to certain points on the line. It is an axiom of the real number system that *every* real number corresponds to a point on the line and vice versa. You can think of real numbers as tags or labels for the points. A line labeled by the real numbers in this way is called a **number line** or a **coordinate line** or a **coordinate axis**.

The relation between real numbers and points is so close that we often think of the real number system *as* a line, and a real number *as* a point. Of course logically, a number and a point are two different things. Still we speak of "the point 6," which is a lot shorter than "the point on the number line corresponding to the real number 6."

Order

Given real numbers a and b, we write

$$a < b \qquad a \text{ is less than } b$$

when a is to the left of b on the number line. For example,

$$3 < 4, \quad 0 < 6, \quad -\tfrac{7}{3} < 0, \quad -18 < 1, \quad -3.5 < -3.4$$

Alternative notation is

$$b > a \qquad b \text{ is greater than } a$$

If a is to the left of b or equal to b, we write

$$a \leq b \quad \text{or} \quad b \geq a$$

We can say that "a is at most b" or "b is at least a." Statements such as $a < b$ or $a \leq b$ are called **inequalities**.

The real numbers greater than 0 are **positive**; those less than 0 are **negative**. In symbols,

$$a > 0 \qquad \text{means} \qquad a \text{ is positive}$$
$$a < 0 \qquad \text{means} \qquad a \text{ is negative}$$

The positive numbers together with 0 are **non-negative**; the negative numbers together with 0 are **non-positive**. In symbols,

$$a \geq 0 \qquad \text{means} \qquad a \text{ is non-negative}$$
$$a \leq 0 \qquad \text{means} \qquad a \text{ is non-positive}$$

Geometric Interpretation of Addition

Adding 1 to any number shifts the corresponding point on the number line one unit to the right. Adding 2 shifts two units to the right, etc. Similarly subtracting 1 shifts one unit to the left, subtracting 2 shifts two units to the left, etc. See Fig. 2.

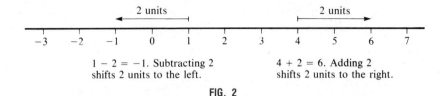

FIG. 2

Now we can express the relations $a < b$ and $a \leq b$ in algebraic language.

$a < b$ if and only if $b - a$ is positive
$a \leq b$ if and only if $b - a$ is non-negative

These statements follow from the relation
$$b = a + (b - a)$$
To get from a to b, we shift by $b - a$. This shift is to the *right* precisely when $b - a$ is positive. Hence b is to the right of a if and only if $b - a$ is positive. Similarly, b is to the right of or equal to a if and only if $b - a \geq 0$.

Signs

The **sign** of a positive number is positive or plus. The sign of a negative number is negative or minus. The number 0 is not given a sign.

Rules of Signs

The sum of two positive numbers is positive. The sum of two negative numbers is negative.

The product and quotient of two numbers is positive if the numbers have the same sign, negative if they have opposite signs.

In particular, the product of a number with itself is positive unless the number is 0. We abbreviate $a \cdot a$ by a^2 ("a squared").

For every real number a,
$$a^2 \geq 0$$
and $a^2 = 0$ only for $a = 0$.

EXAMPLE 1 Show that for every real number c
$$1 + c^2 \geq 1$$
SOLUTION Remember that $a \geq b$ if and only if $a - b \geq 0$. Now
$$(1 + c^2) - 1 = c^2$$

and $c^2 \geq 0$ for every real number. Hence
$$(1 + c^2) - 1 \geq 0$$
which means that $1 + c^2 \geq 1$

EXAMPLE 2 Show that for every real number a
$$(a - 1)^2 + (a - 2)^2 > 0$$
SOLUTION If a is not 1 or 2, then $a - 1 \neq 0$ and $a - 2 \neq 0$, so both $(a - 1)^2 > 0$ and $(a - 2)^2 > 0$. Hence the sum is positive.

If $a = 1$, then $a - 1 = 0$ and $a - 2 = -1$. In this case, the sum is $0^2 + (-1)^2 = 0 + 1 = 1$. Similarly, if $a = 2$, the sum is 1. Thus, $(a - 1)^2 + (a - 2)^2 > 0$ in all cases.

Absolute Value

When someone says that $-1{,}000{,}000$ is a "large negative number," what does that mean? Actually $-1{,}000{,}000$ is *less* than 1 because it lies to the left of 1 on the number line. Still we feel that $-1{,}000{,}000$ is big except for its sign. To measure the size of a real number a regardless of sign, we introduce its **absolute value,** written $|a|$.

Absolute Value For every real number a
$$|a| = \begin{cases} a & \text{if } a \geq 0 \\ -a & \text{if } a < 0 \end{cases}$$

Examples
$$|7| = 7 \qquad |-7| = 7 \qquad |8.34| = 8.34 \qquad |-8.34| = 8.34$$
$$|0| = 0 \qquad |1{,}000{,}000| = 1{,}000{,}000 \qquad |-1{,}000{,}000| = 1{,}000{,}000$$

Note that $|a| > 0$ for every real number a except 0, and that $|-a| = |a|$.

$$|a| > 0 \text{ if } a \neq 0 \qquad |0| = 0 \qquad |-a| = |a|$$

EXAMPLE 3 Find (a) $|0.67|$ (b) $|2 - \pi|$ (c) $|4| - |-5|$
SOLUTION (a) $0.67 > 0$, hence $|0.67| = \underline{0.67}$
(b) $\pi > 3$ so $2 - \pi < 0$. Hence
$$|2 - \pi| = -(2 - \pi) = \underline{\pi - 2}$$
(c) $|4| = 4$ and $|-5| = 5$. Hence
$$|4| - |-5| = 4 - 5 = \underline{-1}$$

Absolute values are useful in measuring distances on the number line. As Fig. 3 illustrates, |a| is the distance between the point a and the point 0.

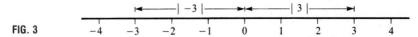

FIG. 3

Now suppose that a and b are any two points. Going from a to b requires a shift of $b - a$. If $b - a > 0$, the distance between the points is $b - a$. But if $b - a < 0$, the distance is $-(b - a)$. In either case, the distance is $|b - a|$. See Fig. 4.

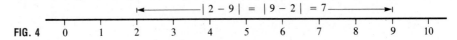

FIG. 4

The absolute value takes care of the sign; we don't have to worry about which point is to the left and which is to the right.

> The distance between points a and b on the number line is $|b - a|$.

With inequalities and absolute values, we can often express geometric statements algebraically or interpret algebraic statements geometrically. Here are a few examples.

ALGEBRAIC STATEMENT	GEOMETRIC STATEMENT
$a < b < c$	b is to the right of a and to the left of c
$b - a = 4$	b is 4 units to the right of a
$|a| < 10$	a is between -10 and 10
$|a - 5| \leq \frac{1}{2}$	a is at most $\frac{1}{2}$ unit from 5

EXAMPLE 4 Express the statement using absolute values and inequalities

(a) The distance between b and -3 is at least 4.

(b) a is closer to 0 than b is.

(c) a and b are between 1 and 2, and a is to the left of b.

SOLUTION (a) The distance between b and -3 is
$$|b - (-3)| = |b + 3|$$
"At least" is translated by $\geq$. Therefore, the statement is
$$|b + 3| \geq 4$$

(b) The distances from a to 0 and from b to 0 are $|a|$ and $|b|$. So the statement is
$$|a| < |b|$$

(c) We have
$$1 < a < 2 \qquad 1 < b < 2 \quad \text{and} \quad a < b$$
We combine these inequalities into one long string:
$$1 < a < b < 2$$

EXERCISES

Express without absolute values
1. $|-4|$
2. $-|-6|$
3. $|3-8|$
4. $|9-4|$
5. $|-2|+|-3|$
6. $|-7|-|-8|$
7. $|a^2+6|$
8. $|-c^2-2|$
9. $(-4)\cdot|-2|$
10. $|-5|-|-6|$
11. $\left|\dfrac{a-b}{b-a}\right|$
12. $\dfrac{-10}{|-10|}$

Find the distance on the number line between the points
13. 7 and 10
14. 8 and 3
15. 4 and -5
16. -6 and 6
17. -2.1 and -2.4
18. $-\tfrac{3}{4}\pi$ and $-\pi$

Express in mathematical notation as briefly as possible
19. a is positive
20. b is negative
21. c is at most 7
22. d is at least 2
23. $\tfrac{22}{7}$ is more than π
24. π^2 is less than 10
25. $(3.019)^2$ is between 9 and 9.12
26. $\tfrac{1}{7}$ is between 0.14 and 0.15
27. x is positive but not more than 5
28. x is non-negative but at most 2.78
29. a is less than b, and b is less than c
30. b is least a and at most c
31. a is at most 1, and b is between 1 and 2
32. a is between 5 and 8, and b is between a and 8
33. the distance from a to b is 3
34. the distance from c to 5 is greater than 1
35. x is more than 2.1 units to the left of y
36. p is more than 4.7 units to the right of q
37. x is closer to -1 than to 6
38. x is twice as far from 0 as from 7

Justify the inequality
39. $a^2+b^2 \geq 0$
40. $-a^2-b^2 \leq 0$
41. $5-a^2-b^2 \leq 5$
42. $4+3a^2 \geq 4$
43. $a^2+(a-3)^2 > 0$
44. $(a-1)^2+3(a+1)^2 > 0$

Show that
45. the product of 3 negative numbers is negative
46. the product of 4 negative numbers is positive
47. a and ab^2 have the same sign ($a \neq 0$, $b \neq 0$)

48. a and $3a(1 + a^2)$ have the same sign ($a \neq 0$)

49. $|a|^2 = a^2$ **50.** $|a - b| = |b - a|$

51. Explain why $a^2 + b^2 + c^2 > 0$ is mathematical shorthand for "at least one of the numbers a, b, c is different from 0."

52. Express the statement in quotes in Exercise 51 using absolute values.

6 ABOUT CALCULATORS

A hand-held calculator is a marvelous tool. Fifteen years ago, a computation such as
$$\frac{(9.8)(154.6) + (7.2)^5}{691.477} \doteq 30.173463$$
would have been drudgery; now it takes about 20 seconds on a calculator. Computations such as
$$3^{12} = 531441 \quad \text{and} \quad \sqrt{233.9} = 15.293789$$
are practically instantaneous. Furthermore, scientific calculators have built-in exponentials and logarithms, which decreases our dependence on tables of these functions.

Of course, a calculator cannot do our thinking for us. But once we have analyzed a problem, it can help us get a numerical answer painlessly, with no extra trouble if the answer does not "come out even."

We recommend a fairly inexpensive scientific calculator with the following types of keys:

(a) $\boxed{+}$ $\boxed{-}$ $\boxed{\times}$ $\boxed{\div}$ $\boxed{+/-}$

(b) $\boxed{x^2}$ $\boxed{\sqrt{}}$ $\boxed{1/x}$ $\boxed{y^x}$

(c) One memory together with the memory-related keys

$\boxed{\text{STO}}$ (also called $\boxed{x \to M}$ or $\boxed{\text{Min}}$)

$\boxed{\text{RCL}}$ (also called $\boxed{\text{MR}}$ $\boxed{\text{RM}}$)

$\boxed{\text{M}+}$ (also called $\boxed{\text{SUM}}$)

(d) Either $\boxed{\log}$ or $\boxed{10^x}$ together with $\boxed{\text{INV}}$ (also called $\boxed{\text{ARC}}$ or $\boxed{\text{F}}$ or $\boxed{\text{2nd}}$)

We will mention some of these keys as we go along. For a detailed discussion of calculator computations, see the Appendix.

REVIEW EXERCISES FOR CHAPTER 1

Compute

1. $2 - 3 \times 4 + 5 \times 6 - 7$ **2.** $1500 \times 36 \div 750 + 1$

3. $6 - \{1 - 3(2 - [4 - 7])\}$

4. $(42 + 44 + 46 + \cdots + 58 + 60) - (41 + 43 + 45 + \cdots + 57 + 59)$

5. $|2 + \pi| + |2 - \pi|$ **6.** $|(-3)(-4)(-5)|$

Express as a sum of terms

7. $(a + 1)(b + 2)$ **8.** $(a + b)(c + d + e)$

Express in algebraic notation, as briefly as possible

9. a and b are negative, and a is less than b

10. the average of c and d is at least 75

11. a is 6 units to the left of b

12. a and b are between 2 and 5, and a is greater than b

13. the distance between a and b exceeds 4

14. b is double the number that is 1 more than a

15. If $a^2b^2 = 0$, show that either $a = 0$ or $b = 0$ or both.

16. Explain why $-(-a) = a$.

17. If $|a| = |b|$, does $a = b$?

18. Show that $\frac{1}{2}(a + |a|) = \begin{cases} a & \text{if } a \geq 0 \\ 0 & \text{if } a < 0 \end{cases}$

19. If $|a| + |b| + |c| = 0$, what can you conclude about a, b, and c?

20. Does subtraction satisfy the commutative law? the associative law?

Fundamentals of Algebra

1 INTEGRAL EXPONENTS

In algebra, there is a standard abbreviation for a number multiplied by itself several times:

$$a^1 = a \qquad a^2 = a \cdot a \qquad a^3 = a \cdot a \cdot a \qquad a^n = \underbrace{a \cdot a \cdot a \cdot a \cdots a}_{n \text{ factors}}$$

Read a^n as "a to the n-th power" or "a to the n." The number n is an **exponent**. The number a^n is the n-th **power** of a.

Examples

$$5^2 = 5 \cdot 5 = 25 \qquad (-2)^3 = (-2) \cdot (-2) \cdot (-2) = -8$$

$$(\tfrac{3}{7})^4 = \tfrac{3}{7} \cdot \tfrac{3}{7} \cdot \tfrac{3}{7} \cdot \tfrac{3}{7} \qquad 10^5 = 10 \cdot 10 \cdot 10 \cdot 10 \cdot 10 = 100{,}000$$

$$10^n = \underbrace{10000 \cdots 00}_{n \text{ zeros}}$$

Look what happens when you multiply a^5 by a^3:

$$a^5 a^3 = \underbrace{(a \cdot a \cdot a \cdot a \cdot a)}_{5 \text{ factors}} \underbrace{(a \cdot a \cdot a)}_{3 \text{ factors}} = \underbrace{a \cdot a \cdot a \cdot a \cdot a \cdot a \cdot a \cdot a}_{8 \text{ factors}} = a^8$$

Thus $a^5 a^3 = a^8$. In the same way, when you multiply m factors of a by n factors of a, you get $m + n$ factors. Therefore,

$$a^m a^n = a^{m+n}$$

Hence, to multiply powers of a number, add exponents.

Now look what happens when you divide a^5 by a^3:

$$\frac{a^5}{a^3} = \frac{a \cdot a \cdot a \cdot a \cdot a}{a \cdot a \cdot a} = \frac{a \cdot a \cdot a}{a \cdot a \cdot a} \cdot \frac{a \cdot a}{1} = a^2$$

Thus, $a^5/a^3 = a^2$. You subtract exponents because the factors in the denominator cancel 3 of the factors in the numerator. In the same way,

$$\frac{a^m}{a^n} = a^{m-n} \qquad \text{when} \qquad m > n$$

For example,

$$\frac{10^7}{10^4} = 10^{7-4} = 10^3 \quad \text{that is,} \quad \frac{10,000,000}{10,000} = 1000$$

What if the denominator contains at least as many factors as the numerator? Can we still subtract exponents? That would require zero and negative exponents. Since $a^n/a^n = 1$, we should have $a^{n-n} = a^0 = 1$. So we define

$$\boxed{a^0 = 1 \qquad a \neq 0}$$

(The symbol 0^0 is undefined.)

Next, we observe that

$$\frac{a^3}{a^5} = \frac{a \cdot a \cdot a}{a \cdot a \cdot a \cdot a \cdot a} = \frac{1}{a \cdot a} = \frac{1}{a^2}$$

so we should have $a^{3-5} = a^{-2} = 1/a^2$. Hence, if n is a positive integer, we define

$$\boxed{a^{-n} = \frac{1}{a^n} \qquad a \neq 0}$$

Examples

$$3^{-1} = \frac{1}{3} \qquad 5^{-2} = \frac{1}{5^2} = \frac{1}{25} \qquad \left(\frac{1}{2}\right)^{-3} = \frac{1}{\left(\frac{1}{2}\right)^3} = \frac{1}{\frac{1}{8}} = 8$$

$$10^{-4} = \frac{1}{10^4} = \frac{1}{10,000} = 0.0001 \qquad 10^{-n} = 0.000 \cdots \underbrace{001}_{n\text{-th place}}$$

We have now defined a^n for every real number a and every integer n, except that $a \neq 0$ when $n \leq 0$. These are the basic rules for working with exponents.

Rules of Exponents If a and b are real numbers and m and n are integers, then

(1) $a^m a^n = a^{m+n}$ \qquad (2) $\dfrac{a^m}{a^n} = a^{m-n}$

(3) $(a^m)^n = a^{mn}$ \qquad (4) $(ab)^n = a^n b^n$

(5) $\left(\dfrac{a}{b}\right)^n = \dfrac{a^n}{b^n}$ \qquad [all denominators non-zero]

These rules hold for *all* integers m and n, not just positive ones.

Examples

Rule (3) $\quad (a^4)^3 = a^4 \cdot a^4 \cdot a^4 = a^{4+4+4} = a^{12} = a^{3 \cdot 4}$

$$(a^4)^{-3} = \frac{1}{(a^4)^3} = \frac{1}{a^{12}} = a^{-12} = a^{4(-3)}$$

26 Fundamentals of Algebra

$$\text{Rule (4)} \quad (ab)^3 = (ab)(ab)(ab) = (aaa)(bbb) = a^3b^3$$

$$(3d)^{-2} = \frac{1}{(3d)^2} = \frac{1}{3^2 d^2} = \frac{1}{3^2} \cdot \frac{1}{d^2} = 3^{-2} d^{-2}$$

$$\text{Rule (5)} \quad \left(\frac{x}{y}\right)^2 = \frac{x}{y} \cdot \frac{x}{y} = \frac{xx}{yy} = \frac{x^2}{y^2}$$

These rules can be extended in various ways. For example,

$$2^2 \cdot 2^3 \cdot 2^4 = 2^{2+3+4} = 2^9 \quad \text{[by Rule (1) extended]}$$
$$(10xy)^3 = 1000 x^3 y^3 \quad \text{[by Rule (4) extended]}$$

EXAMPLE 1 Use the rules of exponents to simplify
    (a) $(cd)^2(c^2d^3)^{-1}$ (b) $(x^2y^{-3})^{-5}$

SOLUTION

(a) $(cd)^2(c^2d^3)^{-1} = (c^2d^2)\dfrac{1}{c^2d^3}$ [Rule (4), definition of a^{-1}]

$$= \frac{c^2}{c^2} \cdot \frac{d^2}{d^3} = \underline{\frac{1}{d}}$$

(b) $(x^2y^{-3})^{-5} = (x^2)^{-5}(y^{-3})^{-5}$ [Rule (4)]

$\qquad\qquad\;\; = x^{2(-5)} y^{(-3)(-5)}$ [Rule (3)]

$\qquad\qquad\;\; = x^{-10} y^{15} = \underline{\dfrac{y^{15}}{x^{10}}}$

EXAMPLE 2 Express $\dfrac{2^{-5} \cdot 8^7}{4^3 \cdot 16}$ as a power of 2.

SOLUTION Replace 8 by 2^3, 4 by 2^2, and 16 by 2^4. Then use the rules of exponents.

$$\frac{2^{-5} \cdot 8^7}{4^3 \cdot 16} = \frac{2^{-5}(2^3)^7}{(2^2)^3(2^4)} = \frac{2^{-5} \cdot 2^{21}}{2^6 \cdot 2^4} \quad \text{[Rule (3)]}$$

$$= \frac{2^{16}}{2^{10}} \quad \text{[Rule (1)]}$$

$$= \underline{2^6} \quad \text{[Rule (2)]}$$

EXAMPLE 3 Compute $\dfrac{2^6 \cdot 5^7}{25 \cdot 10^4}$

SOLUTION Write $25 = 5^2$ and $10 = 2 \cdot 5$. Then everything will be in powers of 2 and 5:

$$\frac{2^6 \cdot 5^7}{25 \cdot 10^4} = \frac{2^6 \cdot 5^7}{5^2(2 \cdot 5)^4} = \frac{2^6 \cdot 5^7}{5^2 \cdot 2^4 \cdot 5^4}$$

$$= \frac{2^6 \cdot 5^7}{2^4 \cdot 5^6}$$

$$= 2^2 \cdot 5^1 = 4 \cdot 5 = \underline{20}$$

Scientific Notation

Exponents provide an efficient way of writing and computing with very large or very small numbers, such as

$$32,000,000,000 \quad 1,876,000 \quad 0.00000\ 00000\ 006$$

Imagine multiplying such numbers as they are written!

In **scientific notation,** each positive number is expressed in the form $c \times 10^n$, where $1 \le c < 10$ and n is an appropriate exponent.

Examples

$$140 = 1.4 \times 10^2 \qquad 0.05 = 5 \times 10^{-2}$$
$$2550 = 2.55 \times 10^3 \qquad 0.0031 = 3.1 \times 10^{-3}$$
$$1,876,000 = 1.876 \times 10^6 \qquad 0.000988 = 9.88 \times 10^{-4}$$

The mass of a neutron is approximately

$$0.00000\ 00000\ 00000\ 00000\ 00016 = 1.6 \times 10^{-24} \text{ gram}$$

A certain fast-food chain has sold

$$37 \text{ billion} = 3.7 \times 10^{10} \text{ hamburgers}$$

EXAMPLE 4 Multiply $(140)(32,000,000,000)(0.00000\ 00000\ 006)$.

SOLUTION Express each factor in scientific notation. Then multiply and group all the powers of 10 together:

$$(1.4 \times 10^2)(3.2 \times 10^{10})(6 \times 10^{-13}) = (1.4)(3.2)(6) \times 10^{2+10-13}$$
$$= 26.88 \times 10^{-1} = \underline{2.688}$$

EXAMPLE 5 Compute $\dfrac{(14,000)(0.00003)(8,800,000)}{(1100)(0.000002)}$

SOLUTION

$$\frac{(1.4 \times 10^4)(3 \times 10^{-5})(8.8 \times 10^6)}{(1.1 \times 10^3)(2 \times 10^{-6})} = \frac{(1.4)(3)(8.8)}{(1.1)(2)} \times 10^{4-5+6-3+6}$$
$$= 16.8 \times 10^8$$

This is the answer. It is not in scientific notation because 16.8 is not between 1 and 10. If you prefer scientific notation, write 16.8 as 1.68×10. Then the answer is

$$16.8 \times 10^8 = (1.68 \times 10) \times 10^8 = 1.68 \times 10^{1+8} = \underline{1.68 \times 10^9}$$

Scientific calculators display very large or very small numbers in scientific notation. For example, the display

$$\boxed{3.940\ 16\ 67}$$

means 3.94016×10^{67}. Your calculator has a key such as $\boxed{\text{EXP}}$ or $\boxed{\text{EE}}$ for entering a power of 10 (usually up to 10^{99} or down to 10^{-99}). To key in the number shown, press

$$3.94016 \boxed{\text{EXP}} 67$$

To key in 1.6×10^{-24}, press

$$1.6 \boxed{\text{EXP}} 24 \boxed{+/-}$$

To key in 10^{12} you must press

$$1 \boxed{\text{EXP}} 12$$

(Without the 1, the calculator treats this number as $0 \times 10^{12} = 0$.)

You do calculations with these numbers in the usual way. To multiply 3.94016×10^{67} by 1.6×10^{-24}, key in

$$3.94016 \boxed{\text{EXP}} 67 \boxed{\times} 1.6 \boxed{\text{EXP}} 24 \boxed{+/-} \boxed{=}$$

(Result: 6.304256×10^{43})

Calculating Powers

Note the key $\boxed{y^x}$ on your calculator. It is used for the (approximate) calculation of powers. To calculate y^x first key in y, then $\boxed{y^x}$, then the exponent x, and finally $\boxed{=}$.

Examples

$(3.52)^7$	$3.52 \boxed{y^x} 7 \boxed{=}$	6695.7409
$(145.3)^{-12}$	$145.3 \boxed{y^x} 12 \boxed{+/-} \boxed{=}$	$1.1293086 \times 10^{-26}$

For a further discussion of $\boxed{y^x}$, see the Appendix.

EXERCISES

Compute

1. 3^3 **2.** $(-3)^3$ **3.** 6^{-2} **4.** 2^{-5} **5.** $(-\frac{2}{3})^5$ **6.** $(-\frac{3}{2})^{-1}$ **7.** $(\frac{1}{3})^0$ **8.** $(\frac{1}{2})^{-2}$

9. $2^{-3}(\frac{1}{3})^2$ **10.** $\dfrac{2^5 \cdot 5^3}{2^6 \cdot 5^{-2}}$ **11.** $2^4 \cdot 3^2 \cdot 6^{-2} \cdot 8^{-1}$ **12.** $10^4 \cdot 25^{-3} \cdot 16^{-1}$

Express as a power of 2

13. 4^2 **14.** 8^{-3} **15.** $(\frac{1}{2})^7$ **16.** $(\frac{1}{64})^2$ **17.** $2 \cdot 4^2 \cdot 8^3$ **18.** $(8^5 \cdot 16^{-2})^3$

19. $\dfrac{64^5}{8^{10}(4 \cdot 32^2)^{-1}}$ **20.** $(\frac{1}{2})^{-8}\left(\dfrac{4^{-4} \cdot 32}{2 \cdot 16^{-1}}\right)$

Express as simply as possible, without negative exponents

21. $(2a^2)(3a^3)$ **22.** $b^4(3b^5)(8b^{10})$ **23.** $\dfrac{10x^5}{2x^2}$

24. $\dfrac{-8x^{10}}{(2x^3)^3}$ **25.** $(-5c)(4c^4)^2$ **26.** $(\frac{1}{2}d)(\frac{4}{9}d^2)(6d^5)$

27. $(3u^{-4})(5u^2)^{-1}$
28. $(2v^4)^{-3}(\tfrac{1}{2}v)^2$
29. $\left(\dfrac{-4x^{-3}}{5}\right)^{-1}$
30. $x^5(x^{-4})^{-2}$
31. $\dfrac{(ab)^6}{ab^2}$
32. $\dfrac{(3a)^3}{(2ab^2)^2}$
33. $(aba^{-4})^2(a^3b)^0$
34. $(-5a^2b^{-3})^{-20}$
35. $(2b^2c^2)^3(-3b^4c)^{-2}$
36. $(ab)^{-5}(2ab^2)(4a^3b)^{-1}$
37. $\left(\dfrac{r^2s}{3rs^3}\right)\left(\dfrac{2r^3s^4}{3r}\right)$
38. $\left(\dfrac{s}{3t^2}\right)^2\left(\dfrac{18s^4t^4}{st^2}\right)$
39. $\dfrac{x^{-2}}{y^{-2}}+\dfrac{y^2}{x^2}$
40. $\left(\dfrac{x}{y}\right)^2\left(\dfrac{y}{z}\right)^3\left(\dfrac{z}{x}\right)^4$
41. $\dfrac{(2pqr)^2}{(p^3q)^2(pr)^{-1}}$
42. $pq^2r^3\left(\dfrac{2p}{qr^3}\right)^{-2}$

Express in scientific notation

43. 62
44. 908
45. 0.0081
46. 0.000076
47. 215,400
48. 55,718,000
49. 2.37
50. 3.005
51. 452,000,000,000,000
52. 0.00000 00004
53. 813×10^9
54. 1295×10^{-18}

Compute and express the result in scientific notation

55. $(180)(30,000,000)(0.00012)$
56. $\dfrac{(20,100)(0.006)}{(0.0000002)(402,000)}$
57. $(0.002)^3(0.00004)(0.000005)(6,000,000,000)$
58. $\dfrac{1}{(800)(200,000)^2(0.00001)^4}$
59. $(200,000)^6$
60. $(10)(200)(3000)(40,000)(500,000)$
61. $\dfrac{(0.0000002)^2}{(5000)^4}$
62. $(8000)(30,000)^5(0.000001)^6$
63. the number of inches in 100 miles
64. the number of cubic centimeters in a cubic kilometer
65. the total mass in grams of 3,000,000 neutrons
66. the length in kilometers of a light-year, the distance light travels in a year. Take the speed of light to be 300,000 km/sec and a year to be 365 days.

Compute on a calculator

67. $(3.5)^{10}$
68. $(1.723)^{-35}$
69. $(9.23 \times 10^{12})(8.41 \times 10^7)$
70. $(4.5 \times 10^{17}) \div (3.18 \times 10^{23})$
71. $(1.5708 \times 10^7)^{-9}$
72. $(2.718 \times 10^{-12})^5$

Determine with a calculator which is larger:

73. 5^{50} or 19^{27}
74. 137^{18} or 3^{81}

2 POLYNOMIALS

In this section we study algebraic expressions such as
$$2x - 3 \qquad x^2 + 6x + 5 \qquad 4x^3 - 11x^2 - 9x + \tfrac{1}{2}$$
Here x represents any real number. We call x a **variable** whose **domain** is the set of all real numbers.

A **monomial** is an expression of the form ax^n, where a is a real number and n is an integer, $n \geq 0$; for example,
$$3 \qquad 2x \qquad x^2 \qquad 4x^3 \qquad (-6)x^4 \qquad \tfrac{1}{8}x^5$$
A monomial ax^n has **degree** n and **coefficient** a. A **polynomial** is a sum of one or more monomials. Usually, we write a polynomial according to descending degrees:
$$a_n x^n + a_{n-1} x^{n-1} + \cdots + a_2 x^2 + a_1 x + a_0 \qquad a_n \neq 0$$
The **degree** of this polynomial is n, the highest degree of the monomials it contains. The numbers $a_n, a_{n-1}, \ldots, a_2, a_1, a_0$ are the **coefficients** of the polynomial. The term $a_n x^n$ is the **leading term**, and a_n is the **leading coefficient**. Thus,
$$9x^6 + x^5 + 3x^2 + 2x$$
is a polynomial of degree 6, with leading term $9x^6$ and leading coefficient 9. (The leading coefficient is always non-zero, but other coefficients may be zero, as are a_4, a_3, and a_0 in this example.)

A non-zero constant is a polynomial of degree zero. The constant 0 is also considered a polynomial but is not assigned a degree. A polynomial with two terms is a **binomial**; one with three terms is a **trinomial**.

When a coefficient of a polynomial is negative, we use a minus sign before that term. For example, instead of
$$x^4 + (-7)x^3 + (-4)x^2 + (-1)x + 3$$
we write
$$x^4 - 7x^3 - 4x^2 - x + 3$$

Polynomial Algebra

Polynomials are sums; we add and subtract them like any other sums.

EXAMPLE 1 Compute (a) $(2x^2 + 6x + 5) + (3x^2 - 8x - 1)$
(b) $(x^4 + 4x^2 + x + 12) - (3x^2 + 7x)$

SOLUTION Use the commutative and associative laws to rearrange and regroup the terms. Then use the distributive law to combine like powers of x:

(a) $(2x^2 + 6x + 5) + (3x^2 - 8x - 1) = (2x^2 + 3x^2) + (6x - 8x) + (5 - 1)$
$$= (2 + 3)x^2 + (6 - 8)x + (5 - 1)$$
$$= 5x^2 - 2x + 4$$

(b) $(x^4 + 4x^2 + x + 12) - (3x^2 + 7x) = x^4 + 4x^2 + x + 12 - 3x^2 - 7x$
$$= x^4 + (4x^2 - 3x^2) + (x - 7x) + 12$$
$$= x^4 + x^2 - 6x + 12$$

> To add or subtract polynomials, add or subtract coefficients of like powers.

We multiply polynomials like any other sums, using the distributive law and the rule $x^m \cdot x^n = x^{m+n}$.

EXAMPLE 2 Compute (a) $(3x + 2)(5x - 1)$ (b) $(x - 2)(4x^2 + 5x + 6)$

SOLUTION

(a) $(3x + 2)(5x - 1) = (3x + 2)(5x) + (3x + 2)(-1)$
$= (3x)(5x) + (2)(5x) + (3x)(-1) + (2)(-1)$
$= 15x^2 + 10x - 3x - 2$
$= \underline{15x^2 + 7x - 2}$

(b) Same technique, only this time you get six terms instead of four:
$(x - 2)(4x^2 + 5x + 6) = (x - 2)(4x^2) + (x - 2)(5x) + (x - 2)(6)$
$= (4x^3 - 8x^2) + (5x^2 - 10x) + (6x - 12)$
$= \underline{4x^3 - 3x^2 - 4x - 12}$

> To multiply polynomials, multiply each term of the first by each term of the second and use the rule of exponents, $x^m \cdot x^n = x^{m+n}$.

For example, the product $(3x - 4)(x^3 + x)$ has 4 terms computed this way:

$$(3x - 4)(x^3 + x) = 3x^4 - 4x^3 + 3x^2 - 4x$$

EXAMPLE 3 Express in the form $a_n x^n + \cdots + a_1 x + a_0$
(a) $(3x - 4)(x^3 + x) + (2x + 1)(3x + 1)$
(b) $(x + 3)(x + 5)(4x - 1)$

SOLUTION (a) Compute each product, then add:
$(3x - 4)(x^3 + x) + (2x + 1)(3x + 1)$
$= (3x^4 - 4x^3 + 3x^2 - 4x) + (6x^2 + 2x + 3x + 1)$
$= 3x^4 - 4x^3 + (3 + 6)x^2 + (-4 + 2 + 3)x + 1$
$= \underline{3x^4 - 4x^3 + 9x^2 + x + 1}$

(b) By the associative law the product is $[(x + 3)(x + 5)](4x - 1)$. Compute the product in brackets, then multiply by $4x - 1$:
$[(x + 3)(x + 5)](4x - 1) = [x^2 + 3x + 5x + 15](4x - 1)$
$= (x^2 + 8x + 15)(4x - 1)$
$= 4x^3 - x^2 + 32x^2 - 8x + 60x - 15$
$= \underline{4x^3 + 31x^2 + 52x - 15}$

EXAMPLE 4 Find the coefficient of x^2 in the product
$(5x^3 + 2x^2 + 7x - 3)(8x^4 - 9x^3 + x^2 + 2x - 12)$

SOLUTION No need to multiply it all out. Just collect all products of terms that produce x^2:

$$(5x^3 + 2x^2 + 7x - 3)(8x^4 - 9x^3 + x^2 + 2x - 12)$$

The desired terms are

$$(2x^2)(-12) + (7x)(2x) + (-3)(x^2) = (-24 + 14 - 3)x^2 = -13x^2$$

Answer -13

EXERCISES

Compute

1. $(x + 4) + (5x - 3)$
2. $(3x - 2) + (x + 6)$
3. $(8x^2 + 5x + 6) + (3x - 4)$
4. $(x^2 + x + 7) - (4x + 1)$
5. $(3x^2 + 7x - 2) - (9x^2 + x + 1)$
6. $(15x^2 - 8x - 11) - (-10x^2 + 4x + 5)$
7. $(\frac{1}{3}x^2 - 2x - 6) + (\frac{4}{3}x^2 - \frac{1}{2})$
8. $(4.1x^2 + 7.05x) - (2x^2 - 1.34x - 0.02)$
9. $(x^3 + 2x^2 - 4x - 5) + (6x^2 - 12x + 1)$
10. $(x^4 + 7x^3 + 8x^2 + 2x + 1) - (3x^3 + 4x^2 + 2x - 1)$
11. $(-x^5 + x + 9) - 2(x^4 + x^3 + x^2 + x + 1)$
12. $3(x^5 - 4x^4 - 6x^3 - 7) - (x^5 + x^4 + 7x^2 + x)$
13. $5(x^8 + 6x^6 + 4x^4 + 2x^2) + 6(x^6 - 5x^5 - 4x^3)$
14. $3(-x^{10} + x + 1) + 4(x^3 - 4x^2 + 2x + 2)$

Multiply

15. $(x + 2)(x - 3)$
16. $(x - 4)(x - 7)$
17. $(2x + 5)(3x + 1)$
18. $(8x - 3)(x + 2)$
19. $(x + 6)(x^2 + x - 1)$
20. $(2x - 1)(x^2 + 3x + 2)$
21. $(x^3 + x)(x^2 + 1)$
22. $(2x^3 - 1)(-x^2 + x)$
23. $(x^5 + \frac{1}{2}x^2)(x^2 - x - \frac{1}{2})$
24. $(x^4 - 3x)(2x^2 + 5x + 1)$
25. $x(x + 1)(x^2 + x + 1)$
26. $(x + 1)(x + 2)(x + 3)$
27. $(x - 1)(x^4 + x^3 + x^2 + x + 1)$
28. $(x + 1)(x^4 - x^3 + x^2 - x + 1)$
29. $(1 + x)(1 + x^2)$
30. (cont.) $(1 + x)(1 + x^2)(1 + x^4)$
31. (cont.) Can you see a pattern?

32. $(1 - x)(1 + x)$

33. (cont.) $(1 - x)(1 + x)(1 + x^2)$

34. (cont.) $(1 - x)(1 + x)(1 + x^2)(1 + x^4)$

35. (cont.) Can you see a pattern?

36. $(1 - x^2)(1 + x + x^2 + x^3 + x^4)$

37. $(x^2 + x + 1)(x^2 - x + 1)$

38. $(x^2 + 2x + 2)(x^2 - 2x + 2)$.

Express as $a_n x^n + \cdots + a_1 x + a_0$

39. $(2x + 3)(x^2 + 1) + (x - 1)(x - 2)$

40. $x(x + 1)(x^2 + 1) - (2x + 1)(2x - 1)$

41. $(x + 2)(4x + 5) + 3(2x - 1)(3x - 1) - 4(x + 1)(x - 3)$

42. $(x + 3)(x^2 - 5x - 4) + 2(x^3 + 1)(x^4 + 1) - x^4(2x + 1)(x^2 + 5)$

43. $(x + a)(x + b)(x + c)$

44. $(x + a)^3$

Find the leading term

45. $(5x^2 + 6x + 7)(2x^4 - x^3 - 3x^2 + x + 12)$

46. $(2x^6 - x^3 - 4)(3x^3 + 2x^2 + x + 1)$

47. $(x^2 + 4x + 1)(3x^3 + 2x + 6)(x^4 + 8x - 7)$

48. $(x + 2)^3(6x - 1)^2$

49. $(x^2 + 3x + 6)(4x^2 - 9x - 1) - (x + 2)(3x^2 + 4) - (2x + 1)(x^3 + x + 5)$

50. $(x^2 + 4)(x^3 + 5) - 2x^3(x + 1)(x - 6) + x(x^2 - 1)(x^2 - 4)$

Find the coefficient of x^3 without computing the whole product

51. $(x^2 + 3x + 1)(2x - 1)$.

52. $(x + 1)(2x - 1)(4x - 1)$

53. $(x^3 - 7x + 6)(x^2 - 4x - 1)$

54. $(2x^4 + 6x^3 - x^2 - x - 7)(x^2 + 3x + 3)$

55. $(1 + x^3)(1 + x^4)(1 + x^5)(1 + x^6)$

56. $(1 + 2x)(1 + 3x^2)(1 + 4x^3)(1 + 5x^4)$

True or false? Why?

57. The product of a polynomial of degree m and a polynomial of degree n is a polynomial of degree $m \cdot n$.

58. Suppose each of two polynomials contain only even powers of x. Then their product is a polynomial with the same property.

59. Suppose n is a positive integer, $n \leq 100$. Show that the number of ways you can make n cents using pennies, nickels, and dimes is the same as the coefficient of x^n in the polynomial

$$(1 + x + x^2 + x^3 + \cdots + x^{100})(1 + x^5 + x^{10} + x^{15} + \cdots + x^{100})$$
$$(1 + x^{10} + x^{20} + x^{30} + \cdots + x^{100})$$

60. Suppose you roll 3 dice of different colors. Show that the number of ways you can make a total of n is the same as the coefficient of x^n in the polynomial

$$(x + x^2 + x^3 + x^4 + x^5 + x^6)^3$$

3 POLYNOMIALS IN SEVERAL VARIABLES

A **monomial** in two variables x and y is an expression of the form $ax^k y^m$, where a is a real number and k and m are non-negative integers. Its **degree** is the sum of the exponents, $k + m$. Thus,

$$4xy \qquad -3y^4 \qquad xy^3 \qquad 2x^5 y^2$$

are monomials of degrees 2, 4, 4, and 7, respectively.

A **polynomial** in two variables x and y is a sum of one or more monomials in x and y. Its **degree** is the highest degree of the monomials it contains. For example,

$$5x - y + 2 \qquad x^2 + xy + 4y^2 + 2x \qquad \tfrac{1}{2}x^4 y^2 - x^3 y - 6x^2 y^2 + 3y + 4$$

are polynomials in two variables of degrees 1, 2, and 6, respectively.

Similarly, a **polynomial** in three variables x, y, and z is a sum of **monomials** of the form $ax^k y^m z^n$, of **degree** $k + m + n$. For example,

$$x + 3y + 2z - 7 \qquad x^2 + y^2 + z^2 - 6xy + y + 2$$
$$y^2 z^2 + 3xyz^2 - xyz - x^2 + 4z$$

are polynomials in three variables of degrees 1, 2, and 4, respectively.

EXAMPLE 1 Compute
(a) $(x^2 + 3y^2 + xy + 2x) + (4x^2 - y^2 + 2xy + 1)$
(b) $(4x^3 + x^2 y + xy^2 + z^2 + xyz) - (4x^3 - 2x^2 y - xyz + 5xz)$

SOLUTION (a) Rearrange and combine similar terms

$(x^2 + 3y^2 + xy + 2x) + (4x^2 - y^2 + 2xy + 1)$
$\qquad = (x^2 + 4x^2) + (3y^2 - y^2) + (xy + 2xy) + 2x + 1$
$\qquad = \underline{5x^2 + 2y^2 + 3xy + 2x + 1}$

(b) $(4x^3 + x^2 y + xy^2 + z^2 + xyz) - (4x^3 - 2x^2 y - xyz + 5xz)$
$\qquad = (4x^3 - 4x^3) + (x^2 y + 2x^2 y) + xy^2 + z^2 + (xyz + xyz) - 5xz$
$\qquad = \underline{3x^2 y + xy^2 + z^2 + 2xyz - 5xz}$

EXAMPLE 2 Multiply
(a) $(2x - y)(x^2 + xy + 3y^2)$ (b) $(xy + z)(xy^2 + y^2 z^2 + y)$

SOLUTION (a) Multiply each term in the first parentheses by each term in the second, use rules of exponents, then collect similar terms:

$(2x - y)(x^2 + xy + 3y^2)$
$\qquad = 2x \cdot x^2 - y \cdot x^2 + 2x \cdot xy - y \cdot xy + 2x \cdot 3y^2 - y \cdot 3y^2$
$\qquad = 2x^3 + (-x^2 y + 2x^2 y) + (-xy^2 + 6xy^2) - 3y^3$
$\qquad = \underline{2x^3 + x^2 y + 5xy^2 - 3y^3}$

(b) $(xy + z)(xy^2 + y^2 z^2 + y)$
$\qquad = xy \cdot xy^2 + z \cdot xy^2 + xy \cdot y^2 z^2 + z \cdot y^2 z^2 + xy \cdot y + z \cdot y$
$\qquad = \underline{x^2 y^3 + xy^2 z + xy^3 z^2 + y^2 z^3 + xy^2 + yz}$

3 Polynomials in Several Variables 35

Perfect Squares

If we square $x + y$, we obtain
$$(x + y)^2 = (x + y)(x + y) = x^2 + xy + yx + y^2 = x^2 + 2xy + y^2$$
Similarly,
$$(x - y)^2 = x^2 - 2xy + y^2$$

$$(x + y)^2 = x^2 + 2xy + y^2$$
$$(x - y)^2 = x^2 - 2xy + y^2$$

These formulas hold for all real numbers x and y. Hence they hold if x and y are replaced by specific real numbers or by any expressions that represent real numbers.

Examples
$$102^2 = (100 + 2)^2 = 100^2 + 2 \cdot 100 \cdot 2 + 2^2 = 10{,}000 + 400 + 4 = 10{,}404$$
$$(a - \tfrac{1}{3})^2 = a^2 - 2 \cdot a \cdot \tfrac{1}{3} + (\tfrac{1}{3})^2 = a^2 - \tfrac{2}{3}a + \tfrac{1}{9}$$
$$(2a + 3b)^2 = (2a)^2 + 2(2a)(3b) + (3b)^2 = 4a^2 + 12ab + 9b^2$$

EXAMPLE 3 Compute (a) $\left(4x + \dfrac{1}{x}\right)^2$ (b) $(x + y + z)^2$

SOLUTION

(a) $\left(4x + \dfrac{1}{x}\right)^2 = (4x)^2 + 2(4x)\left(\dfrac{1}{x}\right) + \left(\dfrac{1}{x}\right)^2 = \underline{16x^2 + 8 + \dfrac{1}{x^2}}$

(b) One approach is to multiply out $(x + y + z)(x + y + z)$. Another is to treat the trinomial $x + y + z$ as a binomial $(x + y) + z$ and square it:
$$[(x + y) + z]^2 = (x + y)^2 + 2(x + y)z + z^2$$
$$= (x^2 + 2xy + y^2) + (2xz + 2yz) + z^2$$
$$= \underline{x^2 + y^2 + z^2 + 2xy + 2xz + 2yz}$$

Recognizing squares is important. For example, spotting that $4a^2 + 12ab + 9b^2$ is really $(2a + 3b)^2$ in sheep's clothing might help to simplify a computation. When you suspect that an expression is the square of a binomial, look for three terms in the form
$$(\text{first})^2 \pm 2(\text{first})(\text{second}) + (\text{second})^2$$

EXAMPLE 4 Express as a perfect square, if possible,
(a) $x^2 - 10x + 25$ (b) $36x^2 + 6xy + \tfrac{1}{4}y^2$ (c) $9x^2 + 8xy + 16y^2$

SOLUTION Try to write the expression as $a^2 \pm 2ab + b^2$.
(a) $x^2 - 10xy + 5^2 = x^2 - 2(x)(5) + 25 = \underline{(x - 5)^2}$
(b) $36x^2 + 6xy + \tfrac{1}{4}y^2 = (6x)^2 + 2(6x)(\tfrac{1}{2}y) + (\tfrac{1}{2}y)^2 = \underline{(6x + \tfrac{1}{2}y)^2}$

(c) $9x^2 + 8xy + 16y^2 = (3x)^2 + 8xy + (4y)^2$

Here the middle term does not fit the pattern, because $8xy \neq 2(3x)(4y)$. Hence the expression is <u>not a perfect square</u>.

Difference of Squares

An especially useful formula is one for the product of the sum and difference of two quantities, $x + y$ and $x - y$:

$$(x + y)(x - y) = x^2 - xy + yx - y^2 = x^2 - y^2$$

Thus, the product is the difference of squares.

$$\boxed{(x + y)(x - y) = x^2 - y^2}$$

EXAMPLE 5 Compute

(a) $(3u^2 + v)(3u^2 - v)$ (b) $103 \cdot 97$ (c) $(x + y + z)(x + y - z)$

SOLUTION (a) Use the formula in the box with $x = 3u^2$ and $y = v$:

$$(3u^2 + v)(3u^2 - v) = (3u^2)^2 - v^2 = \underline{9u^4 - v^2}$$

(b) Note that $103 = 100 + 3$ and $97 = 100 - 3$. Hence

$$103 \cdot 97 = (100 + 3)(100 - 3) = 100^2 - 3^2$$
$$= 10,000 - 9 = \underline{9991}$$

(c) The product is

$$[(x + y) + z][(x + y) - z] = (x + y)^2 - z^2$$
$$= \underline{x^2 + 2xy + y^2 - z^2}$$

EXERCISES

Compute

1. $(x + 2y + 3) + (4x - y - 1)$
2. $(2z - y + 8) + (3x + 2y - 5)$
3. $(-4x + 5y + 6z) - (x - y + z + 2)$
4. $(y + 2z) - (3x + y + z + 1)$
5. $(x^2 + 3y^2) + (x^2 + y^2 - 3y)$
6. $(x^2 + 6xy + 2y^2) - (x^2 + 4xy - y^2 - x - 2)$
7. $(x^2 + y^2 + z^2 + xy + yz + xz) - (3x^2 + y^2 + z^2 + xy + x + y)$
8. $x(3x^2 + xy^3 + yz^4) - y(xz^4 + xy + x^2y^2)$

Multiply

9. $(x + y)(2x + 3y)$
10. $(2x - y)(3x - y)$
11. $(5x + 2y)(5x - 2y)$
12. $(\frac{1}{2}x - \frac{1}{3}y)(2x - 3y)$
13. $(x + y + z)(2x - y)$
14. $(y - z)(3x + 4y + z)$
15. $(x^4 - y^4)(x^4 + y^4)$
16. $(x - y)(x + y)(x^2 + y^2)$

17. $(a + 2b + 3)(a + 2b - 3)$
18. $(a + b + c + d)(a + b - c - d)$
19. $(x^2 + y^2)(x^2 - 3y^2)$
20. $(x^2 + 4)(2x^2 + y^2)$
21. $(r + 1)(s + 1)(t + 1)$
22. $(r + t)(r + 2t)(r + 3t)$
23. $(x - y)(x^2 + xy + y^2)$
24. $(x + y)(x^2 - xy + y^2)$
25. $(x + y)^3$
26. $(x - y)^3$

Find the coefficient of x^2y in the product
27. $(2x^2 + 3x - 4)(y^2 + 8y + 1)$
28. $(x + y)(x^2 + y^2)(x + 1)$
29. $(x + 4y)^3$
30. $(1 + x + y + xy + x^2 + y^2)(1 + 2x + 3y + 2xy + x^2)$

Compute the square
31. $(x - 4y)^2$
32. $(x + 5y)^2$
33. $(6a + 5b)^2$
34. $(7a - 2b)^2$
35. $(u^2 + v^2)^2$
36. $\left(u^2v^2 - \dfrac{3}{uv}\right)^2$
37. $(x + y - z)^2$
38. $(4x - y - 3z)^2$

Mental arithmetic: Compute
39. 105^2
40. 999^2
41. $1003 \cdot 997$
42. $5001 \cdot 4999$

Is the expression a perfect square? If so, of what?
43. $x^2 + 16x + 64$
44. $y^4 + 2y^2 + 1$
45. $z^6 + 4z^3 + 4$
46. $16a^2 + 8a + 1$
47. $x^2y^2 + 3xy + 9$
48. $9c^2 - 30cd + 25d^2$
49. $(x^2 + 1)^2 + 2(x^2 + 1) + 1$
50. $9u^2 + 9u + 1$
51. $a^2b^8 - 4ab^4c^2 + 4c^4$
52. $x^2 + 2xy + y^2 + 4xz + 4yz + 4z^2$

Explain these number games
53. Take a number from 1 to 10. Square the number one larger and square the number one smaller. Subtract. Divide by the number you started with. Now you have 4.
54. Multiply your number by the number 4 larger. Add 4. Take the square root. Subtract the number you started with. Now you have 2.
55. If n is a positive integer, show that $3n^2 + 6n + 5$ is the sum of the squares of three consecutive integers.
56. Consider two consecutive positive integers. Which is bigger, the average of their squares or the square of their average?

4 FACTORING

Factoring is the process of expressing a quantity as a product of factors. When we write
$$105 = 3 \cdot 5 \cdot 7$$

we are factoring the integer 105 into a product of smaller integers. When we write
$$x^2 + 5x + 6 = (x + 2)(x + 3)$$
we are factoring the polynomial $x^2 + 5x + 6$ into a product of polynomials of lower degree.

In this section we discuss factoring of polynomials with integer coefficients into products of polynomials also with integer coefficients. Factoring is useful in simplifying algebraic expressions, in solving equations, and in numerical work.

Common Factors

Recall the distributive law for real numbers:
$$a(b + c) = ab + ac$$
When read from left to right, this is a multiplication formula. But when read from *right to left*, it is a formula for factoring the sum $ab + ac$. It allows us to "factor out" or "remove" the common factor a from $ab + ac$. In a similar way, we can remove common polynomial factors. For example
$$3 + 6x^2 = 3(1 + 2x^2) \qquad x^4 - 6x^3 - 2x^2 = x^2(x^2 - 6x - 2)$$
$$(x - 3)(5x + 6) + 2(x - 3)(x - 4) = (x - 3)[(5x + 6) + 2(x - 4)]$$
$$= (x - 3)(7x - 2)$$

EXAMPLE 1 Factor

(a) $(x + 1)(x + 2)(x + 3) - x(x + 1)(x + 2)$
(b) $2x^5y^3 - 6xy^4 + 4x^2y^2$

SOLUTION (a) A common factor is $x + 1$. Another is $x + 2$. Hence the product $(x + 1)(x + 2)$ is a common factor. Removing it gives
$$(x + 1)(x + 2)(x + 3) - x(x + 1)(x + 2) = (x + 1)(x + 2)[(x + 3) - x]$$
$$= \underline{3(x + 1)(x + 2)}$$

(b) Common factors are 2, x, and y^2, so $2xy^2$ is a common factor. Removing it gives
$$2x^5y^3 - 6xy^4 + 4x^2y^2 = \underline{2xy^2(x^4y - 3y^2 + 2x)}$$

Second Degree Polynomials

If a polynomial $mx^2 + nx + p$ with integer coefficients can be factored, then it can be expressed in the form $(ax + b)(cx + d)$, where a, b, c, d are integers. Since
$$(ax + b)(cx + d) = acx^2 + (ad + bc)x + bd$$
factoring $mx^2 + nx + p$ amounts to finding integers a, b, c, d such that
$$ac = m \qquad ad + bc = n \qquad bd = p$$
This can involve a certain amount of trial and error.

Similarly, if a polynomial $mx^2 + nxy + py^2$ can be factored, then it can be expressed in the form $(ax + by)(cx + dy)$.

EXAMPLE 2 Factor (a) $x^2 + 10x + 9$ (b) $2x^2 - 7x + 3$

SOLUTION (a) The coefficient of x^2 is 1, so we try $(x + b)(x + d)$. Now
$$(x + b)(x + d) = x^2 + (b + d)x + bd$$
so we need integers b and d such that
$$b + d = 10 \qquad bd = 9$$
Clearly $b = 1$ and $d = 9$ works. Therefore,
$$x^2 + 10x + 9 = \underline{(x + 1)(x + 9)}$$
as is easily checked by multiplication.

(b) The coefficient of x^2 is 2, so let us try $(x + b)(2x + d)$. Now,
$$(x + b)(2x + d) = x^2 + (2b + d)x + bd$$
so we need integers b and d such that
$$2b + d = -7 \qquad bd = 3$$
After a little trial and error, we find that $b = -3$ and $d = -1$ do the trick. Hence,
$$2x^2 - 7x + 3 = \underline{(x - 3)(2x - 1)}$$

EXAMPLE 3 Factor (a) $3x^2 - 8xy + 4y^2$ (b) $6x^2 + 19xy + 8y^2$

SOLUTION (a) We try $(3x + by)(x + dy)$. Since
$$(3x + by)(x + dy) = 3x^2 + (b + 3d)xy + bdy^2$$
we need integers b and d such that
$$b + 3d = -8 \qquad bd = 4$$
From $bd = 4$ come the possibilities
$$1, 4 \quad -1, -4 \quad 4, 1 \quad -4, -1 \quad 2, 2 \quad -2, -2$$
The choice $-2, -2$ gives $b + 3d = -8$. Therefore,
$$3x^2 - 8xy + 4y^2 = \underline{(3x - 2y)(x - 2y)}$$

(b) The coefficient of x^2 is 6, so we try either $(x + by)(6x + dy)$ or $(2x + by)(3x + dy)$. With the first choice
$$(x + by)(6x + dy) = 6x^2 + (6b + d)xy + bdy^2$$
so we need integers b, d such that
$$6b + d = 19 \qquad bd = 8$$
From $bd = 8$ come the possibilities
$$1, 8 \quad 8, 1 \quad 2, 4 \quad 4, 2$$
(Negative pairs such as $-1, -8$ can't work because $6b + d > 0$.) We test these pairs and find that none of them gives $6b + d = 19$. Next we try
$$(2x + by)(3x + dy) = 6x^2 + (3b + 2d)xy + bdy^2$$
We need integers b, d such that
$$3b + 2d = 19 \qquad bd = 8$$
Again the possibilities are
$$1, 8 \quad 8, 1 \quad 2, 4 \quad 4, 2$$

Testing, we find that 1, 8 works. Therefore
$$6x^2 + 19xy + 8y^2 = \underline{(2x + y)(3x + 8y)}$$

Factoring Formulas

Every formula for a product of polynomials gives a formula for factoring. As an example, take the product
$$(x + y)(x - y) = x^2 - y^2$$
When read from *right to left*, this is a factorization of the difference of squares. Here are some frequently used formulas:

Factoring Formulas

(1) $x^2 - y^2 = (x + y)(x - y)$
(2) $x^2 \pm 2xy + y^2 = (x \pm y)^2$
(3) $x^3 + y^3 = (x + y)(x^2 - xy + y^2)$
(4) $x^3 - y^3 = (x - y)(x^2 + xy + y^2)$
(5) $x^n - y^n = (x - y)(x^{n-1} + x^{n-2}y + x^{n-3}y^2 + \cdots + xy^{n-2} + y^{n-1})$

Formulas (1) and (2) are familiar; (3), (4), and (5) can be checked by multiplying out their right sides. Each formula holds for all real values of x and y. Therefore, we can replace x and y by any real numbers or any expression that represent real numbers. For example, from (1) follow such factorizations as

$$81a^2 - 4b^2 = (9a)^2 - (2b)^2 = (9a + 2b)(9a - 2b)$$
$$36 - x^4 = 6^2 - (x^2)^2 = (6 + x^2)(6 - x^2)$$
$$(u + v)^2 - \tfrac{1}{49}w^2 = (u + v)^2 - (\tfrac{1}{7}w)^2 = (u + v + \tfrac{1}{7}w)(u + v - \tfrac{1}{7}w)$$

EXAMPLE 4   Factor   (a) $x^5 - 2x^3 + x$ (b) $x^4 - 16y^4$

SOLUTION   (a) First remove the common factor x:
$$x^5 - 2x^3 + x = x(x^4 - 2x^2 + 1)$$
The polynomial in parentheses is a square, $(x^2 - 1)^2$. But wait, we're not done yet because $x^2 - 1 = (x + 1)(x - 1)$. Therefore
$$x^5 - 2x^3 + x = x[(x + 1)(x - 1)]^2$$
$$= \underline{x(x + 1)^2(x - 1)^2}$$

(b) $x^4 - 16y^4 = (x^2)^2 - (4y^2)^2 = (x^2 + 4y^2)(x^2 - 4y^2)$
$\qquad\qquad = (x^2 + 4y^2)[x^2 - (2y)^2]$
$\qquad\qquad = \underline{(x^2 + 4y^2)(x + 2y)(x - 2y)}$

Formulas (3) and (4) allow us to factor the sum and difference of two cubes.

EXAMPLE 5 Factor (a) $x^3 + 8$ (b) $27x^3 - 125y^3$

SOLUTION (a) Since $x^3 + 8 = x^3 + 2^3$, apply Formula (3):
$$x^3 + 8 = x^3 + 2^3 = (x + 2)(x^2 - x \cdot 2 + 2^2)$$
$$= \underline{(x + 2)(x^2 - 2x + 4)}$$

(b) Use Formula (4) for the difference of cubes:
$$27x^3 - 125y^3 = (3x)^3 - (5y)^3 = (3x - 5y)[(3x)^2 + (3x)(5y) + (5y)^2]$$
$$= \underline{(3x - 5y)(9x^2 + 15xy + 25y^2)}$$

Factorization formulas can give information about the factors of positive integers, as the next example shows.

EXAMPLE 6 Show that
 (a) 999,992 is divisible by 98   (b) 999,999,999,998 is divisible by 9998

SOLUTION (a) 999,992 is the difference of cubes:
$$999{,}992 = 1{,}000{,}000 - 8 = (100)^3 - 2^3$$
$$= (100 - 2)(100^2 + 100 \cdot 2 + 2^2)$$
$$= 98(10{,}000 + 200 + 4)$$
$$= \underline{(98)(10{,}204)}$$

(b) $999{,}999{,}999{,}992 = 10^{12} - 8 = (10^4)^3 - 2^3$
$$= (10^4 - 2)[(10^4)^2 + 10^4 \cdot 2 + 2^2]$$
$$= (9998)(100{,}000{,}000 + 20{,}000 + 4)$$
$$= \underline{(9998)(100{,}020{,}004)}$$

Note that part (b) cannot be done on a calculator because a calculator cannot handle a 12-digit number *exactly*.

Factoring by Grouping

Some expressions can be factored by grouping terms suitably.

EXAMPLE 7 Factor
  (a) $xy - 3y + 2x - 6$ (b) $x^5 + x^4 + 2x^3 + 2x^2 + x + 1$

SOLUTION (a) Note that
$$xy - 3y = (x - 3)y \quad \text{and} \quad 2x - 6 = 2(x - 3)$$
Hence $xy - 3y + 2x - 6 = (x - 3)y + 2(x - 3)$
$$= \underline{(x - 3)(y + 2)}$$

(b) $x^5 + x^4 + 2x^3 + 2x^2 + x + 1 = (x^5 + x^4) + (2x^3 + 2x^2) + (x + 1)$
$$= x^4(x + 1) + 2x^2(x + 1) + (x + 1)$$
$$= (x + 1)(x^4 + 2x^2 + 1)$$
$$= \underline{(x + 1)(x^2 + 1)^2}$$

EXERCISES

Factor

1. $3xy - 9xz$
2. $x^2y - 4xy^2$
3. $12a^2bc + 8a^2b^2c^2 + 4a^2c^3$
4. $14a^2b^2 + 21b^2c^2 + 35abc$
5. $(2x + 1)(3x + 1)(4x + 1) - (x + 1)(2x + 1)(3x + 1)$
6. $(4x + 1)(x^2 + 4) + (2x + 5)(x^2 + 4) - 3(x + 2)(x^2 + 4)$
7. $x^2 + 4x + 3$
8. $x^2 + 11x + 30$
9. $x^2 - 9x + 14$
10. $x^2 + x - 6$
11. $3x^2 + 4x + 1$
12. $2x^2 + 11x + 12$
13. $9x^2 - 20x + 4$
14. $6x^2 + 7x - 5$
15. $6x^2 + 25x + 14$
16. $12x^2 - 29x + 10$
17. $x^2 + 5xy - 36y^2$
18. $x^2 - 6xy + 8y^2$
19. $4x^2 + 4xy + y^2$
20. $10x^2 - 17xy + 3y^2$
21. $x^4 + 5x^2 + 6$
22. $x^6 + 7x^3 + 12$
23. $4y^2 - 81$
24. $9z^2 - 100$
25. $a^4 - 16b^2$
26. $b^2c^6 - c^4$
27. $r^2 - 4(s + t)^2$
28. $(2p + 3q)^2 - (p - q)^2$
29. $25 - (x + y + z + 5)^2$
30. $(x + y + z + w)^2 - (x + y + z)^2$
31. $u^6 - 18u^4v^2 + 81u^2v^4$
32. $u^4 - 5u^2v^2 + 4v^4$
33. $(x + 1)(9x^2 - 4) + (x + 2)(3x - 2)$
34. $(x^4 - 1) + (x^3 + x)$
35. $x^4 - 18x^2 + 81$
36. $x^{10} - 2x^6 + x^2$
37. $(a + 2b)^4 - c^4$
38. $a^4 - b^4c^8$
39. $b^3 - 27$
40. $8c^3 + 1$
41. $8x^3 - y^6$
42. $\frac{1}{27}x^3y^6 + 1000$
43. $x^5 - 1$
44. $x^5 - 32$
45. $x^6 - 64$
46. $x^8 - 1$
47. $(x^2 - 1)^2 - 6(x^2 - 1) + 9$
48. $(16x^4 - 1) + (4x^4 - 3x^2 - 1)$
49. $50^3 - 1$
50. $107^2 - 93^2$
51. $x^6 - 9x^3 + 8$
52. $x^6 - 16x^3 + 64$
53. $a^2 + ab + ac + bc$
54. $ab^2 + bc - c^2 - abc$
55. $u^2v - 2v + u^3v^2 - 2uv^2$
56. $3u^2 - uv + 3uw - vw$
57. $x^5 - x^3 - x^2 + 1$
58. $x^6 - 4x^4 - x^2 + 4$
59. $x^2 + x^4 - y^2 - y^4$
60. $x^2 + x - 4y^2 - 2y$
61. Without a calculator, compute $(411.3)^2 - (410.3)^2$.
62. Without dividing, show that 1027 is divisible by 13.
63. Verify Formula (3) by multiplication.
64. Verify Formula (4) by multiplication.

65. Write out Formula (5) for $n = 6$ and verify it by multiplication.

66. Substituting $x = 7$ into the expression $x^4 - 9x^3 + 14x^2$ gives 0. How can you see that without actually computing?

5 RATIONAL EXPRESSIONS

Quotients of polynomials are called **rational expressions.**

Examples

$$\frac{2x + 5}{x^2 - 9} \qquad \frac{x^3 - 7x^2 + x + 4}{5x^4 + 9x + 1} \qquad \frac{y}{x} \qquad \frac{x + 4}{2y - 7} \qquad \frac{x^2 z}{y^2 + z^2 + 1}$$

Let us agree that all denominators are understood to be non-zero. Thus, in the first example above, $x \neq \pm 3$, even though this is not stated explicitly.

Rational expressions represent quotients of real numbers so they obey the usual rules for quotients. For example, the rule

$$\frac{ka}{kb} = \frac{a}{b}$$

allows us to "cancel" a common factor in the numerator and denominator. Recall that an everyday fraction (quotient of integers) is in lowest terms if its numerator and denominator have no common integer factors except ± 1. Similarly, a rational expression is in **lowest terms** if its numerator and denominator have no common *polynomial* factors.

EXAMPLE 1 Express in lowest terms

(a) $\dfrac{6x^2 y^5}{9x^4 y^2}$ (b) $\dfrac{3x - 6}{x^2 - 4x + 4}$ (c) $\dfrac{x^2 - y^2}{x^3 - y^3}$

SOLUTION (a) The numerator and denominator have a common factor $3x^2 y^2$, which we cancel:

$$\frac{6x^2 y^5}{9x^4 y^2} = \frac{(3x^2 y^2)(2y^3)}{(3x^2 y^2)(3x^2)} = \frac{2y^3}{3x^2}$$

(b) Factor the numerator and denominator:

$$\frac{3x - 6}{x^2 - 4x + 4} = \frac{3(x - 2)}{(x - 2)^2} = \frac{3}{x - 2} \qquad \text{[Cancel common factor]}$$

(c) $\dfrac{x^2 - y^2}{x^3 - y^3} = \dfrac{(x - y)(x - y)}{(x - y)(x^2 + xy + y^2)}$ [Factor]

$$= \frac{x + y}{x^2 + xy + y^2} \qquad \text{[Cancel common factor]}$$

To add or subtract fractions, we convert them to equivalent fractions with a common denominator, then add the numerators. Adding and subtracting rational expressions works in the same way.

EXAMPLE 2 Compute

(a) $\dfrac{x}{3x - y} + \dfrac{y}{3x + y}$ (b) $\dfrac{2x + 1}{x^2 - 4} + \dfrac{2}{x - 2} - \dfrac{1}{x - 2}$

SOLUTION (a) A common denominator is $(3x - y)(3x + y)$. Rewrite the expressions so that they have this denominator. Multiply numerator and denominator of the first term by $3x + y$, and those of the second term by $3x - y$, then add:

$$\dfrac{x}{3x - y} + \dfrac{y}{3x + y} = \dfrac{x(3x + y)}{(3x - y)(3x + y)} + \dfrac{(3x - y)y}{(3x - y)(3x + y)}$$

$$= \dfrac{x(3x + y) + (3x - y)y}{(3x - y)(3x + y)}$$

$$= \dfrac{3x^2 + 4xy - y^2}{9x^2 - y^2}$$

(b) A common denominator is $x^2 - 4 = (x + 2)(x - 2)$. The first term is OK. Rewrite the second and third terms. Multiply numerator and denominator of the second term by $x - 2$, and those of the third term by $x + 2$:

$$\dfrac{2x + 1}{x^2 - 4} + \dfrac{2}{x + 2} - \dfrac{1}{x - 2} = \dfrac{2x + 1}{x^2 - 4} + \dfrac{2(x - 2)}{x^2 - 4} - \dfrac{x + 2}{x^2 - 4}$$

$$= \dfrac{(2x + 1) + 2(x - 2) - (x + 2)}{x^2 - 4} = \dfrac{3x - 5}{x^2 - 4}$$

EXAMPLE 3 Compute $\dfrac{1}{x} + \dfrac{1}{x + 1} + \dfrac{1}{x + 2}$

SOLUTION A common denominator is $x(x + 1)(x + 2)$. Multiply numerator and denominator of the first term by $(x + 1)(x + 2)$, those of the second term by $x(x + 2)$, and those of the third term by $x(x + 1)$:

$$\dfrac{1}{x} + \dfrac{1}{x + 1} + \dfrac{1}{x + 2} = \dfrac{(x + 1)(x + 2)}{x(x + 1)(x + 2)} + \dfrac{x(x + 2)}{x(x + 1)(x + 2)} + \dfrac{x(x + 1)}{x(x + 1)(x + 2)}$$

$$= \dfrac{(x^2 + 3x + 2) + (x^2 + 2x) + (x^2 + x)}{x(x + 1)(x + 2)}$$

$$= \dfrac{3x^2 + 6x + 2}{x(x + 1)(x + 2)}$$

To multiply and divide fractions, we use the rules

$$\dfrac{a}{b} \cdot \dfrac{c}{d} = \dfrac{ac}{bd} \qquad \dfrac{\dfrac{a}{b}}{\dfrac{c}{d}} = \dfrac{a}{b} \cdot \dfrac{d}{c} = \dfrac{ad}{bc}$$

These apply to rational expressions as well. To multiply two rational expressions, multiply the numerators and multiply the denominators. To divide, invert the divisor, then multiply.

EXAMPLE 4 Compute and express in lowest terms

(a) $\dfrac{xy}{(x-1)^2} \cdot \dfrac{x^2-1}{y^2+y}$ (b) $\dfrac{\dfrac{x+1}{x^2+1}}{\dfrac{x}{x+3}}$

SOLUTION

(a) $\dfrac{xy}{(x-1)^2} \cdot \dfrac{x^2-1}{y^2+y} = \dfrac{xy(x^2-1)}{(x-1)^2(y^2+y)}$

$= \dfrac{xy(x+1)(x-1)}{(x-1)^2 y(y+1)}$ [Factor]

$= \dfrac{x(x+1)}{(x-1)(y+1)}$ [Cancel common factors]

(b) Invert the divisor and multiply:

$\dfrac{\dfrac{x+1}{x^2+1}}{\dfrac{x}{x+3}} = \dfrac{x+1}{x^2+1} \cdot \dfrac{x+3}{x} = \dfrac{(x+1)(x+3)}{x(x^2+1)}$

EXAMPLE 5 Simplify $\dfrac{\dfrac{1}{x} - \dfrac{1}{x+1}}{1 + \dfrac{1}{x}}$

SOLUTION One way to handle a 4-story fraction like this is to compute the numerator and denominator separately, then to divide the two fractions. A second way, sometimes shorter, is to multiply numerator and denominator by a convenient factor. In this case multiply by $x(x+1)$:

$\dfrac{\dfrac{1}{x} - \dfrac{1}{x+1}}{1 + \dfrac{1}{x}} = \dfrac{x(x+1)\left(\dfrac{1}{x} - \dfrac{1}{x+1}\right)}{x(x+1)\left(1 + \dfrac{1}{x}\right)}$

$= \dfrac{\dfrac{x(x+1)}{x} - \dfrac{x(x+1)}{x+1}}{x(x+1) + \dfrac{x(x+1)}{x}} = \dfrac{(x+1) - x}{x(x+1) + (x+1)}$

$= \dfrac{1}{(x+1)(x+1)} = \dfrac{1}{(x+1)^2}$

EXERCISES

Express in lowest terms

1. $\dfrac{3x^2}{x^3 + 5x}$ 2. $\dfrac{4xy^3}{10x^2y^2}$ 3. $\dfrac{x^2 + x}{(x+1)(x+5)}$

4. $\dfrac{xy - 3y}{(x - 3)(x + y)}$

5. $\dfrac{x - 4}{x^2 - 16}$

6. $\dfrac{x^2 - 25}{x^4 - 5x^3}$

7. $\dfrac{x^2 + 2xy + y^2}{x^2 + 4xy + 3y^2}$

8. $\dfrac{9x^2 - y^2}{9x^2 + 6xy + y^2}$

9. $\dfrac{4x^2 - x - 3}{6x^2 - 7x + 1}$

10. $\dfrac{8x^2 + 41x + 5}{10x^2 + 53x + 15}$

11. $\dfrac{x^3 + x^2}{x^3 + 1}$

12. $\dfrac{x^3 - 1}{x^4 - 1}$

13. $\dfrac{x + y - 2z}{(x + y)^2 - 4z^2}$

14. $\dfrac{2xyz + 3xz^2}{4xy^2 + 12xyz + 9xz^2}$

15. $\dfrac{x^{-1} + y^{-1}}{xy}$

16. $\dfrac{x^{-1} - y^{-1}}{x^{-1} + y^{-1}}$

17. $\dfrac{(x^n)^3}{x^{2n}}$

18. $\dfrac{(xy^2)^{2n}}{(x^n y^n)^3}$

Compute

19. $\dfrac{1}{x - 1} - \dfrac{1}{x}$

20. $\dfrac{3}{x - 1} + \dfrac{4}{x + 1}$

21. $\dfrac{1}{2x + 3} - \dfrac{1}{2x + 5}$

22. $\dfrac{1}{6x - 1} + \dfrac{1}{6x + 1}$

23. $\dfrac{2}{3x - 4} + \dfrac{x}{(3x - 4)^2}$

24. $\dfrac{2x}{(5x + 1)^2} - \dfrac{3}{5x + 1}$

25. $\dfrac{x}{x^2 - 4} + \dfrac{1}{x + 2} + \dfrac{3}{x - 2}$

26. $\dfrac{2x - 3}{x^2 + 4x + 3} - \dfrac{1}{x + 1} - \dfrac{1}{x + 3}$

27. $\dfrac{3x}{x^2 - 4} + \dfrac{1}{(x - 2)^2}$

28. $\dfrac{x}{(x + 5)^2} + \dfrac{x - 2}{x^2 - 25}$

29. $\dfrac{x}{y} + \dfrac{y}{x} + 1$

30. $\dfrac{1}{x} + \dfrac{1}{xy} + \dfrac{1}{xy^2}$

31. $\dfrac{1}{(x + 1)(x + 2)} - \dfrac{1}{x(x + 1)}$

32. $\dfrac{1}{(x - 3)^2} - \dfrac{1}{(x + 3)^2}$

33. $\dfrac{1}{2x - 1} - \dfrac{2}{(2x - 1)^2} + \dfrac{3}{(2x - 1)^3}$

34. $\dfrac{1}{x(x + 1)} + \dfrac{1}{(x + 1)(x + 2)} - \dfrac{2}{(x + 2)(x + 3)}$

Multiply and express in lowest terms

35. $\dfrac{x^2 - 1}{x} \cdot \dfrac{x}{x - 1}$

36. $\dfrac{x^2 - 4}{x + 1} \cdot \dfrac{x^2 + 2x + 1}{x^4 - 16}$

37. $\dfrac{x^2 - 5x + 6}{x^2 + 5x + 4} \cdot \dfrac{x + 4}{x^2 - 6x + 9}$

38. $\dfrac{2x - 1}{x^3 + 8} \cdot \dfrac{x^2 - 5x - 14}{2x^2 - 11x + 5}$

39. $\dfrac{1}{x + y} \cdot \left(\dfrac{1}{x} + \dfrac{1}{y} \right)$

40. $\dfrac{1}{x - y} \cdot \left(\dfrac{y}{x} - \dfrac{x}{y} \right)$

41. $\left(\dfrac{x+y}{x-2y}\right)^2 \cdot \dfrac{x^2 - 3xy + 2y^2}{(x^2 - y^2)^2}$

42. $\dfrac{2x - y}{x^4 - y^4} \cdot \dfrac{(2x^2 + 3xy + y^2)^2}{4x^2 - y^2}$

Compute and express in lowest terms

43. $\dfrac{\dfrac{x+3}{x+4}}{\dfrac{2x}{x+4}}$

44. $\dfrac{\dfrac{x}{x+y}}{\dfrac{2x}{3x+y}}$

45. $\dfrac{\dfrac{1}{a} + \dfrac{1}{b}}{\dfrac{1}{a} - \dfrac{1}{b}}$

46. $\dfrac{\dfrac{b}{a} - \dfrac{a}{b}}{\dfrac{1}{a} + \dfrac{1}{b}}$

47. $\dfrac{\dfrac{z^2 - 1}{z^3 + z}}{\dfrac{z+1}{z^2 + 1}}$

48. $\dfrac{\left(\dfrac{z+w}{z-w}\right)^2}{\dfrac{1}{z^2 - w^2}}$

49. $\dfrac{\dfrac{5}{y+3} - \dfrac{1}{y+1}}{\dfrac{7}{y+3} + \dfrac{2}{y+1}}$

50. $\dfrac{\dfrac{6}{y} + \dfrac{12}{y-4}}{\dfrac{1}{y} + \dfrac{3}{y-4}}$

51. $\dfrac{\dfrac{3}{u+1} + \dfrac{u}{u^2 - 1}}{\dfrac{3}{u+1} - \dfrac{1}{u-1}}$

52. $\dfrac{\dfrac{5u}{u^2 - 4}}{\dfrac{2}{u+2} + \dfrac{3}{u-2}}$

53. $\dfrac{\dfrac{x^2}{y} - \dfrac{y^2}{x}}{x^2 + xy + y^2}$

54. $\dfrac{x^2 + \dfrac{x}{y} - \dfrac{6}{y^2}}{x^2 + \dfrac{4x}{y} + \dfrac{3}{y^2}}$

55. $\dfrac{1}{1 + \dfrac{1}{1 + \dfrac{1}{x}}}$

56. $\dfrac{1 + \dfrac{1}{1 + \dfrac{1}{x}}}{1 - \dfrac{1}{1 - \dfrac{1}{x}}}$

6 ROOTS AND RADICALS

Square Roots

Given a real number a, a number x is a **square root** of a if $x^2 = a$. For example, 3 and -3 are square roots of 9 because $3^2 = 9$ and $(-3)^2 = 9$. Each positive number a has two square roots, one positive and one negative. We denote the positive square root of a by $\sqrt{a}$. Thus

$$\sqrt{9} = 3 \qquad \sqrt{25} = 5 \qquad \sqrt{\tfrac{4}{49}} = \tfrac{2}{7} \qquad \sqrt{0.0016} = 0.04$$

The negative square root of a is $-\sqrt{a}$ since $(-\sqrt{a})^2 = (\sqrt{a})^2 = a$. The number 0 has only one square root and we write $\sqrt{0} = 0$. In the real number system, negative numbers have no square roots because squares are non-negative.

Square roots obey the following algebraic rules:

Fundamentals of Algebra

Rules for Square Roots Let a and b be positive real numbers. Then
(1) $(\sqrt{a})^2 = a$
(2) $\sqrt{a^2} = a$
(3) $\sqrt{ab} = \sqrt{a}\sqrt{b}$
(4) $\sqrt{\dfrac{a}{b}} = \dfrac{\sqrt{a}}{\sqrt{b}}$

Examples $(\sqrt{3})^2 = 3 \qquad \sqrt{15^2} = 15$

$$\sqrt{5 \cdot 14} = \sqrt{5} \cdot \sqrt{14} \qquad \sqrt{\dfrac{7}{12}} = \dfrac{\sqrt{7}}{\sqrt{12}}$$

Rules (1) and (2) come directly from the definition of square roots. Let us check Rule (3). We just check that $\sqrt{a}\sqrt{b}$ is the positive number whose square is ab. Clearly, $\sqrt{a}\sqrt{b} > 0$ because $\sqrt{a} > 0$ and $\sqrt{b} > 0$. Also,

$$(\sqrt{a}\sqrt{b})^2 = (\sqrt{a})^2(\sqrt{b})^2 = ab \qquad \text{[Rule (1)]}$$

(Rule (4) is checked in a similar way.)

Rule (3) helps simplify square roots. For instance,

$$\sqrt{48} = \sqrt{16 \cdot 3} = \sqrt{16}\sqrt{3} = 4\sqrt{3}$$

In general, by Rule (3),

$$\sqrt{r^2 b} = \sqrt{r^2}\sqrt{b} = r\sqrt{b} \qquad r > 0, b > 0$$

EXAMPLE 1 Simplify (a) $\sqrt{36 \cdot 49}$ (b) $\sqrt{500}$ (c) $\sqrt{\dfrac{75}{16}}$

SOLUTION

(a) $\sqrt{36 \cdot 49} = \sqrt{36}\sqrt{49}$ [Rule (3)]
$= 6 \cdot 7 = \underline{42}$

(b) $\sqrt{500} = \sqrt{100 \cdot 5} = \sqrt{100}\sqrt{5}$ [Rule (3)]
$= \underline{10\sqrt{5}}$

(c) $\sqrt{\dfrac{75}{16}} = \dfrac{\sqrt{75}}{\sqrt{16}}$ [Rule (4)]

$= \dfrac{\sqrt{25 \cdot 3}}{4}$

$= \dfrac{\sqrt{25}\sqrt{3}}{4}$ [Rule (3)]

$= \underline{\dfrac{5\sqrt{3}}{4}}$

Note In the following examples and exercises, letters represent positive numbers.

EXAMPLE 2 Simplify (a) $\sqrt{25x^6}$ (b) $\sqrt{225p^4q^7}$ (c) $\sqrt{uv}\sqrt{\dfrac{8}{u^3v}}$

If n is odd, then all real numbers have exactly one n-th root. For instance, -125 has the 3-rd (cube) root -5 since $(-5)^3 = -125$. Nevertheless, we shall stick to n-th roots of positive numbers only.

> **Rules for n-th Roots** Let a and b be positive real numbers. Then
>
> (1) $(\sqrt[n]{a})^n = a$ (2) $\sqrt[n]{a^n} = a$
>
> (3) $\sqrt[n]{ab} = \sqrt[n]{a}\sqrt[n]{b}$ (4) $\sqrt[n]{\dfrac{a}{b}} = \dfrac{\sqrt[n]{a}}{\sqrt[n]{b}}$
>
> (5) $\sqrt[m]{\sqrt[n]{a}} = \sqrt[n]{\sqrt[m]{a}} = \sqrt[nm]{a}$

Rules (1)–(4) are just like the corresponding rules for square roots. Rule (5) is new; it says, for example, that
$$\sqrt[3]{\sqrt{64}} = \sqrt{\sqrt[3]{64}} = \sqrt[6]{64} = 2 \qquad \sqrt{\sqrt[4]{7}} = \sqrt[8]{7}$$
For a justification of Rule (5), see Exercise 61.

EXAMPLE 5 Simplify (a) $\sqrt[4]{810{,}000}$ (b) $\sqrt[5]{0.00064}$

SOLUTION (a) Try to express 810,000 as a product of 4-th powers. In fact,
$$810{,}000 = 81 \times 10{,}000 = 3^4 \times 10^4$$
Therefore $\sqrt[4]{810{,}000} = \sqrt[4]{3^4 \times 10^4} = \sqrt[4]{3^4}\sqrt[4]{10^4}$ [Rule (3)]
$$= 3 \cdot 10 = \underline{30} \qquad \text{[Rule (2)]}$$

(b) Try to express 0.00064 as a product of 5-th powers. In fact,
$$0.00064 = 64 \times 10^{-5} = 64 \times (10^{-1})^5$$
Now $64 = 2 \cdot 2^5$; hence
$$\sqrt[5]{0.00064} = \sqrt[5]{2 \cdot 2^5 \cdot (10^{-1})^5} = \sqrt[5]{2}\sqrt[5]{2^5 \cdot (10^{-1})^5} \qquad \text{[Rule (3)]}$$
$$= \sqrt[5]{2}\sqrt[5]{2^5}\sqrt[5]{(10^{-1})^5} \qquad \text{[Rule (3)]}$$
$$= \sqrt[5]{2} \cdot 2 \cdot 10^{-1} = \underline{\tfrac{1}{5}\sqrt[5]{2}} \qquad \text{[Rule (2)]}$$

Remark We can save a step in solution (b) by using an extension of Rule (3):
$$\sqrt[n]{abc} = \sqrt[n]{a}\sqrt[n]{b}\sqrt[n]{c}$$
and similarly for more factors.

EXAMPLE 6 Simplify (a) $\sqrt[4]{\dfrac{48}{x^{12}y^4}}$ (b) $\sqrt[3]{\sqrt{27}}$

SOLUTION

(a) $\sqrt[4]{\dfrac{48}{x^{12}y^4}} = \dfrac{\sqrt[4]{48}}{\sqrt[4]{x^{12}}\sqrt[4]{y^4}}$ [Rules (3) and (4)]

$\qquad = \dfrac{\sqrt[4]{2^4 \cdot 3}}{\sqrt[4]{(x^3)^4}\sqrt[4]{y^4}} = \dfrac{2\sqrt[4]{3}}{x^3 y}$ [Rule (2)]

52 Fundamentals of Algebra

(b) Since $27 = 3^3$, it is neater to take the cube root first, and then the square root:

$$\sqrt[3]{\sqrt{27}} = \sqrt{\sqrt[3]{27}} = \underline{\sqrt{3}} \qquad [\text{Rule (5)}]$$

EXAMPLE 7 Simplify $\sqrt[4]{\dfrac{4a^2b}{27}} \sqrt[4]{\dfrac{4a^5}{3b^{10}}}$ using at most one radical.

SOLUTION By Rule (3) we can express the product as one big 4-th root:

$$\sqrt[4]{\dfrac{4a^2b}{27}} \sqrt[4]{\dfrac{4a^5}{3b^{10}}} = \sqrt[4]{\dfrac{4a^2b}{27} \cdot \dfrac{4a^5}{3b^{10}}} = \sqrt[4]{\dfrac{16a^7}{81b^9}}$$

$$= \sqrt[4]{\dfrac{2^4 \cdot a^4 \cdot a^3}{3^4 \cdot b^8 \cdot b}}$$

$$= \dfrac{2a}{3b^2} \sqrt[4]{\dfrac{a^3}{b}} \qquad [\text{Rules (2) and (3)}]$$

Calculating Square Roots

The calculator key $\boxed{\sqrt{}}$ calculates (approximately) the square root of the displayed number.

Examples

$\sqrt{1276}$	1 2 7 6 $\boxed{\sqrt{}}$	35.721142
$\sqrt{5.8 \times 10^{27}}$	5 . 8 $\boxed{\text{EXP}}$ 2 7 $\boxed{\sqrt{}}$	7.615773×10^{13}
$\sqrt{1 + \sqrt{3}}$	1 $\boxed{+}$ 3 $\boxed{\sqrt{}}$ $\boxed{=}$ $\boxed{\sqrt{}}$	1.652891

We discuss calculation of *n*-th roots in the next section.

EXERCISES

Simplify

1. $\sqrt{81}$ 2. $\sqrt{144}$ 3. $\sqrt{\tfrac{9}{64}}$ 4. $\sqrt{\tfrac{49}{25}}$ 5. $\sqrt[3]{\tfrac{8}{27}}$ 6. $\sqrt[3]{8{,}000{,}000{,}000}$

7. $\sqrt[4]{0.0256}$ 8. $\sqrt[5]{243}$ 9. $\sqrt{9 \times 10^{12}}$ 10. $\sqrt[4]{4 \times 10^{-20}}$ 11. $\sqrt{36a^8}$

12. $\sqrt{\tfrac{1}{4}b^2}$ 13. $\sqrt[3]{\dfrac{x^6}{125}}$ 14. $\sqrt[4]{\dfrac{y^{12}}{16}}$ 15. $\sqrt{24}$ 16. $\sqrt{75}$ 17. $\sqrt[3]{32}$ 18. $\sqrt[3]{81}$

19. $\sqrt{72c^5}$ 20. $\sqrt{a^6b^7}$ 21. $\sqrt{\dfrac{3a^3b^5}{4c^4}}$ 22. $\sqrt{\dfrac{(a+b)^3}{(c+7)^4}}$ 23. $\sqrt[3]{\dfrac{a^4b^6}{c^3d^9}}$

24. $\sqrt[3]{\dfrac{(a+b+c)^7}{8a^3b^9c^{15}}}$

Compute and simplify

25. $(4 + \sqrt{7})(4 - \sqrt{7})$ 26. $(\sqrt{3} - \sqrt{2})^2$

27. $(a + b\sqrt{2})(a - b\sqrt{2})$ 28. $\sqrt{ab}\left(\sqrt{\dfrac{a}{b}} + \sqrt{\dfrac{b}{a}}\right)$

Rationalize the denominator

29. $\dfrac{6}{\sqrt{6}}$ 30. $\dfrac{1}{\sqrt[3]{4}}$ 31. $\dfrac{\sqrt{2}}{\sqrt{6} + \sqrt{2}}$

32. $\dfrac{5 - \sqrt{3}}{5 + \sqrt{3}}$ 33. $\dfrac{2}{8 - 3\sqrt{5}}$ 34. $\dfrac{1}{7\sqrt{2} - 1}$

35. $\dfrac{\sqrt{a} - \sqrt{b}}{\sqrt{a} + \sqrt{b}}$ 36. $\dfrac{1}{\sqrt{a} + 2\sqrt{b}}$

Verify the formulas

37. $\sqrt{a^2 + x^2} = a\sqrt{1 + \dfrac{x^2}{a^2}}$ 38. $\sqrt{9x^2 - y^2} = y\sqrt{\dfrac{9x^2}{y^2} - 1}$

39. $\sqrt[n]{abc} = \sqrt[n]{a} \sqrt[n]{b} \sqrt[n]{c}$ 40. $\sqrt[4]{a^2} = \sqrt{a}$

Simplify, using at most one radical

41. $\sqrt{16a^3} + \sqrt{25a^5} + \sqrt{49a^7}$ 42. $(\sqrt{3b^5} - \sqrt{2b})^2$

43. $\sqrt[4]{\sqrt[3]{16}}$ 44. $\sqrt{\sqrt{\sqrt{64}}}$ 45. $\sqrt{a\sqrt{a^6}}$

46. $\sqrt[4]{\sqrt{a^8 b^{17}}}$ 47. $\sqrt{2}\sqrt{xy}\sqrt{20yz}$ 48. $\sqrt[4]{2x^3yz}\sqrt[4]{8x^2y^3z^6}$

49. $\sqrt[3]{\dfrac{16xy}{z}}\sqrt[3]{4y^2z^7}$ 50. $\sqrt{\dfrac{5x}{3y}}\sqrt{27x^3y^2}$ 51. $\dfrac{\sqrt{2u^3v^5}}{\sqrt{3u^2v}}$

52. $\dfrac{\sqrt{125u^6v^7w}}{\sqrt{2vw}}$ 53. $\dfrac{5xy}{\sqrt{12x^3y^2}}$ 54. $\dfrac{x\sqrt{50yz}}{\sqrt{98xy^2z^3}}$

Show that

55. $\sqrt{5 + 2\sqrt{6}} = \sqrt{2} + \sqrt{3}$ 56. $\sqrt{9 - 2\sqrt{14}} = \sqrt{7} - \sqrt{2}$

Compute on a calculator

57. $\sqrt{194815}$ 58. $\sqrt{8.75 \times 10^{17}}$

59. $\sqrt{2 + \sqrt{2 + \sqrt{2}}}$ 60. $\sqrt{(4305)^2 + (6952)^2}$

61. Justify Rule (5) for n-th roots this way. Set
$$r = \sqrt[m]{\sqrt[n]{a}} \quad \text{and} \quad s = \sqrt[n]{\sqrt[m]{a}}$$
and compute r^{mn} and s^{nm}.

7 RATIONAL EXPONENTS

In this section we discuss fractional powers of positive numbers. The first question is what a fractional power means. For example, what meaning should we give to $8^{2/3}$? Well, if the rules of exponents hold, the cube of $8^{2/3}$ is 8^2 because
$$(8^{2/3})^3 = 8^{(2/3) \cdot 3} = 8^2$$

Hence $8^{2/3}$ should be the cube root of 8^2:
$$8^{2/3} = \sqrt[3]{8^2} = \sqrt[3]{64} = 4$$
By the same reasoning, the n-th power of $a^{m/n}$ should be a^m, so $a^{m/n}$ should be the n-th root of a^m. Therefore, we define

$$\boxed{a^{m/n} = \sqrt[n]{a^m} \qquad a > 0, \ n > 0}$$

Examples

$$9^{3/2} = \sqrt{9^3} = \sqrt{9^2 \cdot 9}$$
$$= 9\sqrt{9} = 27$$
$$10{,}000^{3/4} = \sqrt[4]{10{,}000^3} = \sqrt[4]{10^{12}}$$
$$= 10^3 = 1000$$

Note that, by definition,
$$a^{1/n} = \sqrt[n]{a^1} = \sqrt[n]{a}$$
Thus, $a^{1/n}$ is our old friend, the n-th root of a. For example,
$$36^{1/2} = \sqrt{36} = 6 \qquad 16^{1/4} = \sqrt[4]{16} = 2$$

Rules of Exponents

With our definition of $a^{m/n}$ all the rules for integer exponents carry over to fractional exponents.

Rules for Rational Exponents If r and s are rational numbers, and if a and b are positive, then

(1) $a^r a^s = a^{r+s}$ (2) $\dfrac{a^r}{a^s} = a^{r-s}$

(3) $(a^r)^s = a^{rs}$ (4) $a^r b^r = (ab)^r$

(5) $\dfrac{a^r}{b^r} = \left(\dfrac{a}{b}\right)^r$

These rules can be proved using properties of integer exponents and roots.

Examples

(1) $3^{1/2} \cdot 3^{1/4} = 3^{1/2 + 1/4} = 3^{3/4}$ (2) $\dfrac{10^2}{10^{4/3}} = 10^{2 - 4/3} = 10^{2/3}$

(3) $a^{-1/6} = (a^{1/6})^{-1} = \dfrac{1}{a^{1/6}}$ (4) $2^{3/5} \cdot 6^{3/5} = (2 \cdot 6)^{3/5} = 12^{3/5}$

(5) $\dfrac{4^{1/3}}{5^{1/3}} = \left(\dfrac{4}{5}\right)^{1/3}$

Rule (3) has a useful consequence. Since
$$a^{m/n} = a^{m \cdot (1/n)} = a^{(1/n) \cdot m}$$

Rule (3) implies
$$a^{m/n} = \sqrt[n]{a^m} = (\sqrt[n]{a})^m$$
Hence

(6) $\quad a^{m/n} = \sqrt[n]{a^m} = (\sqrt[n]{a})^m$

Example
$$64^{5/6} = \sqrt[6]{64^5} = (\sqrt[6]{64})^5$$
It is much easier to compute the third number than the second:
$$64^{5/6} = (\sqrt[6]{64})^5 = 2^5 = 32$$

EXAMPLE 1 Compute (a) $81^{5/4}$ (b) $100^{-7/2}$

SOLUTION

(a) $81^{5/4} = (\sqrt[4]{81})^5 = 3^5 = \underline{243}$ [Rule (6)]

Alternative solution. Since $81 = 3^4$,
$$81^{5/4} = (3^4)^{5/4} = 3^{4(5/4)} = 3^5 = \underline{243} \quad \text{[Rule (3)]}$$

(b) Since $100 = 10^2$,
$$100^{-7/2} = (10^2)^{-7/2} = 10^{2(-7/2)} = \underline{10^{-7}} \quad \text{[Rule (3)]}$$

EXAMPLE 2 Express $\sqrt[6]{8^{5/4} \cdot 4^{-3/2}}$ as a power of 2.

SOLUTION

$\sqrt[6]{8^{5/4} \cdot 4^{-3/2}} = [(2^3)^{5/4} \cdot (2^2)^{-3/2}]^{1/6} = [2^{15/4} \cdot 2^{-3}]^{1/6}$ [Rule (3)]

$\qquad = [2^{(15/4)-3}]^{1/6}$ [Rule (1)]

$\qquad = (2^{3/4})^{1/6}$

$\qquad = 2^{(3/4)(1/6)}$ [Rule (3)]

$\qquad = \underline{2^{1/8}}$

EXAMPLE 3 Express using only one radical (a) $\sqrt[3]{9}\sqrt{\tfrac{1}{3}}$ (b) $\dfrac{\sqrt{a^3b^5}}{\sqrt[4]{a^2b}}$

SOLUTION (a) Convert to fractional exponents
$$\sqrt[3]{9}\sqrt{\tfrac{1}{3}} = \sqrt[3]{3^2}\sqrt{3^{-1}}$$
$$= 3^{2/3} \cdot 3^{-1/2}$$
$$= 3^{2/3-1/2} = 3^{1/6} = \underline{\sqrt[6]{3}}$$

(b) $\dfrac{\sqrt{a^3b^5}}{\sqrt[4]{a^2b}} = (a^3b^5)^{1/2}(a^2b)^{-1/4}$

$\qquad = (a^{3/2}b^{5/2})(a^{-1/2}b^{-1/4})$

$\qquad = a^{3/2-1/2}b^{5/2-1/4}$

$\qquad = ab^{9/4} = ab^{2+1/4} = ab^2 b^{1/4} = \underline{ab^2\sqrt[4]{b}}$

Fundamentals of Algebra

EXAMPLE 4 Simplify and express using only positive exponents

(a) $(4u^6)^{-5/2}$ (b) $\left(\dfrac{p^3}{8q^{-12}}\right)^{2/3}$

SOLUTION

(a) $(4u^6)^{-5/2} = 4^{-5/2}(u^6)^{-5/2}$
$= (2^2)^{-5/2}(u^6)^{-5/2}$
$= 2^{-5}u^{-15} = \dfrac{1}{2^5 u^{15}} = \underline{\dfrac{1}{32u^{15}}}$

(b) $\left(\dfrac{p^3}{8q^{-12}}\right)^{1/3} = \dfrac{(p^3)^{1/3}}{(8q^{-12})^{1/3}}$
$= \dfrac{p}{8^{1/3}(q^{-12})^{1/3}} = \dfrac{p}{2q^{-4}} = \underline{\dfrac{pq^4}{2}}$

Calculating Powers

The calculator key $\boxed{y^x}$ calculates (approximately) powers of positive numbers. To calculate y^x, key in the number y, then press $\boxed{y^x}$, key in x in decimal form, and then press $\boxed{=}$.

Examples $\sqrt[4]{81} = 81^{1/4}$: $\quad 8\ 1\ \boxed{y^x}\ .\ 2\ 5\ \boxed{=} \quad\quad\quad\quad\quad 3$
$(55)^{3/8}$: $\quad 5\ 5\ \boxed{y^x}\ .\ 3\ 7\ 5\ \boxed{=} \quad\quad\quad\quad\quad 4.494030$
$163^{-2/5}$: $\quad 1\ 6\ 3\ \boxed{y^x}\ .\ 4\ \boxed{+/-}\ \boxed{=} \quad\quad\quad\quad 0.130354$

Suppose you want to calculate $5^{3/17}$. You can't use the sequence

$$5\ \boxed{y^x}\ 3\ \boxed{\div}\ 1\ 7\ \boxed{=}$$

because that calculates $5^3 \div 17 = 7.352941$. For $5^{3/17}$ you need two steps: first

$3\ \boxed{\div}\ 1\ 7\ \boxed{=} \quad\quad\quad\quad\quad 0.176471$

then $\quad\quad 5\ \boxed{y^x}\ .\ 1\ 7\ 6\ 4\ 7\ 1\ \boxed{=} \quad\quad\quad\quad\quad 1.328458$

This two-step procedure is awkward. You can streamline the computation if your calculator has either parentheses or a memory. On models with parentheses, this sequence generally works:

$$5\ \boxed{y^x}\ \boxed{(}\ 3\ \boxed{\div}\ 1\ 7\ \boxed{)}\ \boxed{=}$$

(Check this on your calculator.)

On calculators with memory, you can calculate 3/17, store it in the memory, then recall it when needed. The key sequence is

$$3\ \boxed{\div}\ 1\ 7\ \boxed{=}\ \boxed{STO} \quad\quad 5\ \boxed{y^x}\ \boxed{RCL}\ \boxed{=}$$

For details on the memory, see the Appendix.

EXERCISES

[All letters in these exercises represent positive numbers.]

Compute

1. $27^{2/3}$ **2.** $64^{3/2}$ **3.** $25^{-1/2}$

4. $16^{-5/4}$ 5. $(1000)^{5/3}$ 6. $(1{,}000{,}000)^{5/6}$

7. $\left(\frac{4}{49}\right)^{-3/2}$ 8. $(0.001)^{-5/3}$

Express as a power of 2

9. $4\sqrt[3]{2}$ 10. $(\sqrt{2})^{2/3}$ 11. $8(16 \cdot 2^{-7/2})$

12. $\left(\frac{8^{1/4} \cdot 16^{-2}}{4^5}\right)^{1/6}$ 13. $\left(\frac{32}{\sqrt[3]{2}}\right)^{1/6}$ 14. $\sqrt[3]{16}\,\sqrt[4]{\tfrac{1}{4}}$

15. $\dfrac{\sqrt{2^n}}{\sqrt[4]{2^k}}$ 16. $\sqrt{2^n}\,\sqrt[3]{2^k}$

Simplify

17. $(25x^4)^{-3/2}$ 18. $(32y^{10})^{2/5}$ 19. $\left(\dfrac{81u^4}{y^{12}}\right)^{3/4}$

20. $\left(\dfrac{27a^3}{8b^6}\right)^{-4/3}$ 21. $(x^{4/3}y^{-2/3})^3$ 22. $(x^{-3/2}\sqrt{y})^{-2}$

23. $(x^4 y^6 z^{-8})^{5/2}$ 24. $(x^2 y^{-4} z^{-6})^{3/4}$ 25. $(8x\sqrt{x})^{5/3}$

26. $(x^2 \sqrt[3]{x})^{-1/6}$ 27. $(x^{1/2} + x^{3/2})^2$ 28. $x^{1/3}(2x^{2/3} - x^{-1/6})$

Express using at most one radical

29. $\sqrt{2} \cdot \sqrt[3]{2}$ 30. $\dfrac{\sqrt[3]{4}}{\sqrt{2}}$ 31. $\sqrt{b\sqrt{b}}$

32. $\sqrt[3]{\sqrt{b^{1/4}}}$ 33. $\dfrac{\sqrt{ab}}{\sqrt[4]{a^2 b}}$ 34. $\dfrac{\sqrt{3a}}{\sqrt[3]{6a^2}}$

35. $\sqrt{x} \cdot \sqrt[3]{x^2}$ 36. $\sqrt{x} \cdot \sqrt[3]{x} \cdot \sqrt[4]{x}$ 37. $\sqrt{\sqrt{\sqrt{a}}}$

38. $\sqrt{a\sqrt{a\sqrt{a}}}$ 39. $\sqrt[3]{27a^4\sqrt{64a}}$ 40. $\sqrt{\sqrt{a}\,\sqrt[3]{64a^5}}$

Express without radicals using only positive exponents

41. $\left(\dfrac{u^4}{25}\right)^{-3/2}$ 42. $\left(\dfrac{u^{-9}}{27}\right)^{5/3}$ 43. $\left(\dfrac{p^{-3}}{8q^{12}}\right)^{-2/3}$

44. $\left(\dfrac{16}{p^4 q^{-8}}\right)^{-3/4}$ 45. $(\sqrt[3]{ab^2})^{-3/5}$ 46. $\sqrt{\dfrac{a^{-1/3}}{b^{-1/4}}}$

47. $(\sqrt[4]{x^{14} y^{-21/5}})^{-3/7}$ 48. $\left(\dfrac{x}{y}\right)^{1/5}\left(\dfrac{y}{z}\right)^{2/5}\left(\dfrac{z}{x}\right)^{3/5}$

Compute on a calculator

49. $\sqrt[5]{687}$ 50. $\sqrt[6]{9481}$ 51. $\sqrt[4]{2.77 \times 10^{61}}$

52. $\sqrt[3]{1.65 \times 10^{-34}}$ 53. $(2.79)^{4/7}$ 54. $(6.207)^{13/21}$

55. $\sqrt[4]{1 + \sqrt{3}}$ 56. $(6 + \sqrt[5]{7})^{9/4}$

8 AVOIDING COMMON ERRORS

The basic rules of algebra are absolutely essential in this course and all other mathematics courses you will take. Try to master them now so you can avoid mistakes in the future.

Here are some of the most common types of mistakes.

1. *Incorrect use of rules of real numbers*

WRONG	RIGHT
$x/(y + 2) = x/y + 2$	Keep the parentheses: $x/(y + 2)$
$(x + 3)(x + 4) = x + 3 \cdot x + 4$	$(x + 3)(x + 4) = x^2 + 7x + 12$
$(x + y)^2 = x^2 + y^2$	Keep the parentheses: $(x + y)^2$
$\dfrac{1}{x + y} = \dfrac{1}{x} + \dfrac{1}{y}$	Leave as $\dfrac{1}{x + y}$

2. *Poor use of parentheses*

WRONG	RIGHT
$3(x + 1) = 3x + 1$	$3(x + 1) = 3x + 3$
$(\tfrac{1}{2}x)(\tfrac{1}{2}y) = \tfrac{1}{2}xy$	$(\tfrac{1}{2}x)(\tfrac{1}{2}y) = \tfrac{1}{4}xy$
$x - (2y + 1) = x - 2y + 1$	$x - (2y + 1) = x - 2y - 1$

3. *Mistakes with exponents and radicals*

WRONG	RIGHT
$(2x)^3 = 2x^3$	$(2x)^3 = 8x^3$
$(x^3)^4 = x^7$	$(x^3)^4 = x^{12}$
$x^3 x^4 = x^{12}$	$x^3 x^4 = x^7$
$\sqrt{3x} = 3\sqrt{x}$	$\sqrt{3x} = \sqrt{3}\sqrt{x}$
$\sqrt{x + y} = \sqrt{x} + \sqrt{y}$	Leave as $\sqrt{x + y}$

4. *Incorrect use of formulas*

WRONG	RIGHT
$\dfrac{1}{x} + \dfrac{1}{y} = \dfrac{1}{x + y}$	$\dfrac{1}{x} + \dfrac{1}{y} = \dfrac{x + y}{xy}$
$(x + y)^2 = x^2 + y^2$	$(x + y)^2 = x^2 + 2xy + y^2$
$x^3 + y^3 = (x + y)(x^2 + xy + y^2)$	$x^3 + y^3 = (x + y)(x^2 - xy + y^2)$

5. *Sloppy writing*

$\sqrt{x+1}$	Is it $\sqrt{x+1}$ or $\sqrt{x}+1$?
$\dfrac{1}{x+y+3}$	Is it $\dfrac{1}{x+y+3}$ or $\dfrac{1}{x+y}+3$?
x^{-5}	Is it x^{-5} or $x-5$?
$\dfrac{1}{2}x$	Is it $\dfrac{1}{2}x$ or $\dfrac{1}{2x}$ or $\dfrac{1}{2^x}$?

These examples illustrate some, but not all, of the common mistakes made in algebra. Get into the habit of checking your work. Read it over and see if it says what you mean. Are your formulas ambiguous? Are they legible?

EXERCISES

Find the mistake and correct it

1. $x + x + x + x = x^4$
2. $x^3 = 3x$
3. $(2a)(3a) = 6a$
4. $(\tfrac{1}{2}a)^2 = \tfrac{1}{2}a^2$
5. $7(x-1) = 7x - 1$
6. $3x + 6 = 3(x+6)$
7. $(-x)^2 = -x^2$
8. $(-2)(4x-1) = -8x - 2$
9. $\sqrt{x^2+1} = x+1$
10. $\sqrt{4x+4} = 4\sqrt{x+1}$
11. $\sqrt{x} + \sqrt{2x} = \sqrt{3x}$
12. $\sqrt{2x+1} + \sqrt{3x+1} = \sqrt{5x+2}$
13. $\sqrt{36+49} = 6 + 7 = 13$
14. $\sqrt[3]{8+27} = 2 + 3 = 5$
15. $\sqrt{2^9} = 2^3$
16. $\sqrt{4^{10}} = 2^5$
17. $a^{-2}a^{-2} = a^4$
18. $(ab^2)^3 = ab^6$
19. $\sqrt[3]{\sqrt[3]{a}} = \sqrt[6]{a}$
20. $\sqrt{\sqrt[3]{a}} = \sqrt[5]{a}$
21. $\dfrac{1}{x+3} = \dfrac{1}{x} + \dfrac{1}{3}$
22. $\dfrac{a+1}{b+2} = \dfrac{a}{b} + \dfrac{1}{2}$
23. $\dfrac{c+d}{c+e} = 1 + \dfrac{d}{e}$
24. $\dfrac{c^2}{c+d} = \dfrac{c}{1+d}$
25. $(a+b)^3 = a^3 + b^3$
26. $(a+b)^3 = a^3 + 3ab + b^3$
27. $x^{3/2} = \dfrac{x^3}{x^2}$
28. $x^{3/2} = \tfrac{3}{2}x$
29. $\dfrac{4x^3 - 7x^2 + 1}{x} = 4x^2 - 7x + 1$
30. $\dfrac{\tfrac{1}{x}}{\tfrac{1}{y}} = \dfrac{1}{xy}$

REVIEW EXERCISES FOR CHAPTER 2

Express as a power of 3

1. $\left[3^5\left(\frac{1}{3}\right)^2\right]^2\left[27\left(\frac{1}{9}\right)^4\right]$

2. $\dfrac{9^{3/4}\sqrt{27}}{81^{3/2}}$

Express without negative exponents

3. $(4^{-2} + 2^{-3})^{-1}$

4. $\dfrac{(xy^{-1}z^2)^{-3}}{(x^{-2}yz)^2}$

Compute

5. $(2u - \tfrac{1}{2}v)^2$

6. $(x - y)(x + 2y - 3z)$

7. $(1 - x)(1 + x^2)(1 - x^3)$

8. $(2x - 3)(x^3 - 4x^2 + x - 2)$

Factor

9. $16y^4 - 81$

10. $2x - 6y^2 - 4xy + 3y$

11. Express $(50{,}000{,}000)(0.00000\ 00000\ 002)^2$ in scientific notation.

12. Find the coefficient of x^4 in the product
$$(1 - x + x^2 - x^3 + x^4 - x^5)(2 - 3x + 4x^2 - 5x^3)$$

13. Simplify $\sqrt[3]{\dfrac{250x^4}{y^6 z^{10}}}$

14. Express $\dfrac{2 + \sqrt{5}}{5 - 2\sqrt{5}}$ without radicals in the denominator.

Simplify and express in lowest terms

15. $\dfrac{4}{x - 3} - \dfrac{1}{x + 3} - \dfrac{6}{x^2 - 9}$

16. $(x^3 - 3x^2 + 2x)\left(\dfrac{1}{x - 1} + \dfrac{1}{x - 2}\right)$

17. $\dfrac{x^4 - \dfrac{1}{y^4}}{1 - x^2 y^2}$

18. $\dfrac{(xu - yv)^2 + (xv + yu)^2}{x^2 + y^2}$

Find the mistake and correct it

19. $\dfrac{1}{7x} - \dfrac{1}{2x} = \dfrac{1}{5x}$

20. $[\tfrac{1}{2}(a - b)]^2 = \tfrac{1}{2}(a^2 - b^2)$

3

Equations

1 TRANSLATING WORDS INTO MATH

Algebra has many applications in the "real world"—in business, technology, science, etc. Real-life problems do not come expressed in algebra, however; they generally come expressed in words, or part in words and part in numbers. For solutions, we translate them into algebra, solve the corresponding algebra problems, and then translate the answers back into words. Schematically the process looks like this:

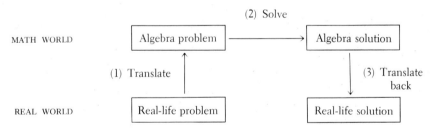

Of the three steps, Step (1) is often the hardest. Once the problem is set up "in math," the actual mathematical solution and its interpretation can be fairly routine. The first two sections of this chapter concentrate on Step (1).

Translating words into algebra is like writing in a new language. In this section, we study the "vocabulary" of algebra. Then, in Section 2, we combine our vocabulary into "sentences," that is, equations.

EXAMPLE 1 Express in algebraic notation

(a) 17% of a number x (b) 50% more than a number y

(c) the average of a, b, and 12

SOLUTION (a) 17% of x means $\frac{17}{100}$ of x,

$$\frac{17}{100}x \quad \text{or} \quad \underline{0.17x}$$

(b) 50% more than y means y plus $\frac{1}{2}$ of itself,

$$y + \tfrac{1}{2}y = \underline{\tfrac{3}{2}y}$$

(c) The average of three numbers is their sum divided by 3. Hence the average of a, b, and 12 is

$$\frac{a + b + 12}{3}$$

Example 1(a) asked for 17% of x. But suppose it asked for "17% of a number." No number is given; we must designate "a number" by a letter. For example, we could say: "Let the number be x. Then 17% of x is $0.17x$."

EXAMPLE 2 Express in algebraic notation
(a) three consecutive integers
(b) the perimeter and area of a square whose side is given
(c) the angles of an isosceles triangle, if one of the equal angles is given

SOLUTION (a) Call the smallest of the three integers n. Then the next integer is $n + 1$, and the one after that is $n + 2$. Hence the integers are $\underline{n, n + 1, n + 2}$.

(b) Call the length of the side s. Then the perimeter of the square is $\underline{4s}$ and the area is $s \cdot s = \underline{s^2}$. See Figure 1.

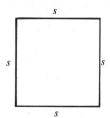
Perimeter = $4s$ Area = s^2
FIG. 1

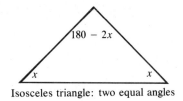
Isosceles triangle: two equal angles
FIG. 2

(c) An isosceles triangle has two equal angles (Fig. 2). Call each of these x (in degrees). Since the sum of all three angles in a triangle is 180 degrees, the third angle is $180 - x - x = 180 - 2x$. Hence the angles are $\underline{x, x, 180 - 2x}$.

EXAMPLE 3 The Audio Shop advertises a sale on stereo equipment: 30% off after the first $50. What is the sale price of equipment whose regular price is x dollars? Assume $x > 50$.

SOLUTION The part of the price after the first $50 is $x - 50$. On this amount the reduction is 30%. Hence the sale price is $50 plus 70% of $x - 50$, that is $50 + 0.7(x - 50)$.

EXAMPLE 4 Express
(a) the distance traveled in t hours at 55 miles per hour
(b) the distance traveled in 7 hours if the first t hours are at 55 mph and the rest of the trip is at 50 mph

SOLUTION (a) At 55 mph, you travel 55 miles in one hour, $2 \cdot 55$ miles in 2 hours, $3 \cdot 55$ miles in 3 hours, etc. Hence, in t hours, the distance is $t \cdot 55$ or $\underline{55t \text{ miles}}$.

(b) The first part of the trip covered $55t$ miles. The rest of the trip took $7 - t$ hours. At 50 mph, the distance covered in this part of the trip was $50(7 - t)$. The total distance was $\underline{55t + 50(7 - t)}$ miles.

Remark The answer to Example 4(b) can be expressed as $350 + 5t$. This makes good sense. For if the entire trip is at 50 mph, then the distance is $7 \cdot 50 = 350$ miles. Each hour driven at 55 mph, rather than 50 mph, increases the distance by 5 miles. Therefore driving t hours at 55 mph increases the 350 miles by $5t$.

EXAMPLE 5 A tank contains 1000 gallons of gasoline. Alcohol is pumped in at the rate of 13 gal/min. After t minutes what fraction of the mixture is alcohol?

SOLUTION In t minutes, $13t$ gallons of alcohol are added. The total volume of the mixture is then $1000 + 13t$. The fraction that is alcohol is

$$\frac{\text{volume of alcohol}}{\text{total volume}} = \underline{\frac{13t}{1000 + 13t}}$$

EXAMPLE 6 A group of people charter a 747 for \$30,000. Express
(a) the cost per person
(b) the cost per person if 70 more people join in

SOLUTION (a) Let n be the number of people in the group. Then

$$\text{cost per person} = \frac{\text{total cost}}{\text{number of people}} = \$\underline{\frac{30{,}000}{n}}$$

(b) Same formula, but now the number of people is $n + 70$. Hence

$$\text{cost per person} = \$\underline{\frac{30{,}000}{n + 70}}$$

EXAMPLE 7 A 40-foot length of rope is cut into two pieces, and then each piece is folded into a square. Express the total area of the two squares.

SOLUTION We are not given the lengths of the pieces. So let us call one length x and the other $40 - x$. See Figure 3.

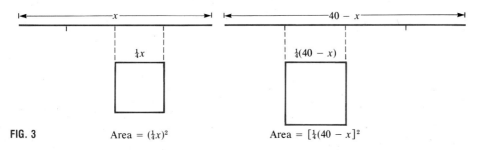

FIG. 3   Area = $(\tfrac{1}{4}x)^2$ Area = $[\tfrac{1}{4}(40 - x)]^2$

64 Equations

The square formed from the first piece has perimeter x, hence side length $\frac{1}{4}x$ and area $(\frac{1}{4}x)^2$. The other square has perimeter $40 - x$, hence side length $\frac{1}{4}(40 - x)$ and area $[\frac{1}{4}(40 - x)]^2$. The total area of the two squares is $\underline{(\frac{1}{4}x)^2 + [\frac{1}{4}(40 - x)]^2}$.

EXERCISES

Express in algebraic notation
1. 15% of x
2. x decreased by 15%
3. the number of inches in y yards
4. the number of seconds in h hours
5. the time of a 200-mile drive at r miles per hour
6. the average speed on a 300-mile trip that took h hours
7. the diagonal of a square of side s
8. the side of a square of area A
9. the cost of renting a car for d days at $80 for the first week, then $20 a day ($d > 7$)
10. the plumber's bill for a job of t hours if he charges $25 for the first hour, and then $10 per hour
11. the value of 100 coins, all nickels and pennies, of which n are nickels
12. the total weight of 50 cans of coffee if n of them are 1-lb cans and the rest are 2-lb cans
13. the perimeter of a right triangle whose legs are 5 and x
14. the height of an equilateral triangle of side s
15. the time for a plane to fly 1000 miles at x miles per hour, the last 300 miles against a headwind of 40 mph
16. the time for a steamboat at speed x miles per hour to go 60 miles downstream aided by a 3-mph current, then return against the current

Assign a letter to a suitable quantity, then express in algebraic notation
17. the sum of a number and its reciprocal
18. the sum of two numbers whose product is 25
19. the average of the squares of two consecutive integers
20. the square of the average of two consecutive integers
21. the area of a triangle whose height is 50% greater than its base
22. the volume of a rectangular box with square base and height triple the side of the base
23. the area of a square inscribed in a given circle
24. the diagonal of a rectangle whose length exceeds its width by 3 inches
25. A 30-foot length of rope is cut into three pieces, two of the same length. Each piece is folded into a square. Express the total area of the three squares.

26. A 30-foot length of rope is cut into a number of equal pieces, and each is folded into a square. Express the total area of these squares.

27. A group of people share the $150 cost of a party. Express the amount each would save if 10 more people joined in.

28. If the price of gas goes up 5 cents a gallon, express how many fewer gallons you could buy for $10.

29. A basketball team has 10 wins and 6 losses. Of the remaining games, half are at home. If they win all the home games and half of the away games, express their winning percentage for the season.

30. A company had twice as many male employees as female. Recently it hired 6 males and 17 females. Express the fraction of employees that are female.

31. The Audio Shop runs a super sale: 40% off after the first $40. Express how much cheaper x dollars worth of equipment (regular price) is now than it was during the sale described in Example 3 of the text.

32. A long convoy of heavy trucks is moving at 20 mph. A motorcycle messenger at 60 mph goes from the rear of the convoy to the front and back again. If the length of the convoy is x miles, how long does it take him? (Hint Consider the motorcycle's speed *relative* to the convoy.)

33. (Just for fun) A rope of length x is cut into three equal pieces. These are folded into a square, an equilateral triangle, and a circle. Express the total perimeter of these three figures.

2 TRANSLATING PROBLEMS INTO ALGEBRA

In this section, we concentrate on interpreting problems as algebraic equations. We practice setting up equations, leaving the actual solution of equations to later sections.

EXAMPLE 1 17% of what number is 68?

SOLUTION Call the unknown number x. We translate "is 68" into mathematical symbols as "= 68." Hence the statement of the problem becomes the equation
$$(17\% \text{ of } x) = 68$$
that is,
$$\tfrac{17}{100}x = 68$$

EXAMPLE 2 One leg of a right triangle is 5, and the perimeter is 20. Find the other leg.

SOLUTION Let x be the length of the second leg. We want an equation involving x. We read the problem and find a disguised equation: "the perimeter is 20," that is,
$$\text{perimeter} = 20$$

The idea now is to express the perimeter in terms of *x*. We draw a sketch (Fig. 1) and label as much as possible using *x*.

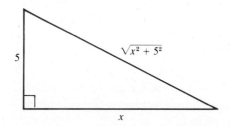

FIG. 1

By the Pythagorean theorem, the hypotenuse is $\sqrt{x^2 + 5^2}$. Therefore
$$\text{perimeter} = x + 5 + \sqrt{x^2 + 25}$$
Hence the desired equation is
$$\underline{x + 5 + \sqrt{x^2 + 25} = 20}$$

Translating word problems into equations takes practice. No single method works in all cases. Still, following some practical guidelines can help.

(1) Read the problem several times. Make sure you know what is given and what is to be found. The unknown quantity is often indicated by words such as "how much," "how far," "when," "what is," "find," etc.

(2) Assign a letter to the unknown. Express as much as you can in terms of this letter.

(3) Draw a diagram if one applies. Label its parts using the letter assigned to the unknown whenever possible.

(4) Find an equation involving the unknown. Often an equation is disguised in the wording of the problem. Be alert to expressions like "the same as," "as much as," or even "is"—they can signal an equation lurking somewhere.

(5) If you get stuck, check that you have used *all* of the data.

EXAMPLE 3 One angle of a triangle is 36°. The second angle is twice as big as the third. Find the second and third angles.

SOLUTION Let the third angle be *x* (in degrees). Then the second angle is 2*x*. Now we need an equation in *x*. We re-read the problem, but it doesn't seem to contain an equation.

Stuck for the moment, we ask whether we have used all the data. No, we have not used the fact that our angles 36, 2*x*, and *x* are the angles of a *triangle*. So far, they could be any old angles. But in a *triangle*, the sum of the angles equals 180°. That's the equation!
$$\underline{36 + 2x + x = 180}$$

EXAMPLE 4 Steve, a slow driver, sets out at a constant speed of 40 mph. I follow one hour later, driving at a steady 55 mph. How long until I catch up?

SOLUTION The unknown is the time required to catch up. Call it t (hours). Now where is an equation? What quantities are equal? Well, "catch up" means that the distances we travel are equal. I'll drive t hours and Steve will drive one hour longer, $t + 1$ hours. So

(my distance in t hours) = (Steve's distance in $t + 1$ hours)

At 55 mph, I cover $55t$ miles in t hours. At 40 mph, he covers $40(t + 1)$ miles in $t + 1$ hours. Hence the equation is

$$55t = 40(t + 1)$$

Suggestion Be consistent with units. Since Example 4 is stated in miles and hours, keep all distances in miles and all times in hours. If I leave 30 *minutes* after Steve the correct equation is

$$55t = 40(t + \tfrac{1}{2}) \quad \text{NOT} \quad 55t = 40(t + 30)$$

EXAMPLE 5 A group of people charter a 747 for \$30,000. Having lots of empty seats, they find a group of 70 people to join them. This lowers the cost per person by \$26.25. How large is the original group?

SOLUTION Let the number of people in the original group be n. The total number of subscribers is $n + 70$.

Now where is an equation? Well, the price is *lowered* by \$26.25. Words such as "lowered" or "decreased" suggest subtraction. Here the *difference* between the two costs *is* \$26.25:

(cost per person with n people) − (cost per person with $n + 70$ people) = 26.25

Now express these costs in terms of n (as in Example 6, page 63):

$$\frac{30{,}000}{n} - \frac{30{,}000}{n + 70} = 26.25$$

EXAMPLE 6 A tank contains 1000 gallons of gasoline. Alcohol is pumped in at the rate of 13 gal/min. How long will it take for the concentration of alcohol in the mixture to reach 10%?

SOLUTION Let the required time be t minutes. Then the equation is

(fraction of alcohol in mixture after t min) = $\dfrac{1}{10}$

As in Example 5, page 63, this fraction is

$$\frac{13t}{1000 + 13t} \quad \text{hence} \quad \frac{13t}{1000 + 13t} = \frac{1}{10}$$

EXAMPLE 7 A table top has area 18 ft². Its shape is a rectangle of length twice the width having semicircular pieces at each of the narrow ends. What is the width of the table?

SOLUTION Let the unknown width be w. Draw a diagram (Fig. 2) and label all important parts in terms of w.

FIG. 2

The equation in this problem is clear:

$$\text{Area} = 18$$

Now express the area in terms of w. The rectangle has area $2w \cdot w = 2w^2$. The two semicircles amount to one circle of radius $\tfrac{1}{2}w$ having area $\pi(\text{radius})^2 = \pi(\tfrac{1}{2}w)^2$. So the total area is $2w^2 + \pi(\tfrac{1}{2}w)^2$. Hence

$$\underline{2w^2 + \pi(\tfrac{1}{2}w)^2 = 18}$$

EXERCISES

Set up the problem as an equation

1. 62 is 12% of what number?
2. 48 is 35% larger than what number?
3. A clock-radio is on sale for $29.75, marked down 30%. What is its regular price?
4. A jacket cost $41.34 including 6% sales tax. What is the price of the jacket?
5. Jane rents a car from agency A at $80 for the first week, and then for $20 per day. Ruth rents a car from agency H at a straight $15 per day. They keep their cars the same length of time and find their rental bills are the same. For how many days did they have the cars?
6. One taxi charges $1 plus 10 cents per $\tfrac{1}{8}$ mile. Another charges 50 cents plus 15 cents per $\tfrac{1}{8}$ mile. For what length of ride are the fares the same?
7. A library discarded 20% of its old books to make space for new purchases. It then acquired 3500 new books, bringing its total collection to 500 more books than it had originally. How many was that?
8. In a certain right triangle one acute angle exceeds the other by 28°. Find these angles.
9. I am playing red and black at a roulette table, betting one chip at a time. After 50 plays, I have lost 6 chips. How many times have I won?
10. The Post Office book rate (1980) was 59¢ for the first pound, 22¢ per pound for the next 6 pounds, and then 13¢ per pound after that. A certain shipment of books went for an average of $19\tfrac{1}{4}$¢ per pound. How much did it weigh?
11. Find two consecutive integers whose squares differ by 97.
12. Find three consecutive odd integers whose squares total 1883.
13. Find a number that exceeds its reciprocal by 4.

14. The difference between a certain positive number and its square root is 12. Find the number.

15. A group of people share the $150 cost of a party. If 10 more people join in, the cost per person decreases by $1.25. How many people were in the original group?

16. If the price of a certain stock would fall by $1, then for $1200 you could buy 2 more shares than you can now. What is the present price per share?

17. A tank contains 1000 gallons of gasoline. A mixture of half gas, half alcohol is pumped in at a constant rate. Find that rate given that the concentration of alcohol reaches 10% in 15 minutes.

18. A tank contains a certain amount of gasoline. A mixture of half gas, half alcohol is pumped in at a constant rate of 50 gallons per minute. The concentration of alcohol reaches 12% in 10 minutes. How much gas did the tank contain at the start?

19. A 40-foot length of rope is cut into two pieces, and then each piece is folded into a square. If the combined area of the squares is 82 ft², how long are the pieces?

20. A 40-foot length of rope is cut into two pieces, and then each piece is folded into a square. If one square has triple the area of the other, how long are the pieces?

21. How can you make a sum of one dollar with pennies and nickels, using 56 coins?

22. A collection of pennies, nickels, and dimes is worth $4. There are 10 more nickels than dimes and 20 more pennies than nickels. How many of each type of coin are there?

23. The number of bacteria in a certain colony doubles every three days. In 15 days there are 128,000 bacteria. How many were there originally?

24. A radioactive chemical decays 50% per day. In how many days will 1 gram of the substance decay to 9.537×10^{-7} grams?

25. The denominator of a certain fraction exceeds its numerator by 3. Increasing both numerator and denominator by 1 adds $\frac{1}{10}$ to the value of the fraction. Find the fraction.

26. Find two consecutive even integers whose reciprocals differ by 1/98.

27. An isosceles triangle has height 3 and perimeter 12. Find its base.

28. Do Example 7 of the text, assuming that the semicircular pieces are placed on the long sides of the rectangle.

29. Find all right triangles whose side lengths are three consecutive integers.

30. Find a right triangle whose longer leg is 1 cm shorter than the hypotenuse and 31 cm longer than the other leg.

31. After lengthening my stride (pace) by 2 inches, I can now walk a mile in 132 fewer paces. What is the length of my new stride?

32. If I increase my average driving speed by 2 mph I will save 5 minutes on a 100-mile drive. What is my usual average speed?

33. I start with a number, square it, subtract 1 and divide by 8. I repeat the process and end up with 55. What number did I start with?

34. I start with a number, add 1, and take the reciprocal. I repeat the operation and end up at the number I started with. Find that number.

3 EQUATIONS; LINEAR EQUATIONS

Now that we have some experience setting up equations, let us start solving them. For example, take the equation
$$55t = 40(t + 1)$$
For most real numbers t, this is not a true statement. If $t = 0$, it says that $0 = 40$, which is false. If $t = 1$, it says that $55 = 80$, again false, and so on. Therefore, we should write
$$55t \stackrel{?}{=} 40(t + 1)$$
because the equation is really a question asking what values of t make it true. Such an equation is called a **conditional equation.** A **solution** or **root** of the equation is a real number that makes it a true statement. **Solving** a conditional equation means finding its solutions.

Let us solve the equation $55t = 40(t + 1)$. We use a sequence of steps:

$$55t = 40(t + 1)$$
$$55t = 40t + 40$$
$$15t = 40 \qquad \text{[subtracting } 40t \text{ from both sides]}$$
$$t = \frac{40}{15} = \frac{8}{3} \qquad \text{[dividing both sides by 15]}$$

Hence the only possible solution is $\frac{8}{3}$. To check that $\frac{8}{3}$ *is* a solution, we substitute the value $t = \frac{8}{3}$ and see whether we get a true statement. We do:

$$55(\tfrac{8}{3}) = 40(\tfrac{8}{3} + 1) \qquad \text{since} \qquad \tfrac{440}{3} = 40(\tfrac{11}{3})$$

Note the strategy used to solve this equation: at each step, we transformed the equation into a simpler equation *having the same solution*. At the last step, the solution was obvious.

Two equations with the same solutions are **equivalent.** In solving equations we use certain operations that transform equations into equivalent ones.

The following operations transform an equation into an equivalent equation:

(1) adding or subtracting the same quantity on both sides

(2) multiplying or dividing both sides by the same non-zero quantity

---

Because of these principles, we really did not have to check the solution $t = \frac{8}{3}$. They guarantee that the equation $t = \frac{8}{3}$ is equivalent to the equation we started with. Nevertheless, checking is a valuable procedure for catching errors.

Identities

Much different from a conditional equation is an **identity** such as
$$x^2 - 1 = (x + 1)(x - 1)$$
This is not a question; it is a true statement for *all* real numbers x. Hence $x^2 - 1$ and $(x + 1)(x - 1)$ are exactly the same thing, but expressed differently.

In general, an identity is an equation that is true for all meaningful values of the variable. For example, the equation
$$\sqrt{4y} = 2\sqrt{y}$$
is an identity; it is true for every $y \geq 0$, the only values for which both sides are defined. Another identity is
$$\frac{1}{u - 2} - \frac{1}{u} = \frac{2}{u(u - 2)}$$
This is true for all values of u except 0 and 2, for which neither side has meaning.

When we talk about "solving an equation," we mean a conditional equation.

Linear Equations

A **linear equation** is an equation of the form
$$ax + b = 0 \qquad a \neq 0$$
or one that can be brought into this form by adding and subtracting quantities on both sides. For example,

$2x + 1 = 4$	can be written as	$2x - 3 = 0$
$x + 6 = 4.93x - 0.85$	can be written as	$3.93x - 6.85 = 0$

The equation $x^2 + 7x = x^2 - 3$ is not linear but is equivalent to the linear equation $7x = -3$.

To solve the linear equation $ax + b = 0$, we subtract b from both sides, and then divide by a:
$$ax + b = 0 \qquad a \neq 0$$
$$ax = -b$$
$$x = -\frac{b}{a}$$

The third equation is equivalent to the first. Hence $-b/a$ is a solution and the only solution.

A linear equation
$$ax + b = 0 \qquad a \neq 0$$
has exactly one solution:
$$x = -\frac{b}{a}$$

EXAMPLE 1 Solve (a) $5x + 1 = 8$ (b) $9(x - 2) = 3(2x + 1) - 6$

SOLUTION
(a)
$$5x + 1 = 8$$
$$5x = 7 \quad \text{[subtracting 1]}$$
$$x = \tfrac{7}{5} \quad \text{[dividing by 5]}$$

(b) First expand both sides:
$$9x - 18 = 6x + 3 - 6$$

Now solve:
$$9x = 6x + 15 \quad \text{[adding 18]}$$
$$3x = 15 \quad \text{[subtracting } 6x\text{]}$$
$$x = \underline{5} \quad \text{[dividing by 3]}$$

EXAMPLE 2 Solve $\dfrac{2x + 1}{x - 4} = \dfrac{3}{5}$

SOLUTION We assume that $x \neq 4$ since that value makes the denominator 0 on the left side. To clear of fractions, we multiply both sides by $5(x - 4)$:
$$5(2x + 1) = 3(x - 4)$$
We expand each side and solve:
$$10x + 5 = 3x - 12$$
$$7x = -17$$
$$x = \underline{-\tfrac{17}{7}}$$

Remark Clearing of fractions in the solution of Example 2 amounts to "cross-multiplying":
$$\frac{a}{b} = \frac{c}{d} \quad \text{is equivalent to} \quad ad = bc$$
provided the denominators are non-zero.

In real-life applications of linear equations, the numbers in the problem may not be so pleasant. Then it helps enormously to use a calculator.

EXAMPLE 3 Solve to 3 decimal places using a calculator
$$53.871x - 12.096 = (3.4)(7.581x + 6.499)$$

SOLUTION It's best to solve for x before calculating:
$$x[53.871 - (3.4)(7.581)] = 12.096 + (3.4)(6.499)$$
$$x = \frac{12.096 + (3.4)(6.499)}{53.871 - (3.4)(7.581)}$$

Now use the calculator: $x = \underline{1.217}$

EXAMPLE 4 When the Celsius temperature is C, the corresponding Fahrenheit temperature is $F = \tfrac{9}{5}C + 32$. Express C in terms of F.

SOLUTION The problem means this: given F, find C. We think of F as a number and C as the unknown. Then we have a linear equation to solve for C:

$$F = \tfrac{9}{5}C + 32 \quad \text{hence} \quad \tfrac{9}{5}C = F - 32$$
$$\underline{C = \tfrac{5}{9}(F - 32)}$$

EXAMPLE 5 Solve for x in terms of y:

$$y = \frac{3x + 5}{2x + 6}$$

SOLUTION It is understood that the denominator $2x + 6$ is non-zero, that is, $x \neq -3$. Multiply both sides by $2x + 6$:

$$y(2x + 6) = 3x + 5$$
$$2yx + 6y = 3x + 5$$

Now think of y as a given number. Then this is a linear equation for the unknown x. Solve for x:

$$2yx - 3x = -6y + 5$$
$$(2y - 3)x = -6y + 5$$
$$x = \underline{\frac{-6y + 5}{2y - 3}} \quad [y \neq \tfrac{3}{2}]$$

EXERCISES

Is the equation a conditional equation or an identity?

1. $x^2 + 7x = 22$
2. $x^3 = 1 + 4\sqrt{x}$
3. $x^3 + 5x = x(x^2 + 5)$
4. $(x + 1)(x - 3) = x^2 - 2x - 3$
5. $4 + \dfrac{1}{x} = \dfrac{4x + 1}{x}$
6. $\dfrac{x^2 + 8x + 2}{x} = x + 8 + \dfrac{2}{x}$
7. $(x + 1)^2 - x^2 = 1$
8. $(x + 1)^2 - x^2 = 2(x + 1) - 1$
9. $4(2x - 3) + 1 = 6(x - 1) + 2x - 5$
10. $\sqrt{x^2 + 9} = x + 3$

Solve for x

11. $4x + 1 = 25$
12. $\tfrac{1}{2}x - 4 = 3$
13. $6x + 3 = 2x - 5$
14. $9x + 7 = 2x + 8$
15. $2(x + 3) = 3(2x - 8)$
16. $5(4x - 3) = 2(x - 12)$
17. $3(5x - 1) + 2(x - 7) = 4x + 3$
18. $3x + 4 - 5(x - 1) = 13x + 10$
19. $\tfrac{1}{2}x + \tfrac{1}{3}(5x + 1) = \tfrac{1}{6}(x - 10)$
20. $2x - \tfrac{1}{4}(x - 7) = \tfrac{1}{3}(2x + 1) + 1$
21. $(x - 3)^2 = x^2 + 2$
22. $(x + 1)^2 = (x - 4)^2 + 6$
23. $(4x + 3)(3x + 1) - 12(x^2 - 2) = 1$
24. $(2x + 1)(3x + 1) = 6x^2 - 8x + 17$

74 Equations

25. $\dfrac{2x-1}{2x+1} = \dfrac{1}{4}$

26. $\dfrac{2x+5}{3x+7} = \dfrac{4}{3}$

27. $\dfrac{1}{x} = \dfrac{4}{3x+1}$

28. $\dfrac{5}{x} = \dfrac{6}{2x-1}$

29. $\dfrac{2x-1}{2x+1} = \dfrac{x+7}{x+8}$

30. $\dfrac{3x+5}{3x+7} = \dfrac{x-1}{x-2}$

31. $\dfrac{x^2-3x}{x^2+4x+7} = 1$

32. $\dfrac{x^2+x+3}{2x^2+4x+19} = \dfrac{1}{2}$

33. $\dfrac{6x}{x+5} = 1 + \dfrac{2x}{x+5}$

34. $\dfrac{x}{4x-1} = \dfrac{1}{4x-1} + 3$

35. $\dfrac{1}{x+3} + \dfrac{4}{x-3} = \dfrac{x}{x^2-9}$

36. $\dfrac{5}{x^2-4} = \dfrac{1}{x-2} + \dfrac{3}{x+2}$

Solve to 3 decimal places, using a calculator

37. $1.83(9.5x + 7.4) = 12.76x - 33.65$

38. $5.47x = 11.3(0.85x + 4.92) - 21.67$

39. $\dfrac{851x + 4961}{2384x + 577} = 1.35$

40. $\dfrac{8806}{65x + 329} = \dfrac{9272}{14x + 101}$

Solve

41. $E = mc^2$ for m

42. $A = \tfrac{1}{2}bh$ for b

43. $\dfrac{1}{f} = \dfrac{1}{f_1} + \dfrac{1}{f_2}$ for f_1

44. $s = s_0 + vt$ for t

45. $\pi r^2 h + 2\pi rh = S$ for h

46. $h = r^2(\pi - h)$ for h

Solve for x in terms of y

47. $\dfrac{x}{a} + \dfrac{y}{b} = 1$

48. $ax - by - c = 0$

49. $x(y + 3) = 4x + 5y + 1$

50. $xy + x + y = 3y - x - 5$

51. $y = \dfrac{2x+1}{3x+5}$

52. $\dfrac{y}{x} = \dfrac{2y+1}{x+3}$

53. $xy^4 - 1 = 2y - x - xy^2$

54. $(2x - y)(y^2 + 1) = xy^2$

Find a number a such that

55. $\dfrac{2x+a}{3x+a} = \dfrac{1}{9}$ has solution $x = 5$

56. $3x + 1 = 2x - 4 + a(5x + 2)$ has no solution

57. $a(x + 4) + 20 = 8x + a(3x - 1)$ is an identity

4 APPLICATIONS OF LINEAR EQUATIONS

In this section, we solve a variety of problems using linear equations.

EXAMPLE 1 17% of what number is 68?

SOLUTION Let the unknown be x. Then x satisfies the equation
$$\frac{17}{100}x = 68$$
To solve for x, multiply both sides by 100:
$$17x = 6800$$
$$x = \frac{6800}{17} = \underline{400}$$

EXAMPLE 2 The Audio Shop advertises a sale on stereo equipment: 30% off after the first $50. If you have $330 to spend, what is the pre-sale price of a system you can now afford?

SOLUTION Let x be the pre-sale price ($x > 50$). The sale price is $50 + 0.7(x - 50)$, as we saw in Example 3, page 62. Hence, for $330 you can buy a system whose pre-sale price satisfies the equation
$$50 + 0.7(x - 50) = 330$$
Solve for x:
$$0.7(x - 50) = 280$$
$$x - 50 = \frac{280}{0.7} = 400$$
$$x = \underline{\$450}$$

EXAMPLE 3 A team has 11 wins and 12 losses. How many consecutive victories will bring their winning percentage up to 60%?

SOLUTION The team now has 11 wins out of 23 games. Suppose they win n straight games. Then the fraction of games won will be
$$\frac{\text{games won}}{\text{games played}} = \frac{11 + n}{23 + n}$$
This is equivalent to 60% provided
$$\frac{11 + n}{23 + n} = \frac{3}{5}$$
Solve for n:
$$5(11 + n) = 3(23 + n)$$
$$55 + 5n = 69 + 3n$$
$$5n - 3n = 69 - 55$$
$$2n = 14$$
$$n = \underline{7}$$

EXAMPLE 4 A large pump can unload an oil tanker in 24 hours. A smaller pump requires 36 hours. How long will it take if both pumps work together?

SOLUTION The key is how much work is done in *one* hour. The big pump does $\frac{1}{24}$ of the job in one hour; the small pump does $\frac{1}{36}$ of the job. Together, in one hour they do

$$\tfrac{1}{24} + \tfrac{1}{36} = \tfrac{3}{72} + \tfrac{2}{72} = \tfrac{5}{72} \text{ of the job.}$$

Suppose it takes t hours to do 1 complete job. Then

$$\tfrac{5}{72}t = 1 \qquad t = \tfrac{72}{5} = \underline{14\tfrac{2}{5} \text{ hours}}$$

EXAMPLE 5 A tank contains 1000 gallons of gasoline. Alcohol is piped in at the rate of 13 gallons per minute. When will the concentration of alcohol in the mixture reach 10%?

SOLUTION We set this one up in Example 6, page 67. Let the required time be t minutes. Then t satisfies

$$\frac{13t}{1000 + 13t} = \frac{1}{10}$$

Solve for t:

$$130t = 1000 + 13t$$
$$117t = 1000$$
$$t = \tfrac{1000}{117} = \underline{8.55 \text{ min}} \qquad [\text{approx.}]$$

EXAMPLE 6 How many liters of a 75% acid solution must be mixed with 15 liters of a 45% acid solution to produce a mixture of 70% acid?

SOLUTION The equation here is

$$\frac{\text{volume of acid in mixture}}{\text{volume of mixture}} = 0.7$$

Let x be the number of liters of the 75% solution to be added. The total volume of the mixture will be $15 + x$ liters. How much acid will it contain?

$$\begin{array}{ll} (0.45)(15) & \text{liters from the given 15 liters} \\ + \ 0.75x & \text{from the } x \text{ liters added} \\ \hline (0.45)(15) + 0.75x = & \text{total volume of acid} \end{array}$$

Therefore, the equation becomes

$$\frac{(0.45)(15) + 0.75x}{15 + x} = 0.7$$

Solve for x:

$$(0.45)(15) + 0.75x = 0.7(15 + x)$$
$$6.75 + 0.75x = 10.5 + 0.7x$$
$$0.75x - 0.7x = 10.5 - 6.75$$
$$0.05x = 3.75$$
$$x = \frac{3.75}{0.05} = \underline{75 \text{ liters}}$$

The distance traveled by a vehicle at constant speed is given by the formula

$$\text{distance} = \text{rate} \times \text{time} \qquad d = rt$$

In many problems you are given two of the three quantities and you must solve for the third. It boils down to applying $d = rt$ or one of the equivalent forms,

$$r = \frac{d}{t} \qquad \text{or} \qquad t = \frac{d}{r}$$

EXAMPLE 7 Two river boats have top speed of 10 mph in still water. They depart at the same time and go at full speed, one upstream against the current, and one downstream with the current. When the first has gone 40 miles, the second has gone 60 miles. How fast is the current?

SOLUTION The unknown is the speed of the current; call it c. Now what quantities are equal? The word "when" in this problem means "at the same time that." Therefore,

$$(\text{time traveled by Boat 1}) = (\text{time traveled by Boat 2})$$

Apply the formula $t = d/r$:

$$\frac{\text{distance traveled by Boat 1}}{\text{speed of Boat 1}} = \frac{\text{distance traveled by Boat 2}}{\text{speed of Boat 2}}$$

The distances are 40 miles for Boat 1 and 60 for Boat 2. The speeds are $10 - c$ for Boat 1 against the current, and $10 + c$ for Boat 2 with the current. Hence the equation becomes

$$\frac{40}{10 - c} = \frac{60}{10 + c}$$

Solve for c:

$$40(10 + c) = 60(10 - c)$$
$$400 + 40c = 600 - 60c$$
$$100c = 200$$
$$c = \underline{2 \text{ mph}}$$

EXAMPLE 8 Express the repeating decimal $0.58585858\cdots$ as a fraction.

SOLUTION Let $0.585858\cdots = x$. Then if we multiply x by 100, we get $58.585858\cdots$, a number whose decimal part is x again! In algebra,

$$100x = 58 + x$$
$$99x = 58$$

Hence

$$x = \underline{\frac{58}{99}}$$

EXERCISES

1. 56 is 7% of what number?   2. 20.7 is 15% larger than what number?

3. A clock-radio is on sale for $29.75, marked down 30%. What is its regular price?

Equations

4. A jacket costs $41.34 including sales tax of 6%. What is the price of the jacket?

5. Jane rents a car from agency A at $80 for the first week, and then for $20 per day. Ruth rents a car from agency H at a straight $15 per day. They keep their cars the same length of time and find their rental bills are the same. For how many days did they have the cars?

6. A bookstore sells a card for $5 that entitles you to a 15% discount on all books. If you have $80 to spend, what is the pre-discount price of books you can now buy?

7. A stockbroker charges a commission of $25 plus 1.5% of the amount of the transaction. A second broker charges $10 plus 1.8% of the amount. What is the amount of a transaction for which the commissions are the same?

8. One plumber charges $19 per call plus $9.60/hour. A second plumber charges a flat $13.60/hour. For a job of how many hours are their charges equal?

9. A team has 11 wins and 9 losses. How many consecutive games must they win to have won $\frac{2}{3}$ of their games?

10. A team has won half of its games. If they win today, their winning percentage will go up to 52%. How many games have they played?

11. In a certain triangle, the second angle is 5° larger than the first, and the third angle is triple the second. Find all three angles.

12. In a certain isosceles triangle, the two equal angles are each 22° larger than the third angle. Find all three angles.

13. The relation between Fahrenheit and Celsius readings is given by $F = \frac{9}{5}C + 32$. For what temperature do both scales give the same reading?

14. For what temperature is the Fahrenheit reading twice the Celsius reading?

15. If your midterm grades are 85, 62, 73, what do you need on the final to average 75? What if the final carries double weight?

16. U.S. government regulations require an average efficiency of at least 27.5 miles per gallon for all cars sold by a manufacturer in 1985. A certain manufacturer projects 1985 sales of 150,000 standard-sized cars averaging 25.6 mpg. How many compacts averaging 29.0 mpg must it sell in 1985 to average exactly 27.5 mpg on all models sold?

17. Find two consecutive integers whose squares differ by 97.

18. If the radius of a certain circle were increased by 1 cm, its area would increase by 23π cm². What is the radius?

19. Al can mow a large lawn in 4 hours, and Bob can do it in 5 hours. Working together, each with a lawnmower, how long will it take them?

20. A large bulldozer can clear a plot of land in 2 hours, and a smaller bulldozer can do it in $3\frac{1}{2}$ hours. Working together, how long will it take them?

21. Suppose an average professor can grade a final exam in 9 hours, and an average assistant can do the same job in 6 hours. If 2 professors and 5 assistants work together, how long will the grading take?

22. If A can plow a certain field in 5 hours, B in 4 hours, and C in 3 hours, then how long will it take them, working together, to plow the field?

23. Working together, Gail and Sue can paint a room in 2 hours. Alone, it takes Gail 6 hours. How long does it take Sue alone?

24. Together, two pumps can empty a tank in 8 hours. Alone, it takes one pump 20 hours. How long does it take the other pump alone?

25. A tank contains 1000 gallons of gasoline. At what constant rate must a mixture of half gas–half alcohol be added so that the concentration of alcohol will reach 10% in 15 minutes?

26. A tank contains a certain amount of pure gasoline. A mixture of half gas–half alcohol is added at the rate of 40 gal/min. After 12 minutes, the concentration of alcohol is 5%. How much gas was in the tank at the start?

27. On a $2\frac{1}{2}$-mile track, how long would it take a race car at 190 mph to "lap" another car at 180 mph?

28. Steve started driving at 40 mph. Later I set out at 50 mph and caught up to him in 2 hours. How long after Steve left did I leave?

29. The current in a river is 2 mph. A boat can go 60 miles downstream in the same time as it can go 36 miles upstream. What is the boat's speed in still water?

30. An eastbound 747 aided by a tailwind of 60 mph flies 1000 miles, while a westbound 747 flying against the same wind flies 800 miles. What is the common speed of these planes in still air?

31. An Olympic athlete averages 4:00 for the mile run; a good college runner averages 4:10. What handicap should the college runner be given to make a fair race? (The Olympic star will run a full mile, the other runner less than a mile.)

32. A hare and a tortoise will run a half-mile race. The hare's speed is 20 miles per hour, and the tortoise's speed is 10 feet per minute. The hare tires easily, so he plans on a one-hour nap along the way. What handicap can he give the tortoise and still tie?

33. How many minutes after 4:00 will the minute hand of a clock first lie directly over the hour hand?

34. At what time between 2:00 and 3:00 will the hands of a clock point in exactly opposite directions?

35. How many grams of 60-40 solder (60% tin, 40% lead) must be mixed with 40-60 solder to produce 500 g of 55-45 solder?

36. How many gallons of 100-proof whiskey (50% alcohol) must be mixed with 80-proof whiskey to produce 50 gal of 95-proof whiskey?

37. Risky stock X pays 10% dividend, and safe bond B pays 6% dividend. How should I invest $10,000 to have an income of $850?

38. How should $20,000 be split between a 5% savings account and a 9% investment fund to yield a 6% annual return?

Express the repeating decimal as a fraction

39. $0.111111\cdots$ **40.** $0.474747\cdots$

41. $0.892892892\cdots$ **42.** $0.508350835083\cdots$

43. $0.75131313\cdots$ **44.** $0.8234234234\cdots$

    (First multiply by 100)   (First multiply by 10)

45. A long convoy of heavy trucks is moving at 20 mph. A messenger on motorcycle at 60 mph can go from the end of the convoy to the front and back again in 6 minutes. How long is the convoy?

46. (Just for fun) If Don can climb a mountain in 6 hours and Mark can climb it in 7 hours, how long will it take them climbing together?

5 QUADRATIC EQUATIONS

A **quadratic equation** is an equation of the form

$$ax^2 + bx + c = 0 \qquad a \neq 0$$

or an equation that can be brought into this form by adding or subtracting terms on both sides. For example,

$$9x^2 - 4x + \tfrac{1}{2} = 0$$

is a quadratic equation. So is

$$3x^2 + 5x - 1 = x^2 + x$$

because subtracting $x^2 + x$ from both sides brings it into the form

$$2x^2 + 4x - 1 = 0$$

Solution by Factoring

We can solve a quadratic equation $ax^2 + bx + c = 0$ whenever we can factor $ax^2 + bx + c$. The method uses a basic property of real numbers: If $rs = 0$, then either $r = 0$ or $s = 0$.

EXAMPLE 1 Solve   (a) $x^2 - 5x = 0$ (b) $x^2 - 3x - 4 = 0$

SOLUTION   (a)   Factor $x^2 - 5x$:

$$x(x - 5) = 0$$

This product is 0 only if either

$$x = 0 \qquad \text{or} \qquad x - 5 = 0$$
$$x = \underline{5}$$

Thus 0 and 5 are the only possible solutions. But they *are* solutions since each one makes the left side 0.

(b)   Factor $x^2 - 3x - 4$:

$$(x - 4)(x + 1) = 0$$

The product is 0 only if either

$$x - 4 = 0 \quad \text{or} \quad x + 1 = 0$$
$$x = \underline{4} \quad\quad\quad\quad x = \underline{-1}$$

EXAMPLE 2 Solve (a) $3x^2 - 7x + 2 = 0$ (b) $x^2 - 6x + 9 = 0$

SOLUTION (a) Factor $3x^2 - 7x + 2$:

$$(3x - 1)(x - 2) = 0$$

Hence either

$$3x - 1 = 0 \quad \text{or} \quad x - 2 = 0$$
$$x = \underline{\tfrac{1}{3}} \quad\quad\quad\quad x = \underline{2}$$

(b) The left side is a perfect square:

$$(x - 3)^2 = 0$$

The only solution is $x = \underline{3}$.

Remark In Example 2(b), the equation can be written as

$$(x - 3)(x - 3) = 0$$

There is only *one root*, 3, but there are *two factors* $x - 3$. We call 3 a **double root** of the equation.

The factoring method for solving quadratic equations is fine when it works. But factoring is generally hopeless unless the coefficients are small integers, and even then it may be hard or impossible. Fortunately, there is a more general method.

Completing the Square

The quadratic equation

$$(x - 2)^2 = 9$$

is easy to solve because the left-hand side is a perfect square. Clearly $x - 2$ must be one of the square roots of 9. Either

$$x - 2 = 3 \quad \text{or} \quad x - 2 = -3$$
$$x = 5 \quad\quad\quad\quad x = -1$$

Now look at the equation

$$x^2 + 6x + 4 = 0$$

The left side is not a perfect square, but something similar is: $x^2 + 6x + 9$. So *make* a perfect square. Add and subtract 9 on the left side:

$$(x^2 + 6x + 9) - 9 + 4 = 0$$
$$(x + 3)^2 - 5 = 0$$
$$(x + 3)^2 = 5$$

Hence either

$$x + 3 = \sqrt{5} \quad \text{or} \quad x + 3 = -\sqrt{5}$$
$$x = -3 + \sqrt{5} \quad\quad\quad\quad x = -3 - \sqrt{5}$$

For short, we write $x = -3 \pm \sqrt{5}$.

We have used a technique called **completing the square**. Any expression $x^2 + bx$ can be written as
$$x^2 + 2(\tfrac{1}{2}b)x$$
These are the first two terms of the perfect square
$$x^2 + 2(\tfrac{1}{2}b)x + (\tfrac{1}{2}b)^2 = (x + \tfrac{1}{2}b)^2$$
So, given $x^2 + bx$, we can create a perfect square by adding and subtracting $(\tfrac{1}{2}b)^2$.

Completing the Square Given any expression $x^2 + bx$ add and subtract $(\tfrac{1}{2}b)^2$:
$$\begin{aligned} x^2 + bx &= x^2 + 2(\tfrac{1}{2}b)x \\ &= x^2 + 2(\tfrac{1}{2}b)x + (\tfrac{1}{2}b)^2 - (\tfrac{1}{2}b)^2 \\ &= (x + \tfrac{1}{2}b)^2 - \tfrac{1}{4}b^2 \end{aligned}$$

Examples

$x^2 + 10x$ Add and subtract $(\tfrac{1}{2} \cdot 10)^2 = 5^2$:
$$\begin{aligned} x^2 + 10x &= x^2 + 2 \cdot 5x + 5^2 - 5^2 \\ &= (x + 5)^2 - 25 \end{aligned}$$

$x^2 - 3x$ Add and subtract $[\tfrac{1}{2}(-3)]^2 = (-\tfrac{3}{2})^2$:
$$\begin{aligned} x^2 - 3x &= x^2 + 2(-\tfrac{3}{2})x + (-\tfrac{3}{2})^2 - (-\tfrac{3}{2})^2 \\ &= (x - \tfrac{3}{2})^2 - \tfrac{9}{4} \end{aligned}$$

EXAMPLE 3 Solve $x^2 - 12x + 29 = 0$ by completing the square.

SOLUTION On the left side, add and subtract $(-6)^2$:
$$\begin{aligned} x^2 + 2(-6)x + (-6)^2 - (-6)^2 + 29 &= 0 \\ (x - 6)^2 - 36 + 29 &= 0 \\ (x - 6)^2 &= 7 \\ x - 6 &= \pm\sqrt{7} \\ x &= \underline{6 \pm \sqrt{7}} \end{aligned}$$

EXAMPLE 4 Solve $x^2 + 5x + 8 = 0$ by completing the square.

SOLUTION Add and subtract $(\tfrac{5}{2})^2$ on the left:
$$\begin{aligned} x^2 + 2 \cdot \tfrac{5}{2}x + (\tfrac{5}{2})^2 - (\tfrac{5}{2})^2 + 8 &= 0 \\ (x + \tfrac{5}{2})^2 &= (\tfrac{5}{2})^2 - 8 \end{aligned}$$
Hence the given equation is equivalent to
$$(x + \tfrac{5}{2})^2 = -\tfrac{7}{4}$$
But this is impossible; the square of a real number cannot be negative. Therefore the equation has no real solutions.

Remark You may prefer to work Example 4 this way. Instead of adding and subtracting $(\tfrac{5}{2})^2$ on the left, just *add* $(\tfrac{5}{2})^2$ on *both* sides:

$$x^2 + 5x + 8 = 0$$
$$x^2 + 2(\tfrac{5}{2})x + (\tfrac{5}{2})^2 + 8 = (\tfrac{5}{2})^2$$
$$(x + \tfrac{5}{2})^2 = (\tfrac{5}{2})^2 - 8$$

We have completed the square to solve equations of the form $x^2 + bx + c = 0$. The method applies also to equations $ax^2 + bx + c = 0$, where $a \neq 1$. We first divide the equation by a:

$$x^2 + \frac{b}{a}x + \frac{c}{a} = 0$$

Then we complete the square as before.

EXAMPLE 5 Solve $3x^2 - 7x - 1 = 0$ by completing the square.

SOLUTION First, divide by 3:

$$x^2 - \tfrac{7}{3}x - \tfrac{1}{3} = 0$$

Now complete the square by adding $(-\tfrac{7}{6})^2$ to both sides:

$$x^2 + 2(-\tfrac{7}{6})x + (-\tfrac{7}{6})^2 - \tfrac{1}{3} = (-\tfrac{7}{6})^2$$
$$(x - \tfrac{7}{6})^2 = (-\tfrac{7}{6})^2 + \tfrac{1}{3} = \tfrac{61}{36}$$
$$x - \tfrac{7}{6} = \pm\sqrt{\tfrac{61}{36}} = \pm\tfrac{1}{6}\sqrt{61}$$

Hence $\quad x = \tfrac{7}{6} \pm \tfrac{1}{6}\sqrt{61} \quad$ or equivalently $\quad x = \underline{\dfrac{7 \pm \sqrt{61}}{6}}$

Completing the square changes a quadratic expression $ax^2 + bx + c$ into the form $a(x + p)^2 + q$. This has uses in algebra other than solving equations.

EXAMPLE 6 Express $2x^2 + 9x + 6$ in the form $a(x + p)^2 + q$.

SOLUTION First take out a factor of 2:

$$2x^2 + 9x + 6 = 2[x^2 + \tfrac{9}{2}x + 3]$$

Now complete the square. Add and subtract $(\tfrac{9}{4})^2$ inside the brackets:

$$2x^2 + 9x + 6 = 2[x^2 + 2 \cdot \tfrac{9}{4}x + (\tfrac{9}{4})^2 - (\tfrac{9}{4})^2 + 3]$$
$$= 2[x^2 + 2 \cdot \tfrac{9}{4}x + (\tfrac{9}{4})^2] + 2[-(\tfrac{9}{4})^2 + 3]$$
$$= 2(x + \tfrac{9}{4})^2 + 2(-\tfrac{81}{16} + \tfrac{48}{16})$$
$$= \underline{2(x + \tfrac{9}{4})^2 - \tfrac{33}{8}}$$

This is the desired form with $a = 2$, $p = \tfrac{9}{4}$, and $q = -\tfrac{33}{8}$.

EXERCISES

Solve by factoring

1. $3x^2 - 4x = 0$ **2.** $5x^2 + 2x = 0$

3. $x^2 - 5x - 14 = 0$ **4.** $x^2 - 7x + 6 = 0$

5. $x^2 + x - 12 = 0$
6. $x^2 - x - 30 = 0$
7. $x^2 + 8x + 16 = 0$
8. $x^2 - 10x + 25 = 0$
9. $2x^2 - 9x + 4 = 0$
10. $2x^2 + 13x + 15 = 0$
11. $6x^2 - 5x + 1 = 0$
12. $9x^2 - 11x + 2 = 0$
13. $8x^2 + 13x - 6 = 0$
14. $12x^2 + 17x + 6 = 0$
15. $12x^2 - 29x + 10 = 0$
16. $6x^2 + 25x + 14 = 0$

Solve by completing the square

17. $x^2 + 4x + 2 = 0$
18. $x^2 + 2x - 5 = 0$
19. $x^2 + 5x + 5 = 0$
20. $x^2 + 7x + 1 = 0$
21. $x^2 - 6x + 11 = 0$
22. $x^2 - 3x + 6 = 0$
23. $3x^2 + 12x + 1 = 0$
24. $2x^2 + 4x - 1 = 0$
25. $\frac{1}{2}x^2 - 4x - \frac{9}{2} = 0$
26. $\frac{1}{2}x^2 + x + 3 = 0$
27. $16x^2 + 8x + 1 = 0$
28. $9x^2 - 6x + 1 = 0$
29. $6(x^2 + x) + 7x = 3x - 1$
30. $3x^2 + 5(x - 1) = 9x - 4$
31. $(2x + 3)^2 = x - 8$
32. $(x + 1)(2x - 3) = x^2 + x + 7$

Express in the form $a(x + p)^2 + q$

33. $2x^2 - 6x + 7$
34. $3x^2 + 12x + 5$
35. $5x^2 + 2x + 1$
36. $4x^2 + 3x + 2$
37. $\frac{1}{3}x^2 - 4x - 2$
38. $\frac{1}{2}x^2 - 6x + 1$
39. $x^2 - 2kx$
40. $x^2 + 8bx$
41. $-3x^2 + 4x$
42. $-2x^2 + 4x + 7$

Verify each statement by completing the square

43. $x^2 - 2x + 6 > 0$ for all real values of x.
44. $x^2 - 3x + 15 > 0$ for all real values of x.
45. The equation $x^2 + 2x + c = 0$ has no real solutions if $c > 1$.
46. The equation $x^2 + cx + 2 = 0$ has no real solutions if $c^2 < 8$.

6 THE QUADRATIC FORMULA; APPLICATIONS

Instead of completing the square each time we solve a quadratic equation, let us do it once and for all. The most general quadratic equation is

$$ax^2 + bx + c = 0 \qquad a \neq 0$$

Dividing by a, we have
$$x^2 + \frac{b}{a}x + \frac{c}{a} = 0$$
which we write as
$$x^2 + 2\left(\frac{b}{2a}\right)x = -\frac{c}{a}$$
Now we complete the square, adding $\left(\frac{b}{2a}\right)^2$ to both sides:
$$x^2 + 2\left(\frac{b}{2a}\right)x + \left(\frac{b}{2a}\right)^2 = \left(\frac{b}{2a}\right)^2 - \frac{c}{a}$$
$$\left(x + \frac{b}{2a}\right)^2 = \frac{b^2}{4a^2} - \frac{c}{a} = \frac{b^2}{4a^2} - \frac{4ac}{4a^2}$$

The result is an equation equivalent to $ax^2 + bx + c = 0$, but in a more convenient form:

(1) $$\left(x + \frac{b}{2a}\right)^2 = \frac{b^2 - 4ac}{4a^2}$$

Let us examine Equation (1). If $b^2 - 4ac$ is positive, then the right side is positive. We take square roots on both sides:
$$x + \frac{b}{2a} = \pm\frac{\sqrt{b^2 - 4ac}}{2a}$$

There are two solutions:

(2) $$x = -\frac{b}{2a} \pm \frac{\sqrt{b^2 - 4ac}}{2a} = \frac{-b \pm \sqrt{b^2 - 4ac}}{2a}$$

If $b^2 - 4ac = 0$, Equation (1) reduces to
$$\left(x + \frac{b}{2a}\right)^2 = 0$$
There is only one solution, $x = -b/2a$ (a double root). This solution is also given by Formula (2) with $b^2 - 4ac = 0$.

Quadratic Formula If $b^2 - 4ac \geq 0$, then the solutions of the quadratic equation
$$ax^2 + bx + c = 0 \qquad a \neq 0$$
are given by the formula
$$x = \frac{-b \pm \sqrt{b^2 - 4ac}}{2a}$$

If $b^2 - 4ac$ is negative, the right side of equation (1) is negative. Then there are no solutions, because the square of a real number cannot be negative. In all cases, the crucial quantity is $b^2 - 4ac$, called the **discriminant** of the quadratic equation. Its sign determines the nature of the solutions.

Solutions of Quadratic Equations The solutions of the quadratic equation
$$ax^2 + bx + c = 0 \qquad a \neq 0$$
depend on the sign of its discriminant $b^2 - 4ac$.

(1) If $b^2 - 4ac > 0$, there are two real solutions,
$$x = \frac{-b \pm \sqrt{b^2 - 4ac}}{2a}$$

(2) If $b^2 - 4ac = 0$, there is one real solution, a double root, $x = -\frac{b}{2a}$.

(3) If $b^2 - 4ac < 0$, there are no real solutions.

Although a quadratic equation may have no real solutions, it always has solutions in a larger number system called the *complex number system*. This will be discussed in Chapter 9.

EXAMPLE 1 Solve (a) $x^2 - 7x + 5 = 0$ (b) $6x^2 + 2x - 3 = 0$

SOLUTION (a) Use the quadratic formula with $a = 1$, $b = -7$, and $c = 5$:

$$x = \frac{-b \pm \sqrt{b^2 - 4ac}}{2a} = \frac{-(-7) \pm \sqrt{(-7)^2 - 4 \cdot 1 \cdot 5}}{2 \cdot 1}$$
$$= \frac{7 \pm \sqrt{49 - 20}}{2}$$
$$= \frac{7 \pm \sqrt{29}}{2}$$

(b) Use the quadratic formula with $a = 6$, $b = 2$, and $c = -3$:

$$x = \frac{-b \pm \sqrt{b^2 - 4ac}}{2a} = \frac{-2 \pm \sqrt{2^2 - 4(6)(-3)}}{2 \cdot 6}$$
$$= \frac{-2 \pm \sqrt{4 + 72}}{12}$$
$$= \frac{-2 \pm \sqrt{76}}{12}$$
$$= \frac{-2 \pm 2\sqrt{19}}{12} = \frac{-1 \pm \sqrt{19}}{6}$$

EXAMPLE 2 Without solving the equation, determine how many real solutions it has

(a) $4x^2 + 20x + 15 = 0$ (b) $2.541x^2 + 1.013x + 4.937 = 0$

SOLUTION Just check the sign of the discriminant.

(a) discriminant = $20^2 - 4 \cdot 4 \cdot 15 = 400 - 240 = 160$

The discriminant is positive. Hence there are <u>two real solutions</u>.

(b) discriminant = $(1.013)^2 - 4(2.541)(4.937)$

No need to compute it exactly; the discriminant is certainly negative. Hence there are <u>no real solutions</u>.

EXAMPLE 3 Solve for x in terms of y:
$$x^2 - xy - 2(y^2 + 1) = 0$$

SOLUTION Think of y as a given number. Then the equation is a quadratic equation for x:
$$ax^2 + bx + c = 0 \quad \text{where} \quad a = 1, \quad b = -y, \quad c = -2(y^2 + 1)$$
By the quadratic formula,
$$x = \frac{y \pm \sqrt{y^2 - 4 \cdot 1 \cdot (-2)(y^2 + 1)}}{2} = \underline{\frac{y \pm \sqrt{9y^2 + 8}}{2}}$$

EXAMPLE 4 Solve $\dfrac{7}{x} + \dfrac{1}{x - 1} = 3$

SOLUTION It is understood that $x \neq 0$ and $x \neq 1$. Multiply both sides by the non-zero quantity $x(x - 1)$:
$$7(x - 1) + x = 3x(x - 1)$$
$$8x - 7 = 3x^2 - 3x$$
$$0 = 3x^2 - 11x + 7$$

By the quadratic formula,
$$x = \frac{11 \pm \sqrt{11^2 - 4 \cdot 3 \cdot 7}}{2 \cdot 3} = \underline{\frac{11 \pm \sqrt{37}}{6}}$$

Applications of Quadratic Equations

EXAMPLE 5 A group of people charter a 747 for $30,000. Having lots of empty seats, they find a group of 70 people to join them. This lowers the cost per person by $26.25. How large is the original group?

SOLUTION We set up this problem in Example 5, page 67. Let n be the number of people in the original group. Then
$$\frac{30,000}{n} - \frac{30,000}{n + 70} = 26.25 = \frac{105}{4}$$

Now solve for n. To clear the equation of fractions, multiply both sides by $4n(n + 70)$:
$$120,000(n + 70) - 120,000n = 105n(n + 70)$$
$$120,000 \cdot 70 = 105n(n + 70)$$

Divide both sides by 105:
$$80,000 = n^2 + 70n$$

Hence n satisfies the quadratic equation
$$n^2 + 70n - 80,000 = 0$$

By the quadratic formula,

$$n = \frac{-70 \pm \sqrt{70^2 - 4(-80{,}000)}}{2} = \frac{-70 \pm \sqrt{324{,}900}}{2} = \frac{-70 \pm 570}{2}$$

Therefore

$$n = \frac{500}{2} = 250 \quad \text{or} \quad n = \frac{-640}{2} = -320$$

But n must be a positive integer. Therefore the answer must be <u>250 people</u>.

EXAMPLE 6 To get from here to the bus stop, I follow the sidewalk to the corner, turn a right angle, then continue along the sidewalk. The total distance is 70 meters. But if I ignore the signs and cut straight across the grass, it's only 50 meters. How far is it from here to the corner?

SOLUTION Let x be the distance from here to the corner (in meters). After turning the corner the remaining distance to the bus stop is $70 - x$ meters (Fig. 1).

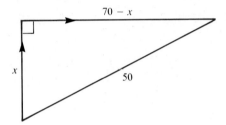

FIG. 1

By the Pythagorean theorem, x must satisfy

$$x^2 + (70 - x)^2 = 50^2$$
$$x^2 + 70^2 - 140x + x^2 = 50^2$$
$$2x^2 - 140x + (4900 - 2500) = 0$$

Dividing by 2, we have

$$x^2 - 70x + 1200 = 0$$

By the quadratic formula,

$$x = \frac{70 \pm \sqrt{4900 - 4(1200)}}{2} = \frac{70 \pm \sqrt{100}}{2} = \frac{70 \pm 10}{2} = 30 \text{ or } 40$$

There seem to be two answers. Why? Well, if $x = 30$, then $70 - x = 40$; and if $x = 40$, then $70 - x = 30$. The two legs of the triangle are 30 m and 40 m, but either one can come first (Fig. 2).

Answer two possibilities, 30 m and 40 m

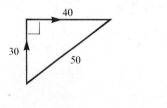 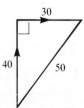

FIG. 2   Two possibilities

EXAMPLE 7 Find two real numbers whose product is 100 and whose sum is
(a) 30 (b) 19

SOLUTION (a) Call the numbers x and $30 - x$. Then
$$x(30 - x) = 100$$
$$30x - x^2 = 100$$
$$0 = x^2 - 30x + 100$$
By the quadratic formula,
$$x = \frac{30 \pm \sqrt{900 - 400}}{2} = \frac{30 \pm 10\sqrt{5}}{2} = 15 \pm 5\sqrt{5}$$
If $x = 15 + 5\sqrt{5}$, then $30 - x = 15 - 5\sqrt{5}$; and if $x = 15 - 5\sqrt{5}$, then $30 - x = 15 + 5\sqrt{5}$. So there is one such pair of numbers: $\underline{15 + 5\sqrt{5} \text{ and } 15 - 5\sqrt{5}}$.

(b) This time call the numbers x and $19 - x$. Then
$$x(19 - x) = 100$$
$$0 = x^2 - 19x + 100$$
But the discriminant is negative: $(-19)^2 - 400 = 361 - 400$. Therefore, the equation has no real solutions. <u>There are no such pairs of real numbers.</u>

EXERCISES

Without solving the equation, determine how many real solutions it has
1. $2x^2 - 8x + 3 = 0$
2. $3x^2 + 7x - 1 = 0$
3. $21x^2 + 100x + 9 = 0$
4. $21x^2 + 9x + 100 = 0$
5. $a^2x^2 - 2x + \frac{1}{a^2} = 0$
6. $x^2 - 2bx + b^2 = 0$
7. $2.012x^2 + 1.854x + 5.791 = 0$
8. $0.86x^2 + 9.49x + 13.72 = 0$

Solve by the quadratic formula
9. $x^2 + 10x + 2 = 0$
10. $x^2 + 5x - 3 = 0$
11. $2x^2 + x - 4 = 0$
12. $3x^2 - 6x - 1 = 0$
13. $5x^2 - 13x + 6 = 0$
14. $6x^2 + 11x + 5 = 0$
15. $10x^2 - 8x + 1 = 0$
16. $4x^2 + 8x - 1 = 0$
17. $9x^2 = 6x + 4$
18. $7x^2 = 17x - 2$
19. $25x^2 - 1 = 10x$
20. $3x^2 - 2x = 1$

Solve for x
21. $x + \frac{1}{x} = 2$
22. $x + \frac{1}{x - 3} = 1$
23. $\frac{1}{x + 1} + \frac{3}{x - 1} = 5$
24. $\frac{3x}{x + 2} = 1 + \frac{1}{x^2 - 4}$
25. $\frac{x}{x^2 + 1} = \frac{1}{3}$
26. $\frac{x - 2}{3x - 8} = \frac{x}{x + 2}$

Solve for x in terms of y
27. $x^2 - 5xy + 2y^2 = 0$
28. $x^2 + 2xy - 5y^2 = 0$

29. $\dfrac{x+y}{x-y} = \dfrac{x}{3x+y}$

30. $\dfrac{x}{y} + \dfrac{y}{x} = 2y$

Give an example of a quadratic equation whose solutions are

31. $\dfrac{3 \pm \sqrt{17}}{2}$

32. $-\tfrac{1}{2} \pm \sqrt{5}$

Given the quadratic equation $ax^2 + bx + c = 0$, with $b^2 - 4ac > 0$, find

33. the sum of the roots

34. the product of the roots

35. Find the length and width of a rectangle with perimeter 40 ft and area 91 ft².

36. Find the length and width of a rectangle of area 100 ft² if the length exceeds the width by 5 ft.

37. Find all numbers that exceed their reciprocals by 4.

38. Find all numbers that are 6 less than their squares.

39. Find all pairs of consecutive positive integers whose cubes differ by 919.

40. Find all triples of three consecutive odd integers whose squares total 1883.

41. A 40-foot length of rope is cut into two pieces, and then each piece is folded into a square. If the combined area of the two squares is 82 ft², find the length of each piece.

42. A 20-foot length of rope is cut into two pieces, and then each piece is folded into a square. If one square has triple the area of the other, find the length of each piece.

43. After lengthening my stride (pace) by 2 inches, I can now walk a mile in 132 fewer paces. What is the length of my new stride?

44. A group of people share the $150 expense of a party. If 10 more people join in, the cost per person goes down by $1.25. How many people were in the original group?

45. Find all right triangles whose three sides have lengths that are consecutive integers.

46. Find a right triangle whose longer leg is 1 cm shorter than the hypotenuse and 31 cm longer than the other leg.

47. I start with a number, take its reciprocal and add 1. I repeat the process and get back to the number I started with. What is that number?

48. Find two numbers whose sum is 48 and whose product is 10 times their difference.

49. Refer to Example 6. Show that it is impossible to save 25 meters by cutting across the grass.

50. If the product of two positive numbers is 100, what is the least their sum can be? (Hint Assume the sum is s and use the technique of Example 7.)

7 OTHER TYPES OF EQUATIONS

Equations of Quadratic Type

An equation of the form
$$ax^4 + bx^2 + c = 0$$
is really a disguised quadratic, since it can be expressed as
$$a(x^2)^2 + bx^2 + c = 0$$
or $\quad ay^2 + by + c = 0 \quad$ where $y = x^2$

Similarly
$$ax^6 + bx^3 + c = 0$$
is a quadratic equation relative to $y = x^3$, and
$$a(x^2 + 1)^2 + b(x^2 + 1) + c = 0$$
is quadratic relative to $y = x^2 + 1$. Such equations are often called equations of **quadratic type.**

EXAMPLE 1 Solve $\quad x^4 - 2x^2 - 2 = 0$

SOLUTION Let $y = x^2$. Then the equation becomes
$$y^2 - 2y - 2 = 0$$
Solve for y by completing the square (or by the quadratic formula if you prefer):
$$y^2 - 2y + 1 = 2 + 1$$
$$(y - 1)^2 = 3$$
$$y = 1 \pm \sqrt{3}$$
But $y = x^2$. Hence each solution x of the original equation satisfies either

$\quad x^2 = 1 + \sqrt{3} \quad\quad$ or $\quad\quad x^2 = 1 - \sqrt{3}$

$\quad x = \underline{\pm \sqrt{1 + \sqrt{3}}} \quad\quad\quad\quad$ (no real solutions because $1 - \sqrt{3} < 0$)

EXAMPLE 2 Solve $\quad x - 10\sqrt{x} + 23 = 0$

SOLUTION Let $y = \sqrt{x}$. The equation becomes
$$y^2 - 10y + 23 = 0$$
By the quadratic formula,
$$y = \frac{10 \pm \sqrt{8}}{2} = 5 \pm \sqrt{2}$$
that is, $\quad\quad \sqrt{x} = 5 \pm \sqrt{2}$

Hence
$$x = (5 \pm \sqrt{2})^2 = 25 \pm 10\sqrt{2} + 2 = \underline{27 \pm 10\sqrt{2}}$$

Polynomial Equations

Solving a polynomial equation
$$a_n x^n + a_{n-1} x^{n-1} + \cdots + a_1 x + a_0 = 0$$
is usually difficult when the degree is more than 2. However, it can be done when we are lucky enough or skillful enough to factor the left-hand side.

EXAMPLE 3 Solve $x^5 - 16x = 0$

SOLUTION Factor the left side:
$$\begin{aligned} x^5 - 16x &= x(x^4 - 16) \\ &= x(x^2 + 4)(x^2 - 4) \\ &= x(x^2 + 4)(x + 2)(x - 2) = 0 \end{aligned}$$
The product is 0 precisely when one of its factors is 0. The factor $x^2 + 4$ is never 0. But each of the other factors can be 0: for the values $\underline{0}$, $\underline{-2}$, and $\underline{2}$, respectively.

EXAMPLE 4 Solve $(x^3 - 1)(2x^2 + 7x + 1) = 0$

SOLUTION The left side can be 0 only if
$$(1) \quad x^3 - 1 = 0 \quad \text{or} \quad (2) \quad 2x^2 + 7x + 1 = 0$$
Therefore, solving the given equation amounts to solving both equations (1) and (2). We solve equation (1) by factoring:
$$x^3 - 1 = (x - 1)(x^2 + x + 1) = 0$$
Hence either
$$\begin{array}{ll} x - 1 = 0 \quad \text{or} & x^2 + x + 1 = 0 \\ x = 1 & \text{(no real solutions;} \\ & \text{discriminant} = -3 < 0 \text{)} \end{array}$$
We solve equation (2) by the quadratic formula

Answer $1, \dfrac{-7 \pm \sqrt{41}}{4}$

Equations Involving Radicals

EXAMPLE 5 Solve $\sqrt{x + 3} = x - 9$

SOLUTION To eliminate the square root, we square both sides:
$$\begin{aligned} x + 3 &= x^2 - 18x + 81 \\ 0 &= x^2 - 19x + 78 \end{aligned}$$
By the quadratic formula,
$$x = \frac{19 \pm \sqrt{361 - 312}}{2} = \frac{19 \pm \sqrt{49}}{2}$$
$$= \frac{19 \pm 7}{2} = 13 \quad \text{or} \quad 6$$

Hence the only possible solutions of the given equation are 13 and 6. Let us check:

$x = 13 \qquad \sqrt{13 + 3} \stackrel{?}{=} 13 - 9$
$\qquad\qquad\qquad 4 = 4 \qquad\qquad$ (correct)

$x = 6 \qquad \sqrt{6 + 3} \stackrel{?}{=} 6 - 9$
$\qquad\qquad\qquad 3 = -3 \qquad\qquad$ (false!)

Thus, 13 is a solution but 6 is not. Hence <u>13 is the only solution.</u>

Example 5 shows that squaring both sides of an equation may introduce **extraneous solutions** (ones that do not satisfy the given equation). For example, take the very simple equation $x = 2$. Squaring both sides gives $x^2 = 4$ with solutions 2 and -2. Obviously, -2 is an extraneous solution.

> Raising both sides of an equation to the same power may introduce extraneous solutions. Therefore, all solutions must be checked.

EXAMPLE 6 Solve $\sqrt{2x + 4} - \sqrt{x + 3} = 1$

SOLUTION We could square both sides, but it's simpler to isolate one of the radicals first (usually the more complicated one):

$$\sqrt{2x + 4} = 1 + \sqrt{x + 3}$$

Squaring, $\qquad 2x + 4 = 1 + 2\sqrt{x + 3} + x + 3$

We'll have to square again, but first we simplify and isolate the radical. Subtracting $x + 4$ from both sides,

$$x = 2\sqrt{x + 3}$$

Now we square again:

$$x^2 = 4(x + 3)$$
$$x^2 - 4x - 12 = 0$$
$$(x - 6)(x + 2) = 0$$

Hence $x = 6$ or $x = -2$. After all that squaring, we certainly must check these values:

$x = 6 \qquad \sqrt{2 \cdot 6 + 4} - \sqrt{6 + 3} \quad \stackrel{?}{=} 1$
$\qquad\qquad\qquad 4 - 3 \qquad = 1 \qquad$ (correct)

$x = -2 \qquad \sqrt{2(-2) + 4} - \sqrt{-2 + 3} \stackrel{?}{=} 1$
$\qquad\qquad\qquad 0 - 1 = 1 \qquad$ (false!)

Therefore <u>6</u> is a solution, but -2 is extraneous.

Some Applications

EXAMPLE 7 One leg of a right triangle is 5, and the perimeter is 20. Find the other leg.

SOLUTION We set up the equation for this problem in Example 2, page 65:

$$x + 5 + \sqrt{x^2 + 25} = 20$$

Subtract $x + 5$ from both sides, then square:
$$\sqrt{x^2 + 25} = 15 - x$$
$$x^2 + 25 = 225 - 30x + x^2$$
$$30x = 200$$
$$x = \frac{20}{3}$$

Check $\frac{20}{3} + 5 + \sqrt{(\frac{20}{3})^2 + 25} \stackrel{?}{=} 20$
$\frac{35}{3} + \sqrt{\frac{625}{9}} \stackrel{?}{=} 20$
$\frac{35}{3} + \frac{25}{3} = 20$ (correct)

EXAMPLE 8 From a cliff, I drop a stone into a lake. Three seconds later, I hear the splash. The stone falls $16t^2$ feet in t seconds, and sound travels 1100 ft/sec in air. How far down is the water?

SOLUTION Let t_1 be the time for the stone to fall and t_2 the time for the sound of the splash to reach me. The equation in this problem is
$$t_1 + t_2 = 3$$
The unknown is the distance from the cliff to the water. Call it d. Let us express t_1 and t_2 in terms of d. Since the stone falls d feet in t_1 seconds,
$$16t_1^2 = d \quad \text{so} \quad t_1 = \sqrt{\frac{d}{16}} = \frac{\sqrt{d}}{4}$$
Since the sound travels d feet in t_2 seconds,
$$1100 t_2 = d \quad \text{so} \quad t_2 = \frac{d}{1100}$$
Now the equation $t_1 + t_2 = 3$ becomes
$$\frac{\sqrt{d}}{4} + \frac{d}{1100} = 3$$
To solve for d, multiply by 1100 and rearrange:
$$275\sqrt{d} + d = 3300$$
$$d + 275\sqrt{d} - 3300 = 0$$
This is an equation of quadratic type:
$$x^2 + 275x - 3300 = 0 \quad \text{where } x = \sqrt{d}$$
By the quadratic formula,
$$x = \frac{-275 \pm \sqrt{275^2 + 4 \cdot 3300}}{2} = \frac{-275 \pm \sqrt{88{,}825}}{2}$$
According to our calculator, $\sqrt{88{,}825} \approx 298.04$. (The symbol $\approx$ means "is approximately equal to.") Hence the positive solution of the quadratic is
$$x \approx \frac{-275 + 298.04}{2} = 11.52$$
Finally,
$$d = x^2 \approx (11.52)^2 \approx \underline{132.7 \text{ ft}}$$

EXERCISES

Solve

1. $x^4 - 7x^2 + 6 = 0$
2. $x^4 - 5x^2 - 14 = 0$
3. $x^4 - 6x^2 + 6 = 0$
4. $x^4 + 2x^2 - 5 = 0$
5. $10x^4 - 7x^2 + 1 = 0$
6. $12x^4 - x^2 - 1 = 0$
7. $x - 5\sqrt{x} + 6 = 0$
8. $x - 9\sqrt{x} + 20 = 0$
9. $x^6 - 2x^3 - 80 = 0$
10. $x^8 - 15x^4 - 16 = 0$
11. $\dfrac{1}{x^4} = \dfrac{6}{x^2} - 8$
12. $\left(\dfrac{x}{4x+1}\right)^2 = \dfrac{x}{4x+1} + 2$
13. $(x^2 + 1)^2 - 3(x^2 + 1) - 1 = 0$
14. $(x^2 - x)^2 - 5(x^2 - x) + 6 = 0$
15. $\dfrac{x^2 + 2}{x^2 + 5} = \dfrac{1}{x^2 - 2}$
16. $\dfrac{x^2}{x^2 + 3} = \dfrac{x^2 + 1}{3x^2 + 1}$
17. $(x^2 - 16)(25x^2 + 9) = 0$
18. $(2x^2 + 1)(x^2 - 36) = 0$
19. $(x^2 - 4x + 4)(x^2 + 3x + 1) = 0$
20. $(x^2 + 5x - 1)(x^2 + 5x - 2) = 0$
21. $x^4 + 5x^3 + 6x^2 = 0$
22. $x^5 - 6x^4 + 5x^3 = 0$
23. $x^2(x - 1)^2 = 4(x - 1)^2$
24. $(2x + 1)^3 = (2x + 1)(x^2 + 9x + 3)$
25. $x^3 = x^6$
26. $x^8 - 1 = 0$
27. $x^7 - x^6 + 4x - 4 = 0$
28. $x^5 + 9x^4 + x + 9 = 0$

Solve

29. $\sqrt{4 - x} = x - 2$
30. $\sqrt{x + 3} = 9 - x$
31. $2\sqrt{x + 5} = x - 3$
32. $5\sqrt{x + 7} = x - 7$
33. $\sqrt{x} + \sqrt{x + 3} = 3$
34. $\sqrt{2x + 1} + \sqrt{5 - x} = 4$
35. $2\sqrt{x} - \sqrt{x - 5} = \sqrt{x + 7}$
36. $2\sqrt{x + 4} + \sqrt{x - 5} = \sqrt{9x + 7}$
37. $\sqrt[4]{8x^2 + 1} = \sqrt{3x - 1}$
38. $\sqrt{x} = \sqrt{10 + 3\sqrt{x}}$

39. Find a number that exceeds its square root by 6.
40. Find a number that is 1 less than twice its square root.
41. Find the base of an isosceles triangle of height 3 and perimeter 12.
42. The longer leg of a right triangle is 1 cm shorter than the hypotenuse. The perimeter is 132 cm. Find the shorter leg.
43. I start with a number, square it, subtract 1, and divide by 8. I then repeat the process and end up with 55. What number did I start with?
44. How long until I hear the splash if I drop a stone into the water from a height of 100 ft? (Refer to Example 8.)

8 SUGGESTIONS FOR SOLVING PROBLEMS

You will spend a lot of time in this course and other math courses solving problems. It pays to develop sound techniques and good habits. We hope you will consider a few suggestions.

Take your time. Before plunging into a problem, take a minute to think. Read the problem several times. Have you seen one like it before?

Examine the data. Be sure you understand what is given and what is to be found. Translate the data into algebraic language. Whenever possible, make a diagram and label it clearly. Look for information hidden in the wording of the problem. If you get stuck, check that you are using *all* the data.

Avoid sloppiness

(1) Avoid sloppiness in writing and in arranging your work. Do calculations in a logical sequence of neat orderly steps. Include all steps except very obvious ones. This will help cut down on errors, or at least make them easier to find.

(2) Avoid sloppiness in algebra, the kind of mistakes discussed in Chapter 2, Section 8. If you analyze a problem correctly and set up the right equation, it's a pity to mess up the solution with sloppy algebra, such as writing $\sqrt{x+3} = \sqrt{x} + \sqrt{3}$.

(3) Avoid sloppiness in units. If you start out in feet, keep all lengths in feet and all areas in square feet. Do not mix feet and inches, minutes and seconds.

(4) Avoid sloppiness in the answer. Be sure to answer the question that is asked. For example, if the problem wants the area of a rectangle, the answer is not the length. If it asks for a time, the answer is not a distance.

Estimate the answer. Make a "ball park" estimate if you can. This will serve as a check against unreasonable answers. For example, if you find that the area of a rectangle is -24 ft^2, you'd better go back and check.

To illustrate some of these suggestions, let us take a very easy problem and consider some "solutions."

EXAMPLE 1 At 50 miles per hour, how long will it take to catch up to Steve who left one hour ago at 40 mph?

DISCUSSION Without any computation, we feel it will take a few hours to catch up, almost certainly more than 1 hour but less than 10 hours.

"SOLUTION" 1 The equation is

$$50t = 40(t + 1)$$
$$50t = 40t + 1 \quad \text{[sloppiness in algebra]}$$
$$10t = 1$$
$$t = \tfrac{1}{10}$$

The answer is that it will take $\tfrac{1}{10}$ hour; that is, 6 minutes to catch up. Do you believe it? If not, go back and find the mistake.

"SOLUTION" 2 $\quad 50t = 400(t + 1)\quad$ [error in writing]
$$-350t = 400$$
$$t = -\frac{40}{35}$$

Hence it takes $-40/35$ hours to catch up. Better check.

"SOLUTION" 3 Since 1 hour is 60 minutes, the equation is
$$50t = 40(t + 60)\quad \text{[sloppiness in units]}$$
$$10t = 2400$$
$$t = 240$$

Hence it takes 240 hours, that is, 10 days to catch up. Again, an unreasonable answer signals a mistake somewhere.

"SOLUTION" 4 $50t = 40(t + 1) = 10t = 40\quad$ [sloppiness in arranging work]
Two steps have been run together and the equal sign has been mauled. *Quantities on opposite sides of an equal sign must be equal.* If you read the "equation" from left to right, you might conclude that $50t = 40$, so $t = \frac{4}{5}$, an unreasonable answer.

"SOLUTION" 5 I catch up in 200 miles because
$$200 = 50 \cdot 4 = 40(4 + 1)$$

Hence the answer is 200 miles.

This solution has merit but shows sloppiness in the answer. The problem asks for the *time* required to catch up. The correct answer is 4 hours.

The Wyatt Earp Principle

Wyatt Earp was a great gunfighter who survived many a shoot-out in the old West. Yet he carried only one gun and used no fancy tricks. His secret? He took an extra split second *to aim*. While the bad guy blazed away wildly with two guns, Earp got his man on the first shot.

Try to face a math exam the way Wyatt Earp faced a gunfight. Instead of calculating wildly with both hands, take a minute to think. You may find the problem is simpler than it looks at first. Certainly you will have a better chance of winning the showdown.

REVIEW EXERCISES FOR CHAPTER 3

Solve the equation

1. $x^2 + 4x + 11 = 0$

2. $6x^2 - 7x + 2 = 0$

3. $x^2 - \frac{4}{3}x + \frac{4}{9} = 0$

4. $\dfrac{x + 1}{x + 4} = \dfrac{2x}{x + 6}$

5. $\dfrac{2x - 3}{5x - 4} = \dfrac{9}{10}$

6. $\dfrac{1}{x} - x = \dfrac{21}{10}$

7. $x^4 - 7x^2 + 12 = 0$

8. $x^{2/3} - 9x^{1/3} + 8 = 0$

Equations

9. $2\sqrt{x} = x - 1$
10. $x = 2 + \sqrt{4x - 3}$
11. $x^5 = 16x$
12. $3x^6 - 13x^4 + 4x^2 = 0$
13. By completing squares, express $x^2 + 2y^2 + x - 8y + 1$ in the form $a(x - p)^2 + b(y - q)^2 + c$.
14. If a, b, c are positive, how many real solutions does the equation $ax^2 + bx - c = 0$ have?
15. A collection of 100 coins, all pennies and nickels, totals $2.76. How many of each coin are there?
16. How much pure salt must be added to 200 gm of 5% salt solution to triple its strength?
17. A team loses 4 straight games, and its winning percentage falls from 50% to 45%. How many wins and losses does it have now?
18. How should $10,000 be invested, partly in a savings account paying 6% yearly interest and partly in bonds paying 12%, in order to earn 7.5% yearly?
19. One leg of a right triangle is 24, and the perimeter is 56. Find the other leg.
20. A 10-mph tailwind allows a small plane to make a 600-mile flight in 15 minutes less than expected. Find the speed of the plane in still air.

Inequalities

1 BASIC PROPERTIES

An inequality is a statement that one quantity is smaller or larger than another. Such statements are very common in real life since we are always comparing sizes or amounts. For example,

> Her bowling average is *over* 160.
>
> The runner's time for the mile is *under* 4 minutes.
>
> A manufacturer produces *more than* 1,000,000 TV sets per year.
>
> A public opinion poll is accurate to *within* 2%.

In this section we discuss basic properties of inequalities. Then, in Sections 2 and 3, we solve inequalities involving a variable much as we solved equations in Chapter 3.

Recall from Chapter 1, Section 5, the notation

$a < b$ a is less than b
$a > b$ a is greater than b

and also

$a \leq b$ a is less than or equal to b
$a \geq b$ a is greater than or equal to b

Algebraically, $a < b$ means that the difference $b - a$ is positive. Geometrically, $a < b$ means the point a is to the left of the point b on the number line.

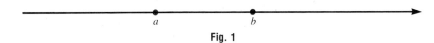

Fig. 1

Properties of Inequalities

If a is to the left of b on the number line, and b is to the left of c, then a is to the left of c. (See Fig. 2.)

Fig. 2

100 Inequalities

In algebraic language this basic property of inequalities says:
$$\text{if } a < b \text{ and } b < c, \text{ then } a < c$$
For example,
$$\sqrt{8} < 3 \text{ and } 3 < \pi, \text{ hence } \sqrt{8} < \pi$$

Given $a < b$, then $a + c < b + c$ for any real number c, for adding c means a shift on the number line. The shifted points $a + c$ and $b + c$ are in the same order as a and b (Fig. 3).

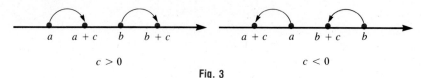

Fig. 3

Thus, we may add or subtract the same number to both sides of an inequality. For example, from the inequality $\sqrt{19} < 4.5$ follow
$$1 + \sqrt{19} < 5.5 \qquad 10 + \sqrt{19} < 14.5 \qquad \sqrt{19} - 3 < 1.5$$

We may multiply both sides of an inequality by the same positive number. For example, from $2 < 3$ follow
$$5 \cdot 2 < 5 \cdot 3 \qquad \tfrac{1}{7} \cdot 2 < \tfrac{1}{7} \cdot 3 \qquad 66 \cdot 2 < 66 \cdot 3$$

But multiplying by a negative number *reverses* the inequality. Thus $2 < 3$, but
$$(-5) \cdot 2 > (-5) \cdot 3 \qquad (-\tfrac{1}{7}) \cdot 2 > (-\tfrac{1}{7}) \cdot 3 \qquad (-66) \cdot 2 > (-66) \cdot 3$$

Let us summarize these facts about inequalities:

Rules for Inequalities
(1) If $a < b$ and $b < c$, then $a < c$.
(2) If $a < b$ and c is any real number, $a + c < b + c$.
(3) If $a < b$ and $c > 0$, then $ca < cb$.
(4) If $a < b$ and $c < 0$, then $ca > cb$.

Rules (3) and (4) include division by c, which is multiplication by $1/c$. Similar rules hold with $\leq$ in place of $<$.

These rules can be checked by subtraction. For example, take Rule (3). If $a < b$, then $b - a$ is positive. If c is positive, then
$$cb - ca = c(b - a) = (\text{positive})(\text{positive}) = \text{positive}$$
Hence $\qquad\qquad\qquad ca < cb$

EXAMPLE 1 Given $1.4 < \sqrt{2} < 1.5$, show that

(a) $1 + \sqrt{2} < 2.5$ \qquad (b) $14 < 10\sqrt{2}$ \qquad (c) $0.08 < \dfrac{\sqrt{2} - 1}{5} < 0.1$

SOLUTION (a) Start with $\sqrt{2} < 1.5$. Add 1 to both sides:
$$1 + \sqrt{2} < 1 + 1.5 = 2.5 \qquad\qquad [\text{Rule (2)}]$$
(b) Start with $1.4 < \sqrt{2}$. Multiply both sides by the positive number 10:

$$14 = 10(1.4) < 10\sqrt{2} \qquad \text{[Rule (3)]}$$

(c) Start with $1.4 < \sqrt{2} < 1.5$. Subtract 1 from all three numbers. (This means subtracting 1 from both sides of $1.4 < \sqrt{2}$ and also $\sqrt{2} < 1.5$.)

$$0.4 < \sqrt{2} - 1 < 0.5 \qquad \text{[Rule (2)]}$$

Now divide by the positive number 5:

$$\frac{0.4}{5} < \frac{\sqrt{2}-1}{5} < \frac{0.5}{5} \qquad \text{[Rule (3)]}$$

$$0.08 < \frac{\sqrt{2}-1}{5} < 0.1$$

What about taking reciprocals on both sides of an inequality? That reverses the inequality provided both numbers have the same sign. For example,

$$2 < 3 \text{ but } \frac{1}{2} > \frac{1}{3} \qquad -4 < -3 \text{ but } \frac{1}{-4} > \frac{1}{-3}$$

---

(5) If $0 < a < b$ or $a < b < 0$, then $\dfrac{1}{a} > \dfrac{1}{b}$

EXAMPLE 2 The ancient geometers had various estimates for the number π. From the estimate $3\frac{10}{71} < \pi < 3\frac{1}{7}$, show that

(a) $\dfrac{3}{\pi} > \dfrac{21}{22}$ (b) $\dfrac{1}{225 - 71\pi} > \dfrac{1}{2}$

SOLUTION (a) Start with $\pi < 3\frac{1}{7} = \frac{22}{7}$

Divide by 3: $\qquad\qquad\qquad \dfrac{\pi}{3} < \dfrac{22}{21} \qquad \text{[Rule (3)]}$

Take reciprocals: $\qquad\qquad \dfrac{3}{\pi} > \dfrac{21}{22} \qquad \text{[Rule (5)]}$

(b) Show that $225 - 71\pi < 2$, and then take reciprocals. Start with

$$\pi > 3\tfrac{10}{71} = \tfrac{223}{71}$$

Multiply by -71: $\qquad -71\pi < -223 \qquad \text{[Rule (4)]}$
Add 225: $\qquad\qquad\quad 225 - 71\pi < 2 \qquad \text{[Rule (2)]}$

Take reciprocals: $\qquad \dfrac{1}{225 - 71\pi} > \dfrac{1}{2} \qquad \text{[Rule (5)]}$

Next, recall the test for equality of fractions (Chapter 1, Section 5):

$$\frac{a}{b} = \frac{c}{d} \quad \text{if and only if} \quad ad = bc$$

There is a similar test for *inequality* of fractions:

> (6) Suppose b and d have the same sign. Then
> $$\frac{a}{b} < \frac{c}{d} \quad \text{if and only if} \quad ad < bc$$

For example,
$$\frac{4}{19} < \frac{5}{23} \quad \text{because} \quad 4 \cdot 23 < 19 \cdot 5, \quad \text{that is,} \quad 92 < 95$$

To justify Rule (6), suppose b and d have the same sign. Then $bd > 0$. If $a/b < c/d$, multiplying both sides by the positive number bd gives $ad < bc$. Conversely, if $ad < bc$, dividing both sides by bd gives $a/b < c/d$.

Note that Rule (5) is a special case of Rule (6). Two other special cases are worth mentioning. For simplicity, we assume all denominators are positive.

> Assume all denominators are positive. Then
>
> (7) $\quad \dfrac{a}{c} < \dfrac{b}{c} \quad$ if and only if $\quad a < b$
>
> (8) $\quad \dfrac{a}{b} < \dfrac{a}{c} \quad$ if and only if $\quad b > c$

Given two fractions with the same positive denominator, Rule (7) says that the larger fraction is the one with the larger numerator. Thus,
$$\frac{3}{5} < \frac{4}{5} \qquad \frac{7}{\sqrt{2}} < \frac{9}{\sqrt{2}} \qquad \frac{1.27\pi}{4.86} < \frac{1.28\pi}{4.86}$$

Given two fractions with the same numerator and positive denominators, Rule (8) says that the larger fraction is the one with the *smaller* denominator. Thus,
$$\frac{7}{6} < \frac{7}{5} \qquad \frac{4}{\sqrt{10}} < \frac{4}{3} \qquad \frac{12.6}{\sqrt{43}} < \frac{12.6}{\sqrt{37}}$$

Estimation

The rules of inequalities help us estimate various quantities, often with a minimum of computation. Especially useful are these two rules, which extend Rules (2) and (3).

> (9) If $\ a < A \ $ and $\ b < B, \quad$ then $\quad a + b < A + B$
> (10) If $\ 0 < a < A \ $ and $\ 0 < b < B, \quad$ then $\quad ab < AB$

For justifications, see Exercises 47 and 48. These rules apply also to three or more

quantities. For instance, suppose $a < A$, $b < B$, and $c < C$. Then
$$a + b + c < A + B + C$$
and
$$abc < ABC \quad \text{if} \quad a > 0, b > 0, c > 0$$

EXAMPLE 3 Verify the estimate using as little arithmetic as possible:
(a) $6 \cdot 8 \cdot 9 \cdot 11 < 5000$ (b) $\pi^2 < 10$
(c) $\frac{5}{9} < \frac{1}{5} + \frac{1}{6} + \frac{1}{7} + \frac{1}{8} + \frac{1}{9} < 1$

SOLUTION (a) Use Rule (10):
$$6 \cdot 8 \cdot 9 \cdot 11 = 48 \cdot 99 < 50 \cdot 100 = 5000$$
(b) Use the estimate $\pi < \frac{22}{7}$ and Rule (10):
$$\pi^2 = \pi \cdot \pi < \frac{22}{7} \cdot \frac{22}{7} = \frac{484}{49} < \frac{490}{49} = 10$$
(c) Each term after the first is less than $\frac{1}{5}$. By Rule (9),
$$\tfrac{1}{5} + \tfrac{1}{6} + \tfrac{1}{7} + \tfrac{1}{8} + \tfrac{1}{9} < \tfrac{1}{5} + \tfrac{1}{5} + \tfrac{1}{5} + \tfrac{1}{5} + \tfrac{1}{5} = 1$$
Similarly, each term but the last is greater than $\frac{1}{9}$. Hence
$$\tfrac{1}{5} + \tfrac{1}{6} + \tfrac{1}{7} + \tfrac{1}{8} + \tfrac{1}{9} > \tfrac{1}{9} + \tfrac{1}{9} + \tfrac{1}{9} + \tfrac{1}{9} + \tfrac{1}{9} = \tfrac{5}{9}$$

EXAMPLE 4 A rectangular box measures 11.83 ft × 4.91 ft × 2.77 ft. Show that its volume is surely less than 200 ft³.

SOLUTION The volume is the product of the three dimensions. By Rule (10),
$$\text{Volume} = (11.83)(4.91)(2.77) < 12 \cdot 5 \cdot 3 = 180$$
The volume is actually less than 180 ft³.

EXAMPLE 5 A triangle is measured to have base 12.3 cm and height 8.5 cm. Both measurements are subject to errors of at most 0.1 cm. What are the possible values of the area?

SOLUTION The area is $\frac{1}{2}bh$, where b is the base and h is the height. Because of possible errors up to 0.1 cm, the true values of b and h are not exact. However, we know that
$$12.2 \leq b \leq 12.4 \quad \text{and} \quad 8.4 \leq h \leq 8.6$$
Therefore, by Rule (10),
$$\tfrac{1}{2}(12.2)(8.4) \leq \tfrac{1}{2}bh \leq \tfrac{1}{2}(12.4)(8.6)$$
$$51.24 \text{ cm}^2 \leq \text{Area} \leq 53.32 \text{ cm}^2$$

EXAMPLE 6 My odometer showed that I drove exactly 300 miles, and I estimate that I used 10 gallons of gas. Hence my estimated gas mileage is 30 miles per gallon. However, my odometer may be off by as much as 2% and my gas estimate may be off by $\frac{1}{2}$ gallon. What values are possible for my actual gas mileage?

SOLUTION The mileage m is the fraction
$$m = \frac{\text{distance driven}}{\text{gallons consumed}} = \frac{d}{g}$$

104 Inequalities

The possible error in *d* is at most 2% of 300; that is, 6 miles. The possible error in *g* is at most $\frac{1}{2}$ gallon. Therefore,
$$294 \le d \le 306 \quad \text{and} \quad 9.5 \le g \le 10.5$$
By Rules (7) and (8),
$$\frac{d}{g} \le \frac{306}{g} \le \frac{306}{9.5} \approx 32.21$$

Thus, the largest possible value of d/g occurs for the largest possible numerator together with the smallest possible denominator. Similarly, the smallest possible value of d/g occurs for the smallest numerator together with the largest denominator:
$$\frac{d}{g} \ge \frac{294}{10.5} = 28$$

Answer
$$28.00 \text{ mpg} \le \text{mileage} \le \frac{306}{9.5} \approx 32.21 \text{ mpg}$$

EXERCISES

Given $1.4 < \sqrt{2} < 1.5$ and $1.7 < \sqrt{3} < 1.8$, show that

1. $\sqrt{2} + 9 < 10.5$
2. $\sqrt{3} - 1 > 0.7$
3. $1 + 4\sqrt{3} > 7.8$
4. $\frac{1}{5}\sqrt{2} + 4 < 4.3$
5. $\frac{4}{\sqrt{3}} > 2$
6. $\frac{7}{\sqrt{2}} < 5$
7. $3.1 < \sqrt{2} + \sqrt{3} < 3.3$
8. $2.38 < \sqrt{6} < 2.7$
9. $\frac{1}{2 - \sqrt{2}} < 2$
10. $\frac{1}{2 - \sqrt{3}} > \frac{8}{3}$

Verify the inequality using as little arithmetic as possible

11. $7854 + 9615 < 7854 + 9761$
12. $\frac{17}{43} + \frac{1}{4} < \frac{17}{43} + \frac{3}{8}$
13. $\frac{16}{47} - \frac{1}{3} > \frac{16}{47} - \frac{1}{2}$
14. $68{,}723 - 11{,}586 > 68{,}723 - 14{,}992$
15. $\frac{3}{\pi} > \frac{3}{4}$
16. $\frac{2}{\pi} < \frac{2}{3}$
17. $9 \cdot 11 \cdot 13 < 1300$
18. $6 \cdot 17 \cdot 101 > 10{,}000$
19. $(2.07)^2(3.026) > 12$
20. $(1.94)(1.87)(3.99) < 16$
21. $\pi\sqrt{2} < 5$
22. $\pi^6 < 10{,}000$
23. $\frac{1}{2} < \frac{1}{11} + \frac{1}{12} + \cdots + \frac{1}{20} < \frac{10}{11}$
24. $\frac{4}{3} < \frac{1}{\sqrt{5}} + \frac{1}{\sqrt{6}} + \frac{1}{\sqrt{7}} + \frac{1}{\sqrt{8}} < 2$

Which is larger?

25. $\dfrac{11}{12}$ or $\dfrac{12}{13}$ **26.** $\dfrac{13}{15}$ or $\dfrac{15}{17}$ **27.** $\dfrac{4}{577}$ or $\dfrac{1}{147}$

28. $\dfrac{4}{13}$ or $\dfrac{11}{36}$ **29.** $\dfrac{\sqrt{2}}{1.5}$ or $\dfrac{1.3}{\sqrt{2}}$ **30.** $\dfrac{\pi}{1.9}$ or $\dfrac{3}{2}$

Which is the best and which is the worst buy?

31. toothpaste; *giant* $5\frac{3}{4}$-oz size at 94¢, *large economy* $6\frac{7}{9}$-oz size at $1.07, or *family* $7\frac{1}{3}$-oz size at $1.20.

32. peanut butter; *small* 4-oz jar at 74¢, *medium* 6.6-oz jar at $1.16, or *large* 12.4-oz jar at $2.08.

33. Bill drove 561 miles in 10 hours. Show that he must have exceeded the 55-mph speed limit at some time.

34. Debbie has grades of 65, 81, and 68 on three tests. Can she raise her average to 85 by doing well on the fourth test?

35. The volume of a sphere of radius r is $\frac{4}{3}\pi r^3$. If the radius is 3.0741 inches, show that the volume exceeds 100 in³.

36. The volume of a rectangular box is at most 1000 in.³. If the length exceeds 20 in. and the width exceeds 16 in., show that the height must be less than $3\frac{1}{8}$ in.

37. A rectangle is measured to be 10 cm × 15 cm. If these measurements are subject to an error of at most 1%, what are the possible values for the area?

38. If your gas mileage varies from 26 mpg to 30 mpg and your gas tank holds between 12 and $12\frac{1}{2}$ gallons, find your maximum and minimum range on a tankful of gas.

39. A group estimates that the cost of a party will run between $500 and $600. The cost will be shared by at least 40 people and at most 45. Find the smallest and largest possible costs per person.

40. A TV poll estimates that 90,000,000 people are watching TV, and of these, 30,000,000 are watching their CBS channel. If these estimates are subject to errors up to 5%, then the percentage of all viewers watching CBS falls in what range?

Let a and b be positive

41. if $\dfrac{a}{b} < 1$ show that $\dfrac{a+1}{b+1} > \dfrac{a}{b}$ **42.** if $\dfrac{a}{b} > 1$ show that $\dfrac{a+1}{b+1} < \dfrac{a}{b}$

43. if $0 < a < b$ show that $a < \sqrt{ab} < b$

44. Show that the average of two numbers is a number between them.

Use subtraction to justify

45. Rule (2) **46.** Rule (4)

47. Justify Rule (9) by proving this string of inequalities

$$a + b < A + b < A + B$$

48. Find a similar justification for Rule (10).

2 SOLVING INEQUALITIES

Let us look at inequalities involving a variable. Statements such as $x^2 + 1 > 0$, which are true for all real numbers x, are called **absolute inequalities.** They correspond to identities in the case of equations. Statements such as $x + 1 > 10$, which are not true for all real numbers x, are called **conditional inequalities.** They correspond to conditional equations.

Solving a conditional inequality means finding the set of all real numbers for which the statement is true. For example, the solution set of $x + 1 > 10$ consists of all real numbers greater than 9. Two conditional inequalities are **equivalent** if they have the same solution set, for example, $x + 1 > 10$ and $x > 9$.

Solving inequalities is similar to solving equations. By a sequence of steps, we transform a given inequality into an equivalent one whose solutions are obvious.

> The following operations transform an inequality into an equivalent inequality:
> (1) adding or subtracting the same quantity on both sides
> (2) multiplying or dividing both sides by the same positive quantity
> (3) multiplying or dividing both sides by the same negative quantity and *reversing* the inequality sign

EXAMPLE 1 Solve the inequality
(a) $3x - 8 < x - 10$ (b) $x(x - 6) \leq x^2 - 9$

SOLUTION (a) Subtract x from both sides:
$$2x - 8 < -10$$
Add 8:
$$2x < -2$$
Divide by 2:
$$\underline{x < -1}$$

(b)
$$x^2 - 6x \leq x^2 - 9$$
Subtract x^2:
$$-6x \leq -9$$
Divide by -6 and reverse the inequality:
$$\underline{x \geq \tfrac{3}{2}}$$

EXAMPLE 2 Solve $-3 \leq \tfrac{1}{2}(x - 4) \leq 1$ and sketch the solution set.

SOLUTION This is really two inequalities, $-3 \leq \tfrac{1}{2}(x - 4)$ and $\tfrac{1}{2}(x - 4) \leq 1$. We can solve both at once:
$$-3 \leq \tfrac{1}{2}(x - 4) \leq 1$$
Multiply by 2:
$$-6 \leq x - 4 \leq 2$$
Add 4:
$$\underline{-2 \leq x \leq 6}$$

On the number line the solutions fill the segment from -2 to 6 inclusive (Fig. 1).

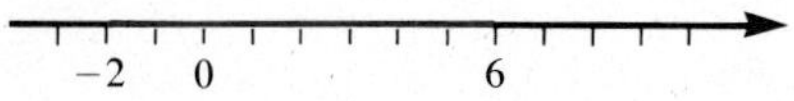

Fig. 1

We denote the segment shown in Fig. 1 by $[-2, 6]$. In general, the set of all real numbers between two given numbers a and b is called an **interval**. If the end points a and b are included, the interval is **closed** and is denoted by $[a, b]$. If a and b are not included, the interval is **open** and is denoted by (a, b). See Figure 2.

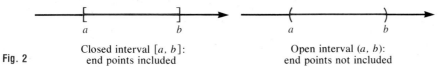

Fig. 2

Closed interval $[a, b]$: end points included

Open interval (a, b): end points not included

An interval that contains one of its end points is **half-open**. We write $[a, b)$ if a is included but b is not, and $(a, b]$ if b is included but a is not (Fig. 3).

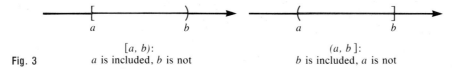

Fig. 3

$[a, b)$: a is included, b is not

$(a, b]$: b is included, a is not

In Example 1(a), the solution $x < -1$ is the set of all points to the left of -1. This set is called an **infinite interval** and denoted by $(-\infty, -1)$. In Example 1(b), the solution $x \geq \frac{3}{2}$ is the set of all points to the right of, or equal to $\frac{3}{2}$. This is an infinite interval denoted by $[\frac{3}{2}, \infty)$. See Figure 4.

Fig. 4 $(-\infty, -1)$ $[\frac{3}{2}, \infty)$

EXAMPLE 3 Solve and sketch the solution set

(a) $(x - 2)(2x + 3) < 0$ (b) $\dfrac{x - 2}{2x + 3} < 0$

SOLUTION (a) The product is negative when the two factors have opposite signs. The factor $x - 2$ is positive when $x > 2$ and negative when $x < 2$. The factor $2x + 3$ is positive when $x > -\frac{3}{2}$ and negative when $x < -\frac{3}{2}$. A diagram helps keep track of the signs (Fig. 5).

```
sign of x − 2    − − − − − − − | − − − | + + + + + + + + + +
sign of 2x + 3   − − − − − − − | + + + | + + + + + + + + + +
                              −3/2    0    2
```
Fig. 5

The signs are opposite between $-\frac{3}{2}$ and 2. Hence $-\frac{3}{2} < x < 2$. Graphically, the solution set is the open interval $(-\frac{3}{2}, 2)$. See Figure 6.

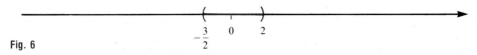

Fig. 6

(b) The quotient is negative when the numerator and denominator have opposite signs. But that is the same problem! Same answer!

Remark Suppose Example 3 had ≤ instead of <. In part (a), $x = 2$ and $x = -\frac{3}{2}$ would also be solutions. So the answer would be the *closed* interval $[-\frac{3}{2}, 2]$. But in part (b), we must exclude $x = -\frac{3}{2}$ to avoid a 0 denominator, so the answer would be the *half-open* interval $(-\frac{3}{2}, 2]$.

EXAMPLE 4 Solve $\dfrac{13x + 16}{2x + 3} < 6$

SOLUTION Subtract 6 from both sides

$$\frac{13x + 16}{2x + 3} - 6 < 0$$

$$\frac{13x + 16 - 6(2x + 3)}{2x + 3} < 0$$

$$\frac{x - 2}{2x + 3} < 0$$

Now solve as in Example 3(b) to obtain $-\frac{3}{2} < x < 2$.

EXAMPLE 5 Solve $\dfrac{5}{x + 1} > \dfrac{4}{x}$

SOLUTION Subtract $4/x$, and then combine the fractions on the left:

$$\frac{5}{x + 1} - \frac{4}{x} > 0$$

$$\frac{5x - 4(x + 1)}{x(x + 1)} > 0$$

$$\frac{x - 4}{x(x + 1)} > 0$$

This fraction is positive when all three factors are positive, or when one is positive and two are negative. Make a drawing to show the signs (Fig. 7).

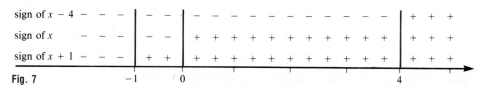

Fig. 7

All signs are positive when $x > 4$. One sign is positive and two are negative when $-1 < x < 0$. Hence the solution set consists of the open interval $(-1, 0)$ together with the infinite interval $(4, \infty)$. See Fig. 8.

Answer $-1 < x < 0$ or $x > 4$

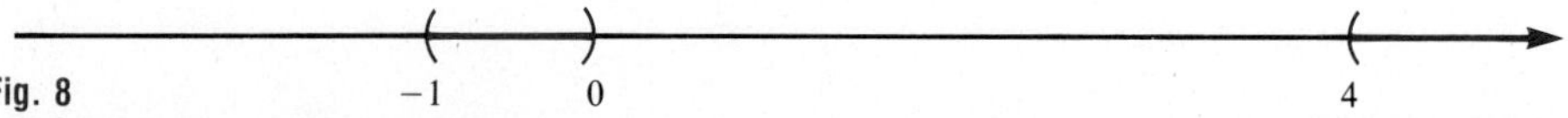

Fig. 8

Terminology The set in Fig. 8 is called the **union** of the intervals $(-1, 0)$ and $(4, \infty)$ and denoted by $(-1, 0) \cup (4, \infty)$. In general, if A and B are two sets of objects, their union $A \cup B$ is the set of all objects that belong either to A or to B.

EXAMPLE 6 Company A will rent a car for $20 per day. Company B will rent a car for $12 per day plus an initial fee of $120. When is it cheaper to rent from Company B?

SOLUTION For d days the rental fees are

Company A: $20d$ Company B: $120 + 12d$

Renting from Company B is cheaper when
$$120 + 12d < 20d$$
$$120 < 8d$$
$$15 < d$$

Answer when the rental period is more than 15 days

EXAMPLE 7 A dietician plans a daily diet of 450 grams consisting of proteins, carbohydrates, and fats. Proteins and carbohydrates produce 4 calories per gram; fats produce 9 calories per gram. If the diet must not exceed 2200 calories, how must the dietician limit the amount of fats?

SOLUTION Let x grams be the amount of fats in the diet. The remaining $450 - x$ grams are proteins and carbohydrates. The calorie values are

fats: $9x$ proteins and carbohydrates: $4(450 - x)$

Total calories must not exceed 2200. Hence
$$9x + 4(450 - x) \le 2200$$
$$5x + 1800 \le 2200$$
$$5x \le 400$$
$$x \le 80$$

Answer Fats must be limited to at most 80 grams.

EXERCISES

Sketch the interval and express it as an inequality in x

1. $[3, 5]$
2. $(0, 4)$
3. $[-1, 2)$
4. $(-3, 1]$
5. $(3, \infty)$
6. $(-\infty, 3]$

Solve the inequality

7. $6x - 1 < 3$
8. $2x + 10 < 7$
9. $10 - x > 4$
10. $6 > 3 - x$
11. $8x - 5 \ge 4x + 1$
12. $3x + 4 \le x - 6$
13. $4 - \tfrac{2}{3}x > x + 2$
14. $\tfrac{1}{2}x + 3 > 9 - 4x$
15. $3(2x - 1) < 4(1 - x)$
16. $5(x - 4) > 2(x + 1) + 3$
17. $-1 < 5x + 9 < 1$
18. $0 < 8x - 1 < 15$
19. $1.8 \le \dfrac{2x - 7}{5} \le 2.6$
20. $-1.03 \le \dfrac{3x + 11}{4} \le 8.15$

Solve the inequality and sketch its solution set

21. $(x - 1)(x - 4) > 0$
22. $(x + 2)(x - 3) < 0$

23. $(2x - 1)(3x - 1) < 0$

24. $(x - 3)(4x - 3) < 0$

25. $x^2(x - 3)(x - 5) \geq 0$

26. $(x^2 + 1)(x + 1)(x - 2) \leq 0$

27. $\dfrac{x}{x - 3} > 0$

28. $\dfrac{2x}{x - 9} < 0$

29. $\dfrac{2x - 7}{x + 5} \leq 0$

30. $\dfrac{x + 1}{4x + 3} \geq 0$

31. $\dfrac{x - 3}{x^2 - 4x} < 0$

32. $\dfrac{x^2 + 5x}{3x + 1} < 0$

33. $\dfrac{1}{x + 5} \geq \dfrac{1}{8}$

34. $\dfrac{4}{x - 3} > 5$

35. $\dfrac{7}{x} < \dfrac{3}{x - 2}$

36. $\dfrac{1}{x + 4} > \dfrac{8}{x}$

37. $\dfrac{x}{2x - 1} \geq 3$

38. $\dfrac{x}{x + 4} \leq \dfrac{1}{6}$

39. $\dfrac{x + 1}{2x + 1} < \dfrac{3}{4}$

40. $\dfrac{1 - 3x}{1 + 3x} < \dfrac{1}{2}$

41. A bookstore sells a card for $5 that entitles you to 15% off on all books. On what amount of purchases will you save over $10 by buying the card?

42. Broker A charges a commission of $25 plus 1.5% of the amount of the transaction. Broker B charges $10 plus 1.8%. For what transactions is A cheaper?

43. Dave's exam scores have been 78, 88, and 92. If the final exam counts double, what score on the final will give him a course average of at least 90?

44. Tank A contains 5000 gallons of water and tank B contains 1000 gallons. Water is pumped from A into B at the rate of 40 gal/min. For how long will A contain more than twice as much water as B?

45. The assets of an established trust fund are $200 million and increasing at the rate of $10 million per year. A new trust fund starts with $100 million and, under aggressive management, its assets increase at the rate of $12 million per year. When will its assets exceed 75% of those of the older trust fund?

46. Mack's has sold 40 billion hamburgers and is selling them at the rate of 2 billion per year. A new chain, Jack's, starts selling hamburgers at the rate of 200 million per year. Assume these rates continue.
(a) How long until Jack's total sales exceed 2% of Mack's?
(b) Will Jack's total sales ever exceed 10% of Mack's?

47. A collection of 100 coins, all pennies and nickels, is worth over $3.75. At least how many nickels are there?

48. A pro football game is a sell-out: 50,000 tickets were sold, some at $12, some at $10. If the total receipts exceed $570,000, at least how many $12 tickets were sold?

49. (See Example 7.) A patient on a restricted diet can have no more than 20% of his daily calories in fats. If he eats 500 grams of food each day, how must the quantity of fats be limited?

50. Find all pairs of consecutive positive integers whose squares differ by at most 15.

3 INEQUALITIES INVOLVING ABSOLUTE VALUES; QUADRATIC INEQUALITIES

Recall the definition of the absolute value of a real number (Chapter 1, Sec. 5):

$$|x| = \begin{cases} x & \text{if } x \geq 0 \\ -x & \text{if } x < 0 \end{cases}$$

On the number line, $|x|$ is the distance from x to 0. Hence if $a > 0$, the inequality $|x| < a$ determines all points whose distance from 0 is less than a. The inequality $|x| > a$ determines all points whose distance from 0 is greater than a. See Figure 1.

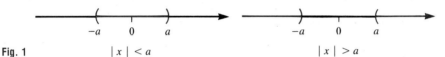

Fig. 1 $|x| < a$ $|x| > a$

Let $a > 0$. Then
$|x| < a$ if and only if $-a < x < a$
$|x| > a$ if and only if $x < -a$ or $x > a$

EXAMPLE 1 Solve the inequality and sketch the solution set

(a) $|x - 3| < 1$ (b) $\left|\dfrac{4x + 1}{3}\right| > 2$

SOLUTION   (a) The inequality means that
$$-1 < x - 3 < 1$$
Add 3:
$$2 < x < 4$$

The solution set is the interval $(2, 4)$. It consists of all points whose distance from 3 is less than 1. See Figure 2a.

(b)
$$\dfrac{4x + 1}{3} < -2 \quad \text{or} \quad \dfrac{4x + 1}{3} > 2$$
$$4x + 1 < -6 \quad\quad\quad 4x + 1 > 6$$
$$4x < -7 \quad\quad\quad\quad 4x > 5$$
$$x < -\dfrac{7}{4} \quad \text{or} \quad x > \dfrac{5}{4}$$

The solution set is $(-\infty, -\tfrac{7}{4}) \cup (\tfrac{5}{4}, \infty)$. See Figure 2b.

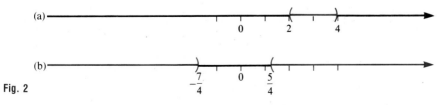

Fig. 2

112 Inequalities

EXAMPLE 2 If I join a tennis club for a $400 annual fee, I can play for only $1 per hour (or part of an hour). If I don't join, it will cost $8 per hour. For how many hours of play yearly will the difference amount to less than $20?

SOLUTION The cost of h hours of play is

$$\text{joining: } 400 + h \qquad \text{not joining: } 8h$$

The difference, in either order, is

$$|8h - (400 + h)| = |7h - 400|$$

The problem requires

$$|7h - 400| < 20$$

Solve for h:

$$-20 < 7h - 400 < 20$$
$$380 < 7h < 420$$
$$54\tfrac{2}{7} < h < 60$$

Therefore, h is an integer between $54\tfrac{2}{7}$ and 60.
Answer
55, 56, 57, 58, or 59 hours

Let us note the following important properties of absolute values:

$$|ab| = |a||b| \qquad \left|\frac{a}{b}\right| = \frac{|a|}{|b|} \qquad b \neq 0$$

For example,

$$|-2x| = |-2||x| = 2|x| \qquad \left|\frac{x}{3}\right| = \frac{|x|}{|3|} = \frac{|x|}{3}$$

These properties can be justified by considering the cases of a and b both positive, both negative, or of opposite signs. See Exercises 41 and 42.

EXAMPLE 3 Solve $\left|\dfrac{2}{x+3}\right| > 1$

SOLUTION Since

$$\left|\frac{2}{x+3}\right| = \frac{|2|}{|x+3|} = \frac{2}{|x+3|}$$

the inequality is

$$\frac{2}{|x+3|} > 1$$

(It is assumed that $x \neq -3$). We multiply both sides by the positive number $|x+3|$:

$$2 > |x+3|$$

which is equivalent to
$$-2 < x + 3 < 2$$
$$-5 < x < -1$$

Answer $-5 < x < -1$, $x \neq -3$; that is, $(-5, -3) \cup (-3, -1)$

Quadratic Inequalities

A quadratic inequality is an inequality of the form
$$ax^2 + bx + c > 0$$
or with $<$ or $\leq$ or $\geq$ in place of $>$.

For the simple quadratic inequalities $x^2 < a$ and $x^2 > a$, where $a > 0$, the solution set is obvious.

> For $a > 0$,
> $x^2 < a$ if and only if $-\sqrt{a} < x < \sqrt{a}$
> $x^2 > a$ if and only if $x < -\sqrt{a}$ or $x > \sqrt{a}$

Corresponding statements hold for $\leq$ and $\geq$.

Methods for solving general quadratic inequalities are factoring (when possible) and completing the square.

EXAMPLE 4 Solve $x^2 + 3 < 4x$

SOLUTION 1 We have $\quad x^2 - 4x + 3 < 0$
Factor: $\quad (x-1)(x-3) < 0$
The factors have opposite signs when $\underline{1 < x < 3}$.

SOLUTION 2 $\quad\quad\quad\quad x^2 - 4x < -3$
Complete the square: $\quad x^2 - 4x + 4 < -3 + 4$
$$(x-2)^2 < 1$$

Hence
$$-1 < x - 2 < 1$$
$$\underline{1 < x < 3}$$

EXAMPLE 5 In t seconds after firing, a projectile is $320t - 16t^2$ feet above the ground. During what period is it higher than 1200 ft?

SOLUTION $\quad\quad 320t - 16t^2 > 1200$
$$0 > 16t^2 - 320t + 1200$$
Divide by 16: $\quad\quad 0 > t^2 - 20t + 75$
Factor: $\quad\quad 0 > (t-5)(t-15)$
The product is negative when $5 < t < 15$.

Answer
$\quad\quad\quad$between 5 and 15 seconds after firing

116 Inequalities

Sketch the solution set

7. $|x + 8| \le 2$

8. $|2x - 3| \ge 1$

Solve the inequality

9. $-1 < 4x + 3 < 8$

10. $\dfrac{2}{5x - 1} > \dfrac{3}{4}$

11. $\dfrac{x}{3x - 8} \ge 0$

12. $\dfrac{2x + 5}{3x + 10} \le \dfrac{3}{7}$

13. $\dfrac{x + 20}{x^2} > 1$

14. $x^4 - 3x^2 + 2 \le 0$

Show that the inequality holds for all real x

15. $x^2 - 3x + 3 \ge 0$

16. $6x \le 9x^2 + 1$

17. A taxi charges $1.25 for the first $\frac{1}{6}$ mile, then 10¢ for each additional $\frac{1}{6}$ mile (or part of $\frac{1}{6}$ mile). For what distances are the fares between $4 and $6?

18. What is the smallest possible perimeter of a rectangle whose area is 80 ft²?

19. The population of New York City is over 7,500,000. Show that at least one day of the year is the birthday of over 20,000 New Yorkers.

20. Thunder was heard 5.0 seconds after a lightning flash. Assume that the speed of sound is 1100 ft/sec but may vary by at most 3% with atmospheric conditions. Also assume that the timing was subject to an error of at most 2%. What are the possible distances from here to the lightning?

5

Coordinates, Lines, Functions

1 COORDINATES IN THE PLANE

In Chapter 1, we labeled each point on a line by a real number. Now we are going to label each point in a plane by an **ordered pair** of real numbers, (a, b). By *ordered*, we mean that a is the first and b is the second number of the pair. Thus $(8, 5)$ and $(5, 8)$ are the same pair, but different ordered pairs.

We draw two perpendicular lines in the plane, one horizontal and one vertical, and label their points as in Figure 1a. The horizontal line is the **x-axis** and the vertical line is the **y-axis.**

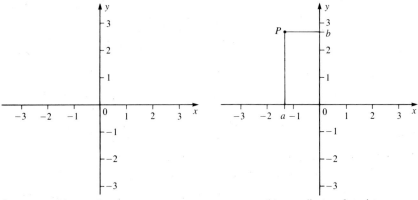

FIG. 1 (a) coordinate axes (b) coordinates of a point

Suppose P is any point in the plane. The vertical line through P meets the x-axis in a point a; the horizontal line through P meets the y-axis in a point b (Figure 1b). We assign to P the ordered pair (a, b). The number a is the **x-coordinate** or **abscissa** of P; the number b is the **y-coordinate** or **ordinate** of P.

Conversely, suppose (a, b) is any ordered pair of real numbers. We draw a vertical line through the point a on the x-axis and a horizontal line through the point b on the y-axis. These intersect in a point P whose coordinates are (a, b). Thus there is a one-to-one correspondence

$$P \leftrightarrow (a, b)$$

118   Coordinates, Lines, Functions

between the set of points in the plane and the set of all ordered pairs of real numbers.

A coordinate system of the type described is called **rectangular** or **cartesian** (for the French mathematician and philosopher Descartes). The point (0, 0) is called the **origin**.

In Figure 2a, we plot some typical points. Note that points of the form $(a, 0)$ lie on the *x*-axis, those of the form $(0, b)$ on the *y*-axis. All other points lie in one of the four quadrants, depending on the signs of their coordinates.

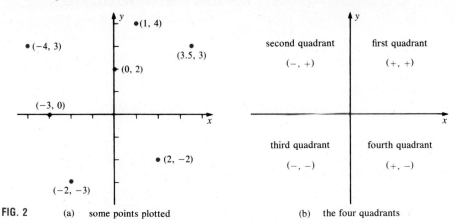

FIG. 2   (a)   some points plotted   (b)   the four quadrants

EXAMPLE 1   Find the coordinates of

(a)   the vertices of the square centered at (3, 0), sides of length 4 and parallel to the axes
(b)   the vertices of the equilateral triangle, sides of length 4, one side on the *x*-axis, one vertex on the positive *y*-axis.

SOLUTION   (a) Draw a figure (Fig. 3a). The vertical sides cross the *x*-axis at 1 and 5. The horizontal sides are at the levels $y = 2$ and $y = -2$. Now you can read off the vertices. Top: (1, 2) and (5, 2). Bottom: (1, −2) and (5, −2).

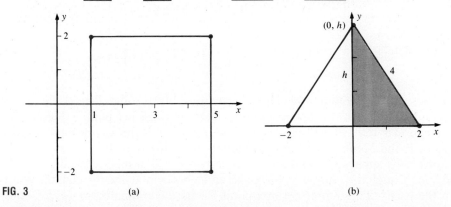

FIG. 3   (a)   (b)

(b) See Figure 3b. By symmetry, two vertices are at (2, 0) and (−2, 0). The third vertex is at a point $(0, h)$, where h is the height of the triangle. To find h, apply the Pythagorean theorem to the shaded right triangle:

$$h^2 = 4^2 - 2^2$$
$$h = \sqrt{12} = 2\sqrt{3}$$

Hence the vertices are $(2, 0)$, $(-2, 0)$, $(0, 2\sqrt{3})$.

EXAMPLE 2 Sketch the set of all points (x, y) for which
(a) $x = 2$ (b) $y = -1$ (c) $x \geq 0$ and $|y| \leq 1$

SOLUTION (a) These are the points $(2, y)$, where y can be any real number. Such points lie 2 units to the right of the y-axis. They form a vertical line (Fig. 4a).

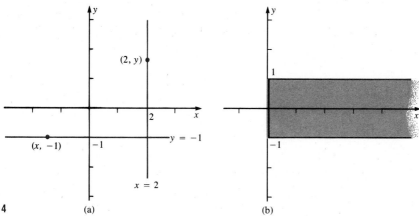

FIG. 4 (a) (b)

(b) All points $(x, -1)$ form the horizontal line one unit below the x-axis (Fig. 4a).

(c) Points (x, y) with $x \geq 0$ lie on or to the right of the y-axis. Points (x, y) with $|y| \leq 1$ lie on and between the levels $y = -1$ and $y = 1$. Points that satisfy *both* conditions fill out the shaded region in Figure 4b.

Distance Formula

Recall that given two points x_1 and x_2 on the x-axis, the distance between them is $|x_2 - x_1|$. For two points (x_1, y_1) and (x_2, y_1) on the same horizontal line in the plane, the distance is also $|x_2 - x_1|$. Similarly, for two points (x_2, y_1) and (x_2, y_2) on the same vertical line, the distance is $|y_2 - y_1|$. See Figure 5a.

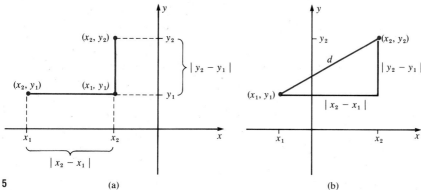

FIG. 5 (a) (b)

Now let us find the distance d between *any* two points (x_1, y_1) and (x_2, y_2). We construct a right triangle as in Fig. 5b. Its legs have length $|x_2 - x_1|$ and $|y_2 - y_1|$. Hence, by the Pythagorean theorem,

$$d^2 = |x_2 - x_1|^2 + |y_2 - y_1|^2 = (x_2 - x_1)^2 + (y_2 - y_1)^2$$

> **Distance Formula** The distance between two points (x_1, y_1) and (x_2, y_2) is
> $$\sqrt{(x_2 - x_1)^2 + (y_2 - y_1)^2}$$

This formula holds even if the points are on the same horizontal or vertical line. Why?

Examples

(1) The distance between $(3, 1)$ and $(7, 4)$ is
$$\sqrt{(7 - 3)^2 + (4 - 1)^2} = \sqrt{4^2 + 3^2} = \sqrt{25} = 5$$

(2) The distance between $(2, -6)$ and $(-3, 1)$ is
$$\sqrt{(-3 - 2)^2 + [1 - (-6)]^2} = \sqrt{(-5)^2 + 7^2} = \sqrt{74}$$

EXAMPLE 3 Show that the points $P = (-6, 5)$, $Q = (4, 1)$ and $R = (8, 11)$ are the vertices of an isosceles right triangle.

SOLUTION Plot the points carefully (Fig. 6). If the triangle is actually isosceles, the figure suggests that the equal sides are $\overline{PQ}$ and $\overline{QR}$. If it is actually a right triangle, the figure suggests that the hypotenuse is $\overline{PR}$.

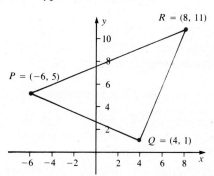

FIG. 6

To confirm this evidence, we just compute the lengths of the three sides and check that
$$\overline{PQ} = \overline{QR} \quad \text{and} \quad (\overline{PR})^2 = (\overline{PQ})^2 + (\overline{QR})^2$$

By the distance formula,
$$(\overline{PQ})^2 = [4 - (-6)]^2 + (1 - 5)^2 = 10^2 + (-4)^2 = 116$$
$$(\overline{QR})^2 = (8 - 4)^2 + (11 - 1)^2 = 4^2 + 10^2 = 116$$
$$(\overline{PR})^2 = [8 - (-6)]^2 + (11 - 5)^2 = 14^2 + 6^2 = 232$$

We see that $\overline{PQ} = \overline{QR} = \sqrt{116}$ and that
$$(\overline{PR})^2 = (\overline{PQ})^2 + (\overline{QR})^2 \quad \text{since} \quad 232 = 116 + 116$$

Hence the triangle is indeed isosceles and a right triangle.

EXAMPLE 4 Let (x, y) be any point on the perpendicular bisector of the segment joining $(0, 5)$ and $(2, -1)$. Show that $y = \frac{1}{3}x + \frac{5}{3}$.

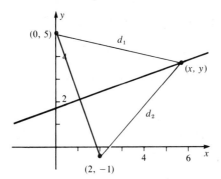

FIG. 7

SOLUTION Points on the perpendicular bisector are equidistant from $(0, 5)$ and $(2, -1)$. Thus, $d_1 = d_2$ in Figure 7. Express d_1 and d_2 using the distance formula, and set $d_1 = d_2$:

$$\sqrt{(x - 0)^2 + (y - 5)^2} = \sqrt{(x - 2)^2 + (y + 1)^2}$$
$$x^2 + y^2 - 10y + 25 = x^2 - 4x + 4 + y^2 + 2y + 1$$
$$-10y + 25 = -4x + 4 + 2y + 1$$
$$4x + 20 = 12y$$

Solve for y:

$$y = \frac{1}{3}x + \frac{5}{3}$$

Midpoint Formula

Let x_1 and x_2 be points on the x-axis. Then the point midway between them is $m = \frac{1}{2}(x_1 + x_2)$. See Fig. 8a. For if $x_1 < x_2$, the distance between them is $x_2 - x_1$. Starting at x_1 and going half way to x_2 brings us to

$$x_1 + \tfrac{1}{2}(x_2 - x_1) = \tfrac{1}{2}(x_1 + x_2)$$

If $x_2 < x_1$, the midpoint is also $\frac{1}{2}(x_1 + x_2)$, by a similar argument.

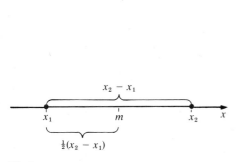
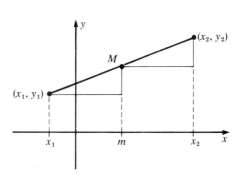

FIG. 8   (a)   (b)

Now let us find the midpoint of the segment joining two points (x_1, y_1) and (x_2, y_2) in the plane. In Figure 8b, the small right triangles are congruent. It follows that m is midway between x_1 and x_2. Therefore, the x-coordinate of M is $\frac{1}{2}(x_1 + x_2)$. Similarly, the y-coordinate of M is $\frac{1}{2}(y_1 + y_2)$.

> **Midpoint Formula** The midpoint of the segment joining (x_1, y_1) and (x_2, y_2) is
> $$\left(\frac{x_1 + x_2}{2}, \frac{y_1 + y_2}{2} \right)$$

For example, the midpoint of the segment joining $(3, 1)$ and $(5, -6)$ is
$$\left(\frac{3 + 5}{2}, \frac{1 - 6}{2} \right) = \left(4, -\frac{5}{2} \right)$$

EXAMPLE 5 The points $(-4, 1)$, $(0, 4)$ and $(10, 3)$ are the vertices of a triangle. Find the length of its shortest median.

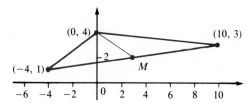

FIG. 9

SOLUTION Recall that a median of a triangle is a line segment from a vertex to the midpoint of the opposite side. Sketch the triangle (Fig. 9). The shortest median is the one from $(0, 4)$ to M, the midpoint of the side connecting $(-4, 1)$ and $(10, 3)$. By the midpoint formula,
$$M = \left(\frac{-4 + 10}{2}, \frac{1 + 3}{2} \right) = (3, 2)$$
The length of the median is the distance from $(0, 4)$ to $(3, 2)$:
$$\sqrt{(3 - 0)^2 + (2 - 4)^2} = \sqrt{3^2 + (-2)^2} = \underline{\sqrt{13}}$$

EXERCISES

Plot and label the points on one graph

1. $(-4, 1), (3, 2), (5, -3), (1, 4)$
2. $(0, -2), (3, 0), (-2, 2), (1, -3)$
3. $(0.2, -0.5), (-0.3, 0), (-1.0, -0.1)$
4. $(75, -10), (-15, 60), (95, 40)$

Find the coordinates (x, y) of the

5. point 3 units to the left of $(5, 6)$
6. point 7 units below $(5, 6)$
7. point $\sqrt{2}$ units "northeast" of $(1, 4)$
8. point 3 units "northwest" of $(0, 0)$
9. vertices of a square centered at $(0, 0)$, sides of length 2 and parallel to the axes

10. vertices of a 3-4-5 right triangle in the first quadrant, right angle at (0, 0), hypotenuse of length 15

11. vertices of a square centered at (1, 3), sides of length 2, at 45° angles with the axes

12. vertices of an equilateral triangle, height 6, one side on the line $x = 2$, one vertex on the negative x-axis

Indicate on a diagram all points (x, y) in the plane for which

13. $x = -3$
14. $y = 2$
15. $1 \leq x \leq 3$
16. $-1 \leq y \leq 2$
17. $x < 3$ and $y > 1$
18. $x > 2$ and $y < 3$
19. $|x| \geq 1$ and $|y| \leq 2$
20. $|x| \geq 2$ and $|y| \geq 2$
21. $xy > 0$ and $|x| \leq 3$
22. $xy < 0$ and $|y| \geq 1$
23. $|x - 3| < 1$ and $y > 0$
24. $|x - 1| < 2$ and $|y| < 1$

Compute the distance between the points

25. (4, 1) and (12, 7)
26. (−9, 6) and (7, −6)
27. (0, 3) and (8, 5)
28. (0, 3) and (0, 11)
29. (−4, −1) and (3, −3)
30. ($\frac{1}{2}$, $\frac{3}{2}$) and (2, 5)

Show that the points are vertices of an isosceles triangle

31. (1, 3), (3, −3), (14, 4)
32. (2, −4), (9, 0), (10, 8)

Show that the points are vertices of a right triangle

33. (2, −3), (−2, 1), (9, 4)
34. (−6, 1), (−2, 9), (4, −4)

Show that the points are vertices of a square

35. (0, 0), (3, 4), (−1, 7), (−4, 3)
36. (−5, −6), (−12, 18), (12, 25), (19, 1)

Compute the lengths of the three medians of the triangle with vertices

37. (0, 0), (2, 6), (10, 4)
38. (0, 2), (−6, 0), (4, 8)

Find an equation that a point (x, y) must satisfy if it lies on the perpendicular bisector of the segment connecting

39. (0, 0) and (4, 6)
40. (2, 4) and (6, 2)

41. Find all points on the x-axis whose distance from the point (2, 7) is 10.

42. Find all points on the line $x = 1$ whose distance from the point (6, 3) is 13.

43. Find all points on the line $y = 1$ that are twice as far from (5, 0) as from the origin.

44. Show that no point on the line $y = 4$ is twice as far from (5, 0) as from the origin.

45. Show that the point (6, 9) is the same distance from the point (0, 1) as from the line $y = -1$.

124 Coordinates, Lines, Functions

46. Show that a point (a, b) has the property described in Exercise 45 if $a^2 = 4b$.

47. Here is a version of the game "Battleship." An enemy battleship is located at a point (m, n), where m and n are integers restricted to the values $0, 1, 2, \ldots, 100$. You have to guess its position.

You take a guess (x_1, y_1), and are then told the error $e_1 = |x_1 - m| + |y_1 - n|$. Then you take a second guess, and are told e_2. You continue this way until you hit. Interpret the numbers $e_1, e_2, \ldots$ geometrically. Show that you can hit on the third try by first guessing $(0, 0)$, then $(0, 100)$.

48. Using the strategy in Exercise 47, express m and n in terms of e_1 and e_2.

49. In Figure 10, the points in the plane with both coordinates *integers* are colored alternately red or black. Is $(21, 37)$ a red or black point?

FIG. 10

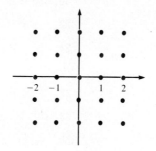

50. [cont.] How can you tell in general whether (m, n) is red or black?

2 EQUATIONS OF LINES

In Chapter 3 we solved equations such as $x^2 - 6x + 2 = 0$ involving one variable. In this chapter and in following chapters, we deal with equations such as

$$y = \tfrac{1}{3}x + \tfrac{5}{3} \qquad x^2 + y^2 = 16 \qquad x = y^2 + 3$$

involving *two* variables. A **solution** is an ordered pair (x, y) of real numbers that makes the equation a true statement. Thus, $(7, 2)$ is a solution of $x = y^2 + 3$, but $(5, 1)$ is not because

$$7 = 2^2 + 3 \qquad \text{but} \qquad 5 \neq 1^2 + 3$$

With each solution (x, y), we associate the corresponding point in the plane. The set of all such points is the **graph** of the equation; it is a picture of the solution set. In this section, we study equations whose graphs are straight lines. Later, we take up equations whose graphs are various curves.

EXAMPLE 1 Sketch the graph of the equation

  (a) $y = x$ (b) $y = 2x$ (c) $y = -\tfrac{1}{2}x$

SOLUTION (a) The graph consists of all points of the form (x, x). Plot a few points, say $(0, 0), (1, 1), (2, 2), (-1, -1), (-2, -2)$. The points lie on a straight line at a 45° angle with the positive x-axis (Fig. 1a).

(b) The graph consists of all points $(x, 2x)$. Plot a few, say, $(0, 0)$, $(1, 2)$, $(2, 4)$, $(-1, -2)$, $(-2, -4)$. The graph is again a straight line through the origin, but steeper than the line in (a). See Figure 1b.

(c) The graph consists of all points $(x, -\frac{1}{2}x)$, for example, $(0, 0)$, $(1, -\frac{1}{2})$, $(2, -1)$, $(-1, \frac{1}{2})$, $(-2, 1)$. It is a line through the origin, going down to the right, and not very steep (Fig. 1c).

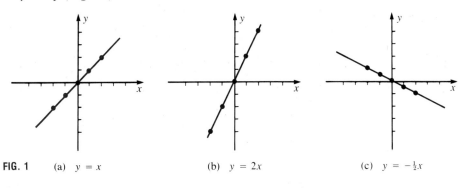

FIG. 1   (a) $y = x$ (b) $y = 2x$ (c) $y = -\frac{1}{2}x$

Slope

Highway engineers measure the steepness of a hill by the ratio of the rise to the run. For instance, the highway sign

indicates that the road rises 1 unit for each 6 units it runs (moves horizontally). See Figure 2.

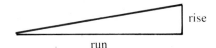

FIG. 2

We use the same idea to measure the steepness of a non-vertical line in the plane. Let (x_0, y_0) and (x_1, y_1) be two distinct points on the line, with $x_0 < x_1$. See Figure 3. The "rise" is $y_1 - y_0$ and the "run" is $x_1 - x_0$. So we define the slope of

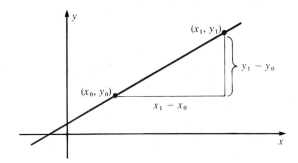

FIG. 3

the line to be the ratio

$$\frac{y_1 - y_0}{x_1 - x_0}$$

Here the denominator $x_1 - x_0$ is positive. The numerator $y_1 - y_0$ is positive if the line rises to the right, negative if it falls to the right, and 0 if the line is horizontal. Therefore,

$$\text{slope} = \frac{y_1 - y_0}{x_1 - x_0} \text{ is } \begin{Bmatrix} \text{positive} \\ \text{negative} \\ 0 \end{Bmatrix} \text{ if the line } \begin{Bmatrix} \text{rises to the right} \\ \text{falls to the right} \\ \text{is horizontal} \end{Bmatrix}$$

We note two features of the definition of slope. First, the slope is unchanged if the two points are reversed:

$$\frac{y_0 - y_1}{x_0 - x_1} = \frac{y_1 - y_0}{x_1 - x_0}$$

Hence it doesn't matter whether $x_1 > x_0$ or $x_1 < x_0$. Second, it also doesn't matter which two points are used to compute the slope of a line. In Figure 4, the right triangles are similar. Hence ratios of corresponding sides are equal.

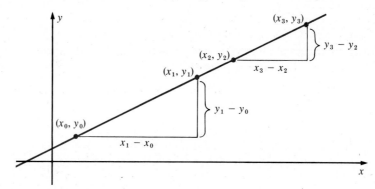

FIG. 4

Let us summarize:

> **Slope** The slope of a non-vertical line in the plane is defined by
> $$\text{slope} = \frac{y_1 - y_0}{x_1 - x_0}$$
> where (x_0, y_0) and (x_1, y_1) are any two distinct points on the line.

EXAMPLE 2 Compute the slope of the line through the given points and sketch the line

(a) $(-2, 1), (7, 3)$ (b) $(2, 5), (4, -1)$ (c) $(-1, -2), (3, -2)$

SOLUTION By the slope formula,

(a) $\text{slope} = \dfrac{3 - 1}{7 - (-2)} = \dfrac{2}{9}$

(b) $\text{slope} = \dfrac{-1 - 5}{4 - 2} = \dfrac{-6}{2} = \underline{-3}$

(c) slope $= \dfrac{-2 - (-2)}{3 - (-1)} = \dfrac{0}{4} = 0$ See Figure 5.

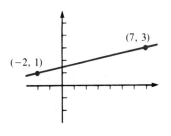
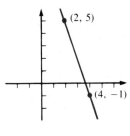
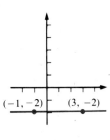

FIG. 5   (a) Slope $\tfrac{2}{9}$ (b) Slope -3 (c) Slope 0

Equations of Lines

Given a line L, let us find an equation whose graph is L. If L is vertical, that's easy: the equation is $x = a$, where a is the point at which L meets the x-axis.

Suppose L is not vertical. Then L is determined by one point it passes through and its slope. For example, take L to be the line through $(8, 5)$ with slope $\tfrac{3}{4}$. A point $(x, y) \neq (8, 5)$ is on L if and only if the slope given by (x, y) and $(8, 5)$ is $\tfrac{3}{4}$.

$$\dfrac{y - 5}{x - 8} = \dfrac{3}{4}$$

That's the equation! Since $x \neq 8$, we can multiply both sides by $x - 8$:

$$y - 5 = \tfrac{3}{4}(x - 8)$$

Even the point $(8, 5)$ satisfies this equation. (It makes both sides 0.) Therefore, the points (x, y) that lie on L are precisely the ones that satisfy the equation.

Similarly, we find an equation for the line through a point (x_0, y_0) with slope m:

> **Point-Slope Form**   The line through the point (x_0, y_0) with slope m is the graph of the equation
>
> $$y - y_0 = m(x - x_0)$$

EXAMPLE 3   Find an equation for the line through $(4, -1)$ with slope -2.

SOLUTION   Apply the point-slope form with $(x_0, y_0) = (4, -1)$ and $m = -2$.
  Answer
$$y + 1 = -2(x - 4)$$

EXAMPLE 4   Find an equation for the line through $(2, 5)$ and $(9, 1)$.

SOLUTION   First find the slope:

$$\text{slope} = \dfrac{1 - 5}{9 - 2} = -\dfrac{4}{7}$$

Now apply the point-slope form with $(x_0, y_0) = (2, 5)$ and $m = -\tfrac{4}{7}$.
  Answer
$$y - 5 = -\tfrac{4}{7}(x - 2)$$

Suppose a line intercepts the y-axis at $(0, b)$ and has slope m. Then the point-slope form gives the equation
$$y - b = m(x - 0) \quad \text{or} \quad y = mx + b$$
This is a particularly useful form for graphing. The number b is the **y-intercept** of the line.

Slope-Intercept Form The line with slope m and y-intercept b is the graph of the equation
$$y = mx + b$$

EXAMPLE 5 Sketch the graph (a) $y = 2x - 1$ (b) $4x + 3y = 5$

SOLUTION (a) The equation has the form $y = mx + b$, where $m = 2$ and $b = -1$. Its graph is the line with slope 2 and y-intercept -1. See Figure 6a.

(b) Convert the equation into the form $y = mx + b$:
$$3y = -4x + 5 \qquad y = -\frac{4}{3}x + \frac{5}{3}$$

The graph is the line with slope $-\frac{4}{3}$ and y-intercept $\frac{5}{3}$. See Figure 6b.

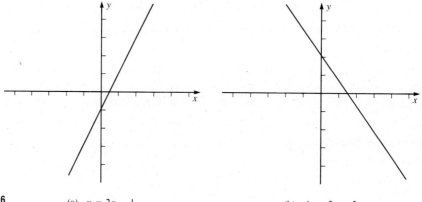

FIG. 6 (a) $y = 2x - 1$ (b) $4x + 3y = 5$

As in Example 5b, we see that the graph of $ax + by = c$ is a line. For
$$by = -ax + c$$
$$y = -\frac{a}{b}x + \frac{c}{b} \qquad b \neq 0$$

This equation represents a line with slope $-a/b$ and y-intercept c/b. If $b = 0$ but $a \neq 0$, the equation reduces to $ax = c$ or $x = c/a$, the equation of a vertical line.

The graph of the equation
$$ax + by = c \qquad a, b \text{ not both } 0$$
is a line.

Parallel and Perpendicular Lines

Two non-vertical lines are parallel if and only if their slopes are equal. That makes sense: equal slopes mean "the same direction." In Figure 7, the lines are parallel if and only if the right triangles shown are similar. But then

$$\frac{y_1 - y_0}{x_1 - x_0} = \frac{y_3 - y_2}{x_3 - x_2}$$

that is, the slopes are equal.

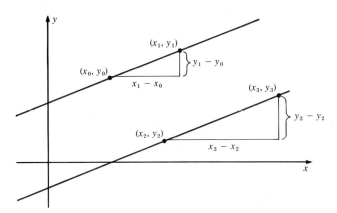

FIG. 7

Two non-vertical lines with slopes m_1 and m_2 are perpendicular if and only if $m_1 = -1/m_2$ or equivalently $m_1 m_2 = -1$. (For a justification, see Exercise 46.)

Parallel and Perpendicular Lines Two non-vertical lines with slopes m_1 and m_2 are

$$\left.\begin{array}{l}\text{parallel}\\ \text{perpendicular}\end{array}\right\} \text{ if and only if } \begin{cases} m_1 = m_2 \\ m_1 m_2 = -1 \end{cases}$$

Examples
(1) The lines $y = 4x + 2$ and $y = 4x - 7$ are parallel: $m_1 = m_2 = 4$.
(2) The lines $y = \frac{2}{3}x - 1$ and $y = -\frac{3}{2}x + 14$ are perpendicular:

$$m_1 m_2 = (\tfrac{2}{3})(-\tfrac{3}{2}) = -1$$

EXAMPLE 6 Find an equation for the line through $(-2, 3)$ and
 (a) parallel to the line $3x - 8y = 1$
 (b) perpendicular to the line $y = 4x + 11$

SOLUTION In both parts, the answer is $y - 3 = m(x + 2)$, where m is the proper slope.
 (a) Write the line $3x - 8y = 1$ as

$$8y = 3x - 1$$
$$y = \frac{3}{8}x - \frac{1}{8}$$

130 Coordinates, Lines, Functions

Hence its slope is $\frac{3}{8}$. A parallel line must have the same slope. Set $m = \frac{3}{8}$.

(b) The slope of $y = 4x + 11$ is 4. A perpendicular line must have slope $m = -\frac{1}{4}$.

Answer

(a) $y - 3 = \frac{3}{8}(x + 2)$ (b) $y - 3 = -\frac{1}{4}(x + 2)$

EXERCISES

Find the slope of the line through the given points

1. (0, 0), (3, 4)
2. (0, 0), (6, 2)
3. (−1, 2), (4, 2)
4. (6, 5), (9, 5)
5. (−8, 13), (2, 7)
6. (1, −4), (3, −6)

By computing slopes, determine whether the points lie on a straight line

7. (3, 1), (10, 8), (14, 12)
8. (−1, −6), (0, −5), (3, −2)
9. $(0, \frac{1}{2})$, $(1, \frac{3}{2})$, $(5, \frac{11}{2})$
10. (9, 7), (5, 6), (−7, 4)

Sketch the graph

11. $y = 2x - 3$
12. $y = \frac{1}{2}x + 3$
13. $y = -\frac{1}{4}x + 1$
14. $y = -3x - \frac{1}{2}$
15. $y - 3 = 0$
16. $3y = 5$
17. $2x - 3y = 6$
18. $4x - 3y = 9$
19. $\frac{3}{4}x + y - 2 = 0$
20. $x - \frac{1}{2}y + 2 = 0$

Plot on the same set of coordinate axes

21. $y = -\frac{1}{2}x + b$ $b = -1, 0, 1, 2$
22. $y = mx + 1$ $m = -1, -\frac{1}{3}, 1, 2$
23. $y - b = 2(x - 1)$ $b = -2, 0, 1, 2$
24. $y - 4 = 3(x - a)$ $a = -1, 0, 1, 2$
25. $x - y = c$ $c = -2, -1, 0, 1, 2$
26. $x + y = d$ $d = -2, -1, 0, 1, 2$

Verify that the points are vertices of a parallelogram

27. (0, 0), (0, 5), (3, 9), (3, 4)
28. (−4, −5), (3, 3), (9, 6), (2, −2)

Find an equation for the line

29. through (4, 2), slope 1
30. through (1, 5), slope $\frac{1}{2}$
31. through (−7, 4), vertical
32. through (10, 9), horizontal
33. through (−5, 6) and (2, 4)
34. through (7, 5) and (8, 3)
35. through $(\frac{1}{2}, -\frac{1}{2})$ and (0, 3)
36. through $(\frac{2}{3}, -\frac{1}{3})$ and $(1, \frac{4}{3})$

37. through $(4, -3)$, parallel to $2x - y = 19$

38. through $(1, 6)$, perpendicular to $8x - 5y = 12$

Using slopes, show that the points are vertices of a right triangle

39. $(-3, 5), (-1, 3), (4, 12)$ 40. $(-6, 15), (3, 2), (6, 6)$

41. Show that the parallelogram in Exercise 27 is a rhombus (all sides equal).

42. Let $(x_1, y_1), (x_2, y_2), (x_3, y_3)$, and (x_4, y_4) be the vertices of any quadrilateral. Show that the midpoints of its sides are the vertices of a parallelogram.

43. Verify that (a, b) and $(-b, -a)$ are symmetric with respect to the line $y = -x$.

44. Verify that $(0, 4)$ and $(\frac{16}{5}, \frac{12}{5})$ are symmetric with respect to the line $y = 2x$.

45. A steep mountain road climbs with a constant grade of 1:7 (slope $\frac{1}{7}$). It is an 18-mile drive to the summit (Fig. 9). What is the elevation of the summit in feet?

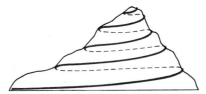

FIG. 9

46. Figure 10 shows the lines $y = m_1 x$ and $y = m_2 x$. They are perpendicular if and only if $d_1^2 + d_2^2 = d_3^2$. Use the distance formula to show that this condition boils down to $m_1 m_2 = -1$.

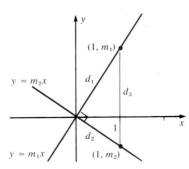

FIG. 10

Exercises 47–50 refer to Figure 10, page 124. The points in the figure are called **lattice points**.

47. If n is an integer, show that the line $y = x + n$ contains only lattice points of the same color. How does the color depend on n?

48. Show that the lines $y = x$, $y = 3x$, $y = 5x$, etc., contain only red lattice points, but the lines $y = 2x$, $y = 4x$, $y = 6x$, etc., contain both red and black lattice points.

49. Show that $y = mx + n$ contains lattice points of only one color if m is odd, both colors if m is even.

50. The line $y = \sqrt{2}x$ contains no lattice points besides $(0, 0)$. Why?

51. Show that the points (a, b) and (b, a) are symmetric in the line $y = x$; that is, $y = x$ is the perpendicular bisector of the segment connecting them.

3 FUNCTIONS

A question that arises all the time in mathematics is how one quantity depends on another; for example, how the price of an item depends on the demand, how the density of traffic depends on the time of day, how the area of a circle depends on the radius, etc. Such relationships are expressed by functions.

> **Definition of Function** A **function** is a rule that assigns to each member of a set **D** a unique member of a set **Y**.

Examples
(1) The rule that assigns to each state its capital city. Here **D** is the set of U.S. states, and **Y** is the set of U.S. cities.
(2) The rule rule that assigns to each day in the year 1982, the number of cars built on that day by the General Motors Corp. Here **D** is the set of 365 days in the year and **Y** is the set of non-negative integers.
(3) The rule that assigns to each real number its square. Here **D** is the set of all real numbers and **Y** is the set of all non-negative real numbers.

For us, Example 3 is the most important. We will deal only with functions where both **D** and **Y** are the sets of real numbers. The set **D** is the **domain** of the function. To each number x in **D**, the function assigns a number y. We call x the **independent variable** and y the **dependent variable**. The set of all assigned numbers y is the **range** of the function.

Sometimes we will describe domains and ranges by shorthand set notation such as $\{x > 2\}$, or $\{-1 < x \leq 3\}$ or $\{$all real $x\}$.

To picture a function, think of a machine or black box that converts inputs into outputs (Fig. 1). The set of all acceptable inputs is the domain; the set of all possible outputs is the range.

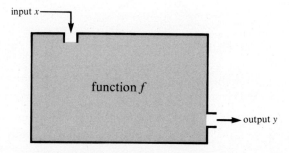

Functions are usually denoted by letters such as f, g, or h. Given a function f, the number y that f assigns to x is denoted $f(x)$ (read "f of x" or "f at x"). In terms of machines, $f(x)$ is the output x is when the input.

Usually a function f is given by a formula showing how to find $f(x)$ from x. For example, the formula

$$f(x) = 2x + 5$$

means that f assigns to x the number $2x + 5$. Thus, f assigns to 4 the number $2 \cdot 4 + 5 = 13$. In **functional notation**, $f(4) = 13$.

EXAMPLE 1 Let $f(x) = 2x + 5$. Compute

    (a) $f(0)$ (b) $f(3)$ (c) $f(-\sqrt{7})$ (d) $f(9.02)$

SOLUTION (a) $f(0) = 2 \cdot 0 + 5 = \underline{5}$

    (b) $f(3) = 2 \cdot 3 + 5 = \underline{11}$

    (c) $f(-\sqrt{7}) = 2(-\sqrt{7}) + 5 = \underline{-2\sqrt{7} + 5}$

    (d) $f(9.02) = 2(9.02) + 5 = \underline{23.04}$

Remark It is not necessary to use the letter x in the formula $f(x) = 2x + 5$. The function f can be described just as well by

    $f(t) = 2t + 5$ $f(u) = 2u + 5$ or even $f(\Box) = 2\Box + 5$

EXAMPLE 2 Let $g(u) = u^2 - 1$. Compute

    (a) $g(-u)$ (b) $g(\sqrt{x})$ (c) $g\left(\dfrac{1}{t}\right)$ (d) $\dfrac{g(3 + h) - g(3)}{h}$

SOLUTION Don't worry that the inputs are letters. The point is that $g(\text{input}) = (\text{input})^2 - 1$ as long as the input represents a real number.

    (a) $g(-u) = (-u)^2 - 1 = \underline{u^2 - 1}$

    (b) $g(\sqrt{x}) = (\sqrt{x})^2 - 1 = \underline{x - 1}$

    (c) $g\left(\dfrac{1}{t}\right) = \left(\dfrac{1}{t}\right)^2 - 1$ $t \neq 0$

    (d) $\dfrac{g(3+h) - g(3)}{h} = \dfrac{[(3+h)^2 - 1] - (3^2 - 1)}{h}$

                         $= \dfrac{(9 + 6h + h^2 - 1) - (9 - 1)}{h}$

                         $= \dfrac{6h + h^2}{h} = \underline{6 + h}$ $h \neq 0$

EXAMPLE 3 Let $h(x) = 3x$. Which of these statements is true?

    (a) $h(x + 1) = h(x) + 1$ (b) $h(2x) = 2h(x)$

    (c) $h(x_1 + x_2) = h(x_1) + h(x_2)$ (d) $h(x^2) = [h(x)]^2$

SOLUTION

    (a) $h(x + 1) = 3(x + 1) + 1 = 3x + 4$

        and $h(x) + 1 = 3x + 1$ [not the same]

(b) $h(2x) = 3(2x) = 6x$
and $2h(x) = 2(3x) = 6x$ [the same]
(c) $h(x_1 + x_2) = 3(x_1 + x_2) = 3x_1 + 3x_2$
and $h(x_1) + h(x_2) = 3x_1 + 3x_2$ [the same]
(d) $h(x^2) = 3x^2$
and $[h(x)]^2 = (3x)^2 = 9x^2$ [not the same]

Answer (b) and (c)

When a function is given by a formula, its domain is understood to be the set of all real numbers for which the formula makes sense. For example, if $f(x) = x^2$, then the domain of f is {all real x}; if $g(x) = \sqrt{x}$, then the domain of $g(x)$ is $\{x \geq 0\}$.

EXAMPLE 4 Find the domain of the function given by

(a) $f(x) = \dfrac{1}{\sqrt{x-2}}$ (b) $g(x) = \sqrt{x} + \sqrt{1-x^2}$

SOLUTION The square root $\sqrt{x-2}$ is defined for $x \geq 2$. Since $\sqrt{x-2}$ occurs in the denominator, we must exclude $x = 2$. Hence the domain of f is $\underline{\{x > 2\}}$.

(b) The domain is the set of real numbers x for which both terms are defined. Now, $\sqrt{x}$ is defined for $x \geq 0$, and $\sqrt{1-x^2}$ is defined for $-1 \leq x \leq 1$. Both terms are defined on the common part of these sets, $\underline{\{0 \leq x \leq 1\}}$.

Not every function is given by a formula. For example,
(1) $f(t)$ is the temperature at O'Hare airport t hours after midnight on January 1, 1981
(2) $f(w)$ is the first-class postage on a letter of weight w ounces
(3) $p(t)$ is a patient's blood pressure t seconds after a monitoring instrument begins recording it

EXAMPLE 5 Let $f(t)$ be the temperature at O'Hare airport t hours after midnight. Express, using functional notation,
(a) the temperature at 1:45 AM
(b) the increase in temperature from 8 AM to 10 AM
(c) the average hourly temperature for readings taken at
1 AM, 2 AM, . . . , 11 AM, 12 noon

SOLUTION (a) Since 1:45 AM is $1\frac{3}{4}$ hours after midnight, it corresponds to $t = 1.75$. At that time the temperature is $\underline{f(1.75)}$.

(b) The temperatures at 8 AM and 10 AM are $f(8)$ and $f(10)$, respectively. The increase from 8 AM to 10 AM is the difference $\underline{f(10) - f(8)}$.

(c) The 12 readings are $f(1), f(2), \ldots, f(12)$. The average of these numbers is

$$\dfrac{f(1) + f(2) + \cdots + f(12)}{12}$$

Terminology It is very common to speak of "the function $f(x)$" or "the function $y = f(x)$" or even "the function x^2." Strictly speaking, these are incorrect. The *function* is f. It assigns to x the *number* $f(x)$ or y or x^2. Still, these expressions are used all the time and are generally harmless.

Notation The letters f and x are the all-time favorite symbols for a function and its independent variable. In applications, however, we often use symbols that suggest the quantities involved. For example, $f(x) = \pi x^2$ gives the area of a circle of radius x, but a more natural notation is $A(r) = \pi r^2$.

EXERCISES

Compute the values of the function

1. $f(x) = 3x - 1$
   (a) $f(1)$ (b) $f(-1)$ (c) $f(0.15)$ (d) $f(-12)$

2. $f(x) = |x - 4|$
   (a) $f(-2)$ (b) $f(2)$ (c) $f(4 - \sqrt{7})$ (d) $f(\sqrt{7} - 4)$

3. $f(x) = \dfrac{1}{x}$
   (a) $f(5)$ (b) $f(-3)$ (c) $f(\tfrac{1}{4})$ (d) $f(0.001)$

4. $f(x) = x^2 + x - 1$
   (a) $f(3)$ (b) $f(-1)$ (c) $f(-10)$ (d) $f\left(\dfrac{-1 + \sqrt{5}}{2}\right)$

5. $f(x) = \dfrac{x}{3x + 2}$
   (a) $f(0)$ (b) $f(-34)$ (c) $f(\tfrac{5}{8})$ (d) $f(-0.67)$

6. $f(x) = \sqrt{x + 4}$
   (a) $f(0)$ (b) $f(-3)$ (c) $f(0.41)$ (d) $f(57^2 + 4 \cdot 57)$

What function is computed by the sequence of calculator keys?

7. $x \;\boxed{\times}\; 4 \;\boxed{+}\; 3 \;\boxed{=}\;$

8. $x \;\boxed{-}\; 2 \;\boxed{=}\; \boxed{x^2}$

9. $x \;\boxed{-}\; 5 \;\boxed{=}\; \boxed{1/x}$

10. $x \;\boxed{\sqrt{}}\; \boxed{\sqrt{}}$

Find the domain of f where $f(x) =$

11. $\dfrac{x}{(x - 1)^3}$

12. $\dfrac{1}{7x^2 + 3}$

13. $\dfrac{x}{x^2 - 4x + 3}$

14. $\dfrac{x + 1}{x^2 - 9}$

15. $\sqrt{4 - x^2}$

16. $\sqrt{x^2 - \tfrac{1}{4}}$

17. $\dfrac{1}{\sqrt{x + 5}}$

18. $\dfrac{1}{(x^2 + 2)\sqrt{x - 1}}$

19. $\sqrt{6 - x} + 2\sqrt{x}$

20. $\sqrt{x} + \sqrt{x - 1} + \sqrt{2 - x}$ 21. $\sqrt{3 - \sqrt{x}}$ 22. $\sqrt{4 + \sqrt{x^2 + 5}}$

Find a function $f(x)$ that describes

23. the number of seconds in x hours

24. the Celsius reading corresponding to $x°$ Fahrenheit

25. the diagonal of a square of side x

26. the time required for a plane to fly 1000 miles at x miles/hr if the last 300 miles are against a 40-mph headwind

27. the value after one year of a $10,000 investment if x dollars are invested at 6% yearly interest and the rest at 9%

28. the one-day rental of a car at $20 per day plus 20 cents per mile

29. Suppose the function $h(t)$ describes the elevation of a rocket t seconds after launch. Express, using functional notation, (a) its elevation after 3 sec (b) the distance it rises between times $t = 1$ and $t = 2$ (c) its average speed over the first 5 sec.

30. After a shut-down, a steel plant starts up again slowly. Suppose $f(x)$ is the time required until it can produce x tons of steel. Express, using functional notation
(a) the time needed to produce the first 100 tons
(b) the time needed to produce the next 100 tons
(c) the time required to produce p pounds

Express the equation in words

31. $f(t + 8) = 2f(t)$, where $f(t)$ is the value of an investment after t years

32. $m(t + 35) = \frac{1}{2}m(t)$, where $m(t)$ is the mass of a sample of radioactive material after t days

33. $d(t + 1) = d(t) - \frac{1}{3}$, where $d(t)$ is the depth (in feet) of oil in a tank t minutes after a valve is opened

34. $h(2v) = 4h(v)$, where $h(v)$ is the maximum height reached by a projectile fired with initial velocity v

Compute $f(x - 1)$, $f(-x)$, $f(2x)$, $f(x^2)$

35. $f(x) = 8x + 5$ **36.** $f(x) = 3x^2 + 1$ **37.** $f(x) = \dfrac{x}{x + 1}$ **38.** $f(x) = \dfrac{1}{x^3}$

The following table gives some values of the function f. (Assume $f(x) = 0$ if $|x| > 3$.)

x	-5	-4	-3	-2	-1	0	1	2	3	4	5
$f(x)$	0	0	1	8	9	6	7	7	4	0	0

Make a similar table for

39. $f(-x)$ **40.** $f(x - 1)$

In Exercises 41–2 let $f(x) = \begin{cases} 1 \text{ if } x > 0 \\ 0 \text{ if } x = 0 \\ -1 \text{ if } x < 0 \end{cases}$

Compute

41. $f(2)$, $f(-5)$, $f(2x)$ **42.** $f(-4.7)$, $f(t^2 + 1)$, $f(x) + f(-x)$

43. Which functions satisfy $f(3x) = 3f(x)$?
    (a) $f(x) = x$ (b) $f(x) = 7x$ (c) $f(x) = x + 8$ (d) $f(x) = x^3$

44. Which functions satisfy $f(x_1 + x_2) = f(x_1) + f(x_2)$?
    (a) $f(x) = ax$ (b) $f(x) = x^2$ (c) $f(x) = \dfrac{1}{x}$ (d) $f(x) = \sqrt{x}$

45. Show that $f(x) = x^n$, where n is any integer, satisfies $f(x_1 x_2) = f(x_1) f(x_2)$. Give an example of a function that does not satisfy this equation.

46. Find an example of a function that satisfies $f(x^3) = [f(x)]^3$ and an example of a function that does not.

4 GRAPHS OF FUNCTIONS

Each function has a graph showing its "life history" at a glance. The **graph** of f is the set of all points $(x, f(x))$, where x is in the domain of f. We can picture $f(x)$ as the "height" of the graph at x, above the x-axis if $f(x) > 0$, below if $f(x) < 0$. See Figure 1.

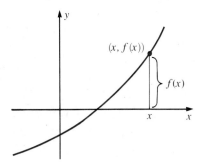

FIG. 1

The graph of a function shows the domain and range (Fig. 2). The domain is the part of the x-axis directly above or below the graph. The range is the set of all y values or "heights" of the graph. Note that a vertical line can intersect the graph at only one point, because a function assigns exactly one value of y to each x in its domain.

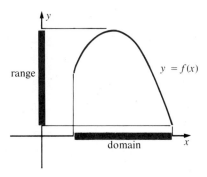

FIG. 2

The graph of a *function* f consists of all points (x, y) with $y = f(x)$. Therefore, it is also the graph of the *equation* $y = f(x)$.

Coordinates, Lines, Functions

EXAMPLE 1 Sketch the graph of the function
(a) $f(x) = \frac{1}{2}x - 3$ (b) $f(x) = |x|$

SOLUTION The graph of f is the graph of the equation $y = \frac{1}{2}x - 3$. It is a straight line (Fig. 3a) with slope $\frac{1}{2}$ and y-intercept -3.

(b) By definition, $|x| = \begin{cases} x \text{ if } x \geq 0 \\ -x \text{ if } x < 0 \end{cases}$

Hence the graph of $y = |x|$ coincides with the line $y = x$ for $x \geq 0$ and with the line $y = -x$ for $x < 0$. See Fig. 3b.

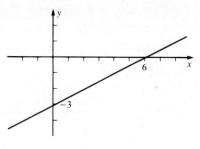

FIG. 3   (a) Graph of $f(x) = \frac{1}{2}x - 3$ (b) Graph of $f(x) = |x|$

Terminology It is a mouthful to say "the graph of the function f, where $f(x) = \frac{1}{2}x - 3$." We often use shorter expressions such as "the graph of $y = \frac{1}{2}x - 3$" or "the graph of $f(x) = \frac{1}{2}x - 3$" or even "the graph of $\frac{1}{2}x - 3$."

EXAMPLE 2 Sketch the graph of $f(x) = \sqrt{x}$.

SOLUTION The domain is $\{x \geq 0\}$. Clearly $\sqrt{x}$ is 0 at $x = 0$ and increases as x increases. Hence the graph starts at $(0, 0)$ and rises as x moves to the right.

For more precise information, we plot some points estimating the values of $\sqrt{x}$ on a calculator.

x	0	0.5	1	2	3	4	5	6	7	8	9
$\sqrt{x}$	0.00	0.71	1.00	1.41	1.73	2.00	2.24	2.45	2.65	2.83	3.00

As expected, the graph rises (Fig. 4). Notice how slowly, however. It will not reach the height 10 until $x = 100$, the height 20 until $x = 400$, etc.

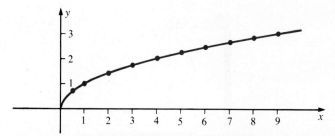

FIG. 4

Sometimes we graph functions not given by formulas. One example is the greatest integer function:

$$f(x) = [x] = \text{the greatest integer } y \text{ such that } y \leq x$$

For example, $[3.1416] = 3$, $[6] = 6$, $[-1.7] = -2$. The graph (Fig. 5a) jumps at each integer value of x. Another example is the saw-tooth function $s(x)$, whose graph is shown in Figure 5b.

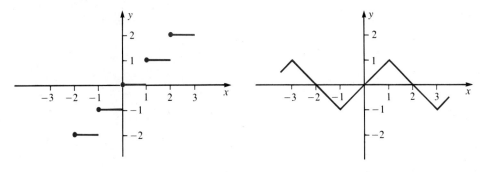

(a) Greatest integer function $[x]$ (b) Saw-tooth function $s(x)$

FIG. 5

Some Graphing Principles

Given the graph of $y = f(x)$, it is easy to construct graphs of related functions: $y = f(x) \pm k$, $y = -f(x)$, and $y = cf(x)$.

First we consider the graph of $y = f(x) + k$ for $k > 0$. Each of its points look like $(x, y + k)$, where (x, y) is on the graph of $y = f(x)$. But $(x, y + k)$ is k units higher than (x, y). Hence the graph of $y = f(x) + k$ is k units higher than that of $y = f(x)$. Similarly, the graph of $y = f(x) - k$ is k units lower (Fig. 6).

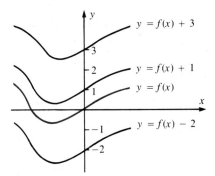

FIG. 6

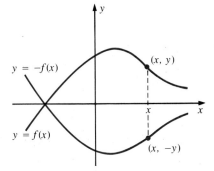

FIG. 7

> Let $k > 0$. Then the graph of
> $$\left.\begin{array}{l} y = f(x) + k \\ y = f(x) - k \end{array}\right\} \text{ is the graph of } y = f(x) \text{ shifted } k \text{ units } \left\{\begin{array}{l} \text{upward} \\ \text{downward} \end{array}\right.$$

Next, the graph of $y = -f(x)$. Each of its points is $(x, -y)$, where (x, y) is on the graph of $y = f(x)$. But $(x, -y)$ is the reflection of (x, y) in the x-axis. Therefore, the graph of $y = -f(x)$ is the reflection of the graph of $y = f(x)$. See Figure 7.

> The graph of $y = -f(x)$ is the reflection of the graph of $y = f(x)$ in the x-axis.

EXAMPLE 3 Graph $y = 1 - \sqrt{x}$

SOLUTION Fig. 4 shows the graph of $y = \sqrt{x}$. Reflect it in the x-axis to obtain (Figure 8a) the graph of $y = -\sqrt{x}$. Then shift upward one unit (Fig. 8b) to obtain the graph of $y = -\sqrt{x} + 1$.

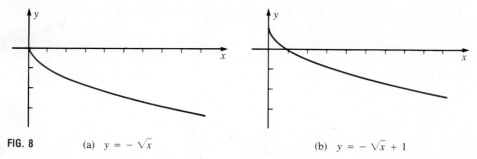

FIG. 8   (a) $y = -\sqrt{x}$ (b) $y = -\sqrt{x} + 1$

Next, the graph of $y = 2f(x)$. Each of its points is $(x, 2y)$, where (x, y) is on the graph of $y = f(x)$. But $(x, 2y)$ is "twice as high" as (x, y). Therefore, the graph of $y = 2f(x)$ is the graph of $y = f(x)$ stretched vertically by a factor of 2. Similarly, the graph of $y = cf(x)$ is the graph of $y = f(x)$ stretched vertically by a factor of c, provided $c > 0$. When $0 < c < 1$, "stretched" must be interpreted as "shrunk." (See Figure 9.)

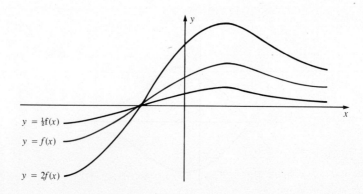

FIG. 9 $y = \tfrac{1}{2}f(x)$
            $y = f(x)$
            $y = 2f(x)$

Let $c > 0$. The graph of $y = cf(x)$ is the graph of $y = f(x)$ stretched vertically by a factor of c.

EXAMPLE 4 Graph $y = \frac{1}{2}s(x) - 2$, where $s(x)$ is the saw-tooth function in Figure 5b, page 139.

SOLUTION First shrink the graph of $y = s(x)$ vertically by a factor of $\frac{1}{2}$. That gives the graph (Fig. 10a) of $y = \frac{1}{2}s(x)$. Then shift downward 2 units. That gives the graph (Fig. 10b) of $y = \frac{1}{2}s(x) - 2$.

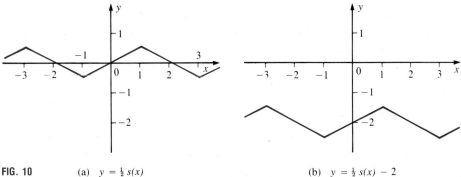

FIG. 10   (a) $y = \frac{1}{2} s(x)$ (b) $y = \frac{1}{2} s(x) - 2$

Even and Odd Symmetry

Certain functions have symmetry properties that are useful in graphing. A function f is an **even function** if $f(-x) = f(x)$. For example, $f(x) = x^2$ is an even function because $(-x)^2 = x^2$.

An even function takes the same values at x and at $-x$. Hence whenever (x, y) is on its graph, so is $(-x, y)$. But the points (x, y) and $(-x, y)$ are symmetric in the y-axis. Therefore, the graph is symmetric in the y-axis (Fig. 11a).

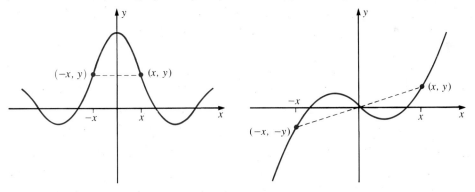

(a)   Even function: symmetry in the y-axis   (b)   Odd function: symmetry in the origin

FIG. 11

A function f is an **odd function** if $f(-x) = -f(x)$. For example, $f(x) = x^3$ is an odd function because $(-x)^3 = -x^3$. An odd function takes values of opposite sign at x and $-x$. Hence whenever (x, y) is on its graph, so is $(-x, -y)$. But the points (x, y) and $(-x, -y)$ are symmetric in the origin, as shown in Fig. 11b. Therefore, the graph is symmetric in the origin.

> A function f is an **even function** if
> $$f(-x) = f(x)$$
> The graph of an even function is symmetric in the y-axis.
> A function f is an **odd function** if
> $$f(-x) = -f(x)$$
> The graph of an odd function is symmetric in the origin.

Even and odd functions are convenient to graph because you only have to do half the work. First graph the function for $x \geq 0$. Then get the rest of the graph by symmetry.

EXAMPLE 5 Graph (a) $f(x) = x^2$ (b) $f(x) = \frac{1}{2}x^3$

SOLUTION (a) The function is even, so we graph it first for $x \geq 0$. The values of x^2 increase rapidly as x increases. The graph would soon go off the top of the page unless we take only small values of x. Let us tabulate a few:

x	0	0.5	1.0	1.5	2.0	2.5
x^2	0	0.25	1.0	2.25	4.0	6.25

We plot the corresponding points and connect them by a smooth curve. This is the graph for $x \geq 0$. Then we get the rest of the graph free: we just reflect in the y-axis (Fig. 12a).

(b) The function is odd, so the graph is symmetric in the origin. The values of

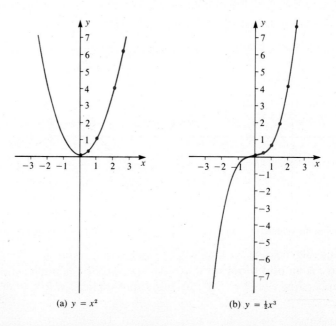

FIG. 12 (a) $y = x^2$ (b) $y = \frac{1}{2}x^3$

$\tfrac{1}{2}x^3$ increase very rapidly, so again we tabulate only for some small positive values of x. (We round off our calculator results to 2 places.)

x	0	0.5	1.0	1.5	2.0	2.5
$\tfrac{1}{2}x^3$	0	0.06	0.05	1.69	4.00	7.81

We plot the corresponding points and connect them by a smooth curve. Then we reflect in the origin to get the rest of the graph (Fig. 12b).

EXERCISES

Graph

1. $f(x) = \tfrac{1}{2}x + 4$
2. $f(x) = 3x - 1$
3. $f(x) = -3$
4. $f(x) = 2$
5. $f(x) = |x - 1|$
6. $f(x) = |2x|$
7. $f(x) = 2|x| - 1$
8. $f(x) = \tfrac{1}{4}|x| + 1$
9. $f(x) = 2\sqrt{x}$
10. $f(x) = -2\sqrt{x}$
11. $f(x) = \sqrt{x - 4}$
12. $f(x) = \sqrt{x + 1}$
13. $f(x) = \begin{cases} 0 & x \le 0 \\ 2x & x > 0 \end{cases}$
14. $f(x) = \begin{cases} x - 1 & x \le 3 \\ 2 & x > 3 \end{cases}$

Recall the saw-tooth function $s(x)$ in Figure 5b. Graph

15. $f(x) = -s(x)$
16. $f(x) = 3s(x)$
17. $f(x) = |s(x)|$
18. $f(x) = 1 - 2s(x)$

Graph

19. $f(x) = \tfrac{1}{4}x^2$
20. $f(x) = -\tfrac{1}{3}x^2$
21. $f(x) = -\tfrac{1}{2}x^3$
22. $f(x) = \tfrac{1}{10}x^3$
23. $f(x) = 9 - x^2$
24. $f(x) = x^2 - 4$

25. A taxi charges 75¢ as soon as you enter, and then 10¢ after each $\tfrac{1}{6}$ mile. Graph $f(x)$, the cost of riding x miles in this cab.

26. A parking lot charges $2 for the first hour (or part of an hour), and then $1 per hour. Graph $f(x)$, the cost of parking for x hours.

27. First-class postage (1981 rate) is 18¢ per ounce or part of an ounce. Graph $f(x)$, the first-class postage on a letter of x ounces.

28. A 120-yard high hurdle race has 10 hurdles. The first is 15 yards after the start, and all the rest are spaced at 10-yard intervals. Graph $f(x)$, the number of hurdles remaining after a runner has covered x yards.

29. A dump truck with 10,000-lb capacity hauls away dirt from a building site. It takes a half hour to load up, then a half hour to haul the dirt away and return for another load. Graph $f(t)$, the weight of dirt removed in t hours after the first arrival of the truck.

30. (cont.) Suppose the contractor calls in a second truck with capacity 20,000 lb, but requiring one hour to load up, then another hour to haul the dirt away and return. Both trucks arrive together in the morning. Now graph $f(t)$, the combined weight removed in t hours.

31. Graph $f(x)$, the distance from x to the next integer.

32. Graph $f(x)$, the distance from x to the nearest integer.

Determine whether $f(x)$ is odd or even or neither. $f(x) =$

33. $x^7 - 4x^3$ **34.** $x^6 + 5x^2$ **35.** $2x - 1$

36. $x^3 - x^2$ **37.** $\dfrac{x^3}{x^2 + 5}$ **38.** $\dfrac{7}{x^3}$

39. $\dfrac{x^3 - 1}{x^3 + 2}$ **40.** $\dfrac{x^5}{x^4 + 6x^2 + 1}$ **41.** $\sqrt{1 - 5x^2}$

42. $(|x| - 1)^3$ **43.** $(x^2 + 5)^9$ **44.** $(x^3 + x)^6$

45. Show that $f(x) = x|x|$ is an odd function and sketch its graph.

46. Show that $f(x) = \sqrt{|x|}$ is an even function and sketch its graph.

5 OPERATIONS ON FUNCTIONS; INVERSE FUNCTIONS

There are several ways of building new functions out of old ones. Given functions f and g, we define functions $f \pm g$, fg, and f/g:

$$[f + g](x) = f(x) + g(x) \qquad [f - g](x) = f(x) - g(x)$$

$$[fg](x) = f(x)g(x) \qquad [f/g](x) = \frac{f(x)}{g(x)}$$

The sum, difference, and product are defined wherever both $f(x)$ and $g(x)$ are defined (the intersection of their domains). The quotient requires also that $g(x) \neq 0$.

Examples

(1) $f(x) = 2x - 3 \qquad g(x) = x^2 - 4$

Then

$$[f + g](x) = (2x - 3) + (x^2 - 4) = x^2 - 2x - 7 \qquad \{\text{all } x\}$$
$$[fg](x) = (2x - 3)(x^2 - 4) \qquad \{\text{all } x\}$$
$$[f/g](x) = \frac{2x - 3}{x^2 - 4} \qquad \{x \neq \pm 2\}$$

(2) $f(x) = \sqrt{x - 2} \qquad g(x) = \sqrt{3 - x}$

The domain of f is $\{x \geq 2\}$. The domain of g is $\{x \leq 3\}$. The common part (intersection) of these domains is $\{2 \leq x \leq 3\}$.

$$[f + g](x) = \sqrt{x - 2} + \sqrt{3 - x} \qquad \{2 \leq x \leq 3\}$$
$$[fg](x) = \sqrt{(x - 2)(3 - x)} \qquad \{2 \leq x \leq 3\}$$
$$[f/g](x) = \sqrt{\frac{x - 2}{3 - x}} \qquad \{2 \leq x < 3\}$$

Note that $x = 3$ is excluded from the domain of $[f/g]$.

5 Operations on Functions; Inverse Functions

Composition of Functions

Given functions f and g, we define their **composition** $f \circ g$ (also called the **composite function** of f and g) by

$$[f \circ g](x) = f[g(x)]$$

Think of substituting one function into the other, or replacing the variable of f by the function g.

Examples

(1) $f(x) = x^2 + 2x \qquad g(x) = -3x$

Then
$$\begin{aligned}[f \circ g](x) = f[g(x)] &= [g(x)]^2 + 2[g(x)] \\ &= (-3x)^2 + 2(-3x) \\ &= 9x^2 - 6x\end{aligned}$$

(2) $f(x) = 3x - 4 \qquad g(x) = 2x^2 - x + 1$

Then
$$\begin{aligned}[f \circ g](x) = f[g(x)] &= 3g(x) - 4 \\ &= 3(2x^2 - x + 1) - 4 \\ &= 6x^2 - 3x - 1\end{aligned}$$

Figure 1 shows a black box picture of $f \circ g$. The boxes f and g are connected "in series"; the output from g is the input to f. Note that x must be an acceptable input for g, and $g(x)$ must be an acceptable input for f.

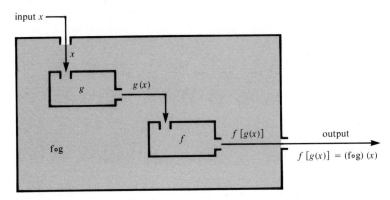

FIG. 1

EXAMPLE 1 Let $f(x) = x - 5$ and $g(x) = \sqrt{x}$. Compute the composite function and find its domain

(a) $f \circ g$ \qquad (b) $g \circ f$

SOLUTION (a) $[f \circ g](x) = f[g(x)] = g(x) - 5 = \sqrt{x} - 5$ Clearly the domain is $\{x \geq 0\}$. This checks with Fig. 1, since g accepts only inputs $x \geq 0$, while f accepts any input $g(x)$.

(b) $[g \circ f](x = g[f(x)] = \sqrt{f(x)} = \sqrt{x - 5}$. The domain is $\{x \geq 5\}$.

Inverse Functions

An important technique in mathematics is turning problems around and looking at "inverse problems." Here are a few examples:

(1) *Problem:* Given the radius of a circle, find its area.
Inverse problem: Given the area of a circle, find its radius.
(2) *Problem:* A metal rod weighs 8 grams per cm of length. Given its length, find its weight.
Inverse problem: Given the weight of the rod, find its length.
(3) *Problem:* One gram of radioactive carbon-14 disintegrates, losing half its mass in 5668 years. Find its mass after t years.
Inverse problem: Given the mass, find t.

A closely related technique is turning *functions* around to find *inverse functions*. Let us start with an easy example, the function $f(x) = 2x$. Given x, it assigns the number $y = 2x$. Inversely, given y, there corresponds $x = \frac{1}{2}y$. The function $g(y) = \frac{1}{2}y$ is the inverse function of $f(x) = 2x$.

For a graphical interpretation, see Figure 2a. The arrows indicate how each x determines a unique y. Thus, $y = f(x) = 2x$. Now reverse the arrows (Fig. 2b). Then each y determines a unique x. Thus, $x = g(y) = \frac{1}{2}y$.

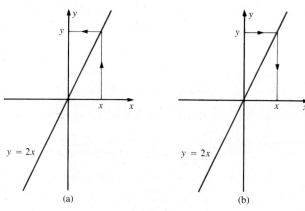

FIG. 2

The function f takes you from x to y. Then g takes you from this y back to x. Thus,
$$g[f(x)] = \tfrac{1}{2}f(x) = \tfrac{1}{2}(2x) = x$$
Similarly, g takes you from y to x. Then f takes you from x back to y:
$$f[g(y)] = 2g(y) = 2(\tfrac{1}{2}y) = y$$
Thus, the actions of f and g neutralize each other; apply one after the other and you are back where you started. In the language of composite functions,
$$(g \circ f)(x) = x \qquad (f \circ g)(y) = y$$

> **Inverse Functions** Suppose f and g are functions such that
> $$g[f(x)] = x \quad \text{for all } x \text{ in the domain of } f$$
> $$f[g(y)] = y \quad \text{for all } y \text{ in the domain of } g$$
> Then g is the **inverse function** of f, and f is the **inverse function** of g.

EXAMPLE 2 Find the inverse function $g(y)$
  (a) $f(x) = 2x - 5$ (b) $f(x) = x^3$

SOLUTION (a) Given y, the inverse $g(y)$ is that value of x for which $2x - 5 = y$. Solve for x:
$$2x - 5 = y$$
$$x = \tfrac{1}{2}(y + 5)$$
Hence
$$g(y) = \tfrac{1}{2}(y + 5)$$

Check
$$g[f(x)] = \tfrac{1}{2}[f(x) + 5] = \tfrac{1}{2}[(2x - 5) + 5] = x$$
$$f[g(y)] = 2g(y) - 5 = 2[\tfrac{1}{2}(y + 5)] - 5 = y$$

(b) Solve $y = x^3$ for x. Thus, $x = y^{1/3}$; that is, $g(y) = y^{1/3}$.

Check
$$g[f(x)] = [f(x)]^{1/3} = (x^3)^{1/3} = x$$
$$f[g(y)] = [g(y)]^3 = (y^{1/3})^3 = y$$

Existence of Inverse Functions

Not every function has an inverse function. For example, take $f(x) = x^2$. Given $y = 4$, *either* $x = 2$ *or* $x = -2$. Thus, given y, we cannot assign to it a unique x.

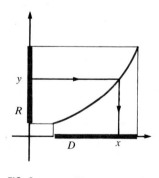
(a)

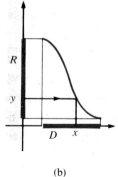
(b)

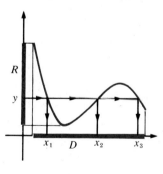
(c)

FIG. 3

For most functions, the graph shows whether an inverse function exists. Look at Fig. 3. Each of the three functions has domain **D** and range **R**. In (a) and (b) an inverse function g exists; to each y in **R** corresponds a single x in **D**. The domain of g is **R**; the range of g is **D**. In (c), however, some values of y correspond to several points x. Therefore, no inverse function exists. The trouble is that the graph wiggles. That does not happen in (a) or (b), where the graph increases steadily or decreases steadily.

> If the graph of f is either increasing or decreasing, then f has an inverse function g. If f has domain **D** and range **R**, then g has domain **R** and range **D**.

Graphs of Inverse Functions

Let f and g be a pair of inverse functions. Suppose (a, b) is a point on the graph of $y = f(x)$. Then $g(b) = a$. Hence (b, a) is a point on the graph of $y = g(x)$. But (a, b) and (b, a) are symmetric in the line $y = x$ (Exercise 51, page 132). Therefore, the graphs are also symmetric in this line (Fig. 4).

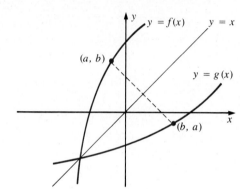

FIG. 4

In Chapter 7 we deal with an important function whose graph is of the type shown in Figure 5.

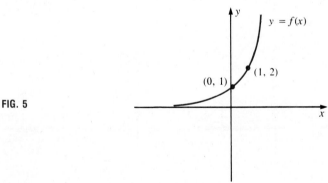

FIG. 5

EXAMPLE 4 Figure 5 shows the graph of a function f. Sketch the graph of its inverse function g. Find the domain and range of g, and the values $g(0)$ and $g(1)$.

SOLUTION The graph of f is increasing, so an inverse function g exists. To graph it, reflect the graph of f in the line $y = x$. See Fig. 6. The domain of g is $\{x > 0\}$;

the range is {all y}. By reflection, the points (1, 0) and (2, 1) lie on the graph of $y = g(x)$. Therefore, $g(1) = 0$ and $g(2) = 1$.

Answer

(Fig. 6) $g(1) = 0, g(2) = 1$ Domain = {$x > 0$}, Range = {all y}

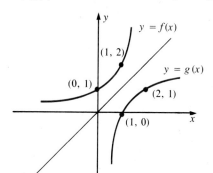

FIG. 6

Remark As Example 4 shows, we can talk about an inverse function even though we have no formula for it. For example, the graph of $f(x) = x^7 + 4x^3$ is increasing, so an inverse function $x = g(y)$ exists. But finding an explicit formula for it would require solving $x^7 + 4x^3 = y$ for x, which is hopeless.

EXERCISES

Find $[f + g](x)$, $[fg](x)$ and $[f/g](x)$ and their domains

1. $f(x) = 3x + 1$ $g(x) = x - 4$ 2. $f(x) = 2x - 1$ $g(x) = 2x + 3$
3. $f(x) = x^2 + 3x$ $g(x) = x^2 - 1$ 4. $f(x) = x^3$ $g(x) = x^2 - x$
5. $f(x) = \sqrt{1 - x}$ $g(x) = \sqrt{x - 2}$ 6. $f(x) = \sqrt{x}$ $g(x) = \sqrt{25 - x^2}$
7. $f(x) = x^2 + 3$ $g(x) = \dfrac{1}{x}$ 8. $f(x) = \dfrac{1}{x^2 - 4}$ $g(x) = 5x$

Compute $[f \circ g](x)$ and $[g \circ f](x)$, and give their domains

9. $f(x) = 3x + 1$ $g(x) = x - 2$ 10. $f(x) = \tfrac{1}{2}x + 5$ $g(x) = 4x - 1$
11. $f(x) = (x - 1)^2$ $g(x) = 2x - 1$ 12. $f(x) = x^2 + 3x$ $g(x) = \dfrac{1}{x}$
13. $f(x) = \sqrt{x}$ $g(x) = x^2 - 9$ 14. $f(x) = \sqrt{x}$ $g(x) = |x|$
15. $f(x) = \tfrac{1}{3}(x - 7)$ $g(x) = 3x + 7$ 16. $f(x) = x^5$ $g(x) = x^{1/5}$
17. $f(x) = \dfrac{5x + 6}{4x + 5}$ $g(x) = \dfrac{5x - 4}{-6x + 5}$ 18. $f(x) = \dfrac{2x + 3}{3x + 5}$ $g(x) = \dfrac{5x - 3}{-3x + 2}$

Compute $[f \circ g](x)$

19. $f(x) = x$ and $g(x)$ is any function   20. $g(x) = x$ and $f(x)$ is any function

How do the calculator sequences represent composition of functions? How do the composite functions differ?

21. $\boxed{x^2}\ \boxed{\sqrt{\ }}$ and $\boxed{\sqrt{\ }}\ \boxed{x^2}$ **22.** $\boxed{+}\ \boxed{1}\ \boxed{=}\ \boxed{x^2}$ and $\boxed{x^2}\ \boxed{+}\ \boxed{1}\ \boxed{=}$

Find the inverse function

23. $f(x) = \tfrac{1}{2}x + 6$
24. $f(x) = -3x + 2$
25. $f(x) = 8x^3$
26. $f(x) = x^2 - 4,\ x \geq 0$
27. $f(x) = x^4 + 1,\ x \geq 0$
28. $f(x) = \dfrac{1}{x^2},\ x > 0$
29. $f(x) = \sqrt{x + 3},\ x \geq -3$
30. $f(x) = 1 - \sqrt{x},\ x \geq 0$
31. $f(x) = (x - 2)^3$
32. $f(x) = \sqrt[3]{5 - x}$
33. $f(x) = \begin{cases} 2x, & x \geq 0 \\ x, & x < 0 \end{cases}$
34. $f(x) = \begin{cases} x + 2, & x \geq 0 \\ 3x + 2, & x < 0 \end{cases}$

Express the statement as an explicit function and find its inverse function

35. For a child between the ages of 6 and 12, multiply the age by 2.5 and add 30 to get the ideal height in inches

36. To each Fahrenheit reading of temperature corresponds a Celsius reading

37. To each length s corresponds the surface area of a cube with side length s

38. To each length r corresponds the area of a circle of radius r

Express the meaning of the inverse function in words

39. At elevation x feet above sea level the atmospheric pressure is $y = f(x)$ lb/in²

40. After driving t hours, I have traveled $s = f(t)$ miles

41. After grading n papers, my steadily rising blood pressure is $B = f(n)$

42. When the depth of water in a certain irregularly shaped tank is x feet, the tank contains $y = f(x)$ cubic feet of water

Plot the function and its inverse function on the same set of axes

43. $f(x) = 2x + 1$
44. $f(x) = -\tfrac{1}{2}x + 3$
45. $f(x) = \sqrt{x + 2}$
46. $f(x) = \begin{cases} 3x, & x \geq 0 \\ \tfrac{1}{3}x, & x < 0 \end{cases}$
47. the function in Figure 7
48. the function in Figure 8 (next page)

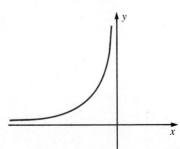

FIG. 7

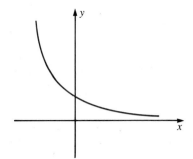

FIG. 8

A function f is called **strictly increasing** if whenever x_1 and x_2 are in the domain of f and $x_1 < x_2$, then $f(x_1) < f(x_2)$. Show that

49. the sum of two strictly increasing functions is strictly increasing

50. the composite of two strictly increasing functions is strictly increasing

REVIEW EXERCISES FOR CHAPTER 5

Given the points $(-2, -3)$ and $(4, 7)$, find
   1. the slope of the segment connecting them
   2. the distance between them

Find an equation for the line
   3. through $(-1, 4)$, slope $\tfrac{2}{3}$
   4. through $(5, 0)$, parallel to $x + y = 7$
   5. through $(6, 3)$ and $(10, -5)$
   6. through $(9, -4)$ and perpendicular to $y = 2x - 9$
   7. perpendicular bisector of the segment from $(1, -5)$ to $(3, 6)$
   8. Find the range and domain of the function $f(x) = \dfrac{1}{x^2}$.

Compute $f(5)$, $f(-x)$, $f\!\left(\dfrac{1}{t}\right)$ and $\dfrac{f(x+h) - f(x)}{h}$ for

   9. $f(x) = 3x^2 - x$ **10.** $f(x) = 1 - \dfrac{1}{x}$

   11. Is the function even, odd, or neither?
      (a) $f(x) = |x - 2|$ (b) $g(x) = \left(x - \dfrac{1}{x}\right)^3$ (c) $f(x) = \sqrt{25 - x^2}$

   12. If $f(x)$ is an odd function, show that $f(0) = 0$.

Sketch the graph

13. $y = \frac{1}{2}(x + 5)$

14. $2x + 3y = 12$

15. $\dfrac{x}{5} + \dfrac{y}{3} = 1$

16. $y = \sqrt{1 - x}$

17. $y = -x^2$

18. $y = 1 - \frac{1}{2}|x|$

19. If $f(x) = x - 4$ and $g(x) = 1/x$, compute $[f \circ g](x)$ and $[g \circ f](x)$ and find their domains.

20. The value of an account after t years is $f(t)$. Suppose $f(t + 1) = (1.08)f(t)$. Express this equation in words.

Find the inverse function and its domain

21. $f(x) = \frac{3}{4}x - 8$

22. $f(x) = 5 - 2\sqrt{x}$

Functions and Graphs

1 LINEAR AND QUADRATIC FUNCTIONS

This chapter deals with some useful functions and equations, and their graphs. We begin with linear and quadratic functions.

Linear Functions

A function f is **linear** if
$$f(x) = ax + b \qquad a \neq 0$$
The graph of $y = f(x)$ is a straight line with slope a and y-intercept b.

A linear function is defined for all x. In many applications, however, quantities x and y are related by a linear equation $y = ax + b$, but x is restricted somehow. In such cases we still say that y is a linear function of x (on some domain), or that there is a **linear relation** between x and y.

Examples

(1) Some doctors use this rule of thumb. For children from 6 to 12 years old, the average height y (in inches) is related to the age by $y = 3x + 20$. Thus y is a linear function of x on the domain $\{6 \leq x \leq 12\}$.

(2) According to a formula for an ideal marriage, the age B of the bride should be related to the age G of the groom by $B = \frac{3}{4}G + 4$. Thus B is a linear function of G. The domain must be restricted, however, Otherwise, a groom of 4 should marry a bride of 7, and a groom of 200 should marry a bride of 154. A reasonable domain here is $\{20 \leq G \leq 40\}$.

(3) The circumference C of a circle is a linear function of the radius r given by $C = 2\pi r$. Since r is a positive length, the domain is $\{r > 0\}$.

Suppose it is known that two quantities are connected by a linear relation. How do we find the formula for that relation? Well, the graph of that formula is a line, so we can use information about equations for lines. An example is the relation between Fahrenheit and Celsius.

EXAMPLE 1 Water freezes at 0°C or 32°F, and boils at 100°C or 212°F. Express Fahrenheit as a linear function of Celsius.

SOLUTION For suitable constants a and b,
$$F = aC + b$$
The graph of this linear equation is a line with slope a and F-intercept b. Now $F = 32$ when $C = 0$, and $F = 212$ when $C = 100$. Hence the line passes through the points $(0, 32)$ and $(100, 212)$. See Figure 1.

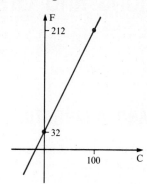

FIG. 1

$$\text{slope} = a = \frac{212 - 32}{100 - 0} = \frac{180}{100} = \frac{9}{5} \qquad F\text{-intercept} = b = 32$$

Hence
$$F = \tfrac{9}{5}C + 32$$

The slope $\tfrac{9}{5}$ means that F increases by $\tfrac{9}{5}$ when C increases by 1. In other words, $\tfrac{9}{5}°F = 1°C$.

EXAMPLE 2 An investor buys an apartment house for $500,000. For tax purposes, its value V depreciates linearly to 0 in 25 years. Find a formula for V after t years $(0 \le t \le 25)$.

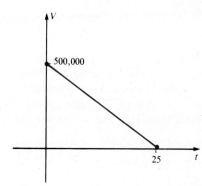

FIG. 2

SOLUTION 1 Linear or straight-line depreciation means that V is a decreasing linear function of t. Thus
$$V = at + b$$

where $a < 0$. The graph of this equation is a straight line with slope a and V-intercept b. We are given two points on this line, $(0, 500{,}000)$ and $(25, 0)$. See Figure 2. Therefore,

$$\text{slope} = a = \frac{0 - 500{,}000}{25 - 0} = -20{,}000 \qquad \text{V-intercept} = b = 500{,}000$$

Hence,

$$V = -20{,}000t + 500{,}000 = 20{,}000(25 - t)$$

SOLUTION 2 In one year, V loses $\frac{1}{25}$ of its value, that is, \$20,000. Hence, in t years, it loses $20{,}000t$. Since V starts at 500,000, in t years the value decreases to

$$V = 500{,}000 - 20{,}000t = 20{,}000(25 - t)$$

Quadratic Functions

A function f is quadratic if

$$f(x) = ax^2 + bx + c \qquad a \neq 0$$

To study the graphs of quadratic functions, we start with some simple ones and build up to the general quadratic.

The graph of $f(x) = x^2$ is familar (Fig. 3a). From it we obtain the graph of $f(x) = ax^2$ by vertical stretching or shrinking, and by reflection in the x-axis when $a < 0$. The curves are called **parabolas** (Fig. 3b,c).

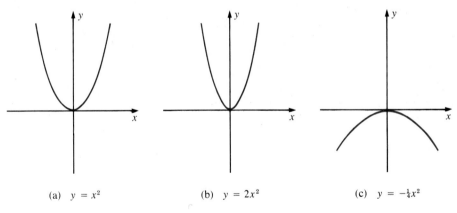

(a) $y = x^2$ (b) $y = 2x^2$ (c) $y = -\frac{1}{4}x^2$

FIG. 3

EXAMPLE 3 Graph (a) $f(x) = (x - 1)^2$ (b) $f(x) = (x + 2)^2$

SOLUTION (a) Tabulate some values

x	-4	-3	-2	-1	0	1	2	3	4	5	6
$(x-1)^2$	25	16	9	4	1	0	1	4	9	16	25

The values are the same as for $f(x) = x^2$, but shifted one unit to the right. Hence the graph of $f(x) = (x - 1)^2$ is the graph of $f(x) = x^2$ shifted one unit to the right See Figure 4a, next page.

(b) Tabulate some values:

x	-5	-4	-3	-2	-1	0	1	2	3	4	5
$(x+2)^2$	9	4	1	0	1	4	9	16	25	36	49

The values are the same as for $f(x) = x^2$, but shifted two units to the *left*. The same is true of the graphs (Fig. 4b).

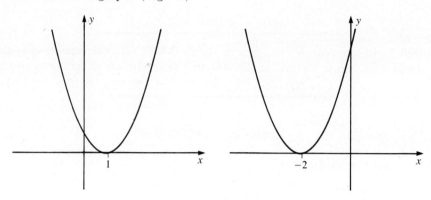

FIG. 4 (a) $y = (x-1)^2$ (b) $y = (x+2)^2$

The reasoning of Example 3 leads to a general graphing principle:

Horizontal shifts Let $h > 0$. The graph of
$$\left. \begin{array}{l} y = f(x-h) \\ y = f(x+h) \end{array} \right\} \text{ is the graph of } y = f(x) \text{ shifted } h \text{ units } \begin{cases} \text{to the right} \\ \text{to the left} \end{cases}$$

Let us apply this principle to graph the general quadratic function
$$f(x) = ax^2 + bx + c$$
By completing the square, we can express the function as
$$f(x) = a(x-h)^2 + k$$
Hence the graph of $f(x)$ is just the parabola $y = ax^2$ shifted horizontally $|h|$ units and vertically $|k|$ units.

EXAMPLE 4 Graph (a) $f(x) = 2x^2 + 12x + 19$ (b) $f(x) = -x^2 + 2x + 3$

SOLUTION (a) Factor a 2 out of the first two terms, then complete the square:
$$\begin{aligned} f(x) &= 2(x^2 + 6x) + 19 \\ &= 2(x^2 + 6x + 9) - 18 + 19 \\ &= 2(x+3)^2 + 1 \end{aligned}$$

For the desired graph, shift the curve $y = 2x^2$ three units to the left and one unit up (Fig. 5a).

(b) Complete the square:
$$\begin{aligned} f(x) &= -(x^2 - 2x) + 3 \\ &= -(x^2 - 2x + 1) + 1 + 3 \\ &= -(x-1)^2 + 4 \end{aligned}$$

Shift the parabola $y = -x^2$ one unit to the right and four units up (Fig. 5b).

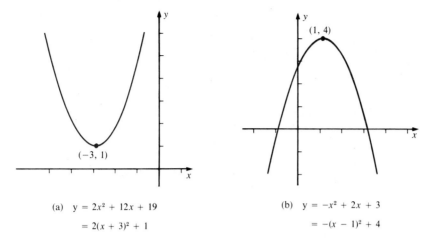

(a) $y = 2x^2 + 12x + 19$
  $= 2(x + 3)^2 + 1$

(b) $y = -x^2 + 2x + 3$
  $= -(x - 1)^2 + 4$

FIG. 5

Optimization Problems

Various problems require the maximum or minimum value of a quadratic function. Again, our strategy is completing the square.

EXAMPLE 5 A projectile is fired with muzzle velocity v_0 at a 30° angle with the ground. After t seconds its height above the ground is $y = \frac{1}{2}v_0 t - 16t^2$ feet. If $v_0 = 960$ ft/sec, find the greatest height the projectile reaches. When does it reach that height?

SOLUTION Complete the square:
$$y = 480t - 16t^2 = -16(t^2 - 30t)$$
$$= -16(t^2 - 30t + 15^2) + 16 \cdot 15^2$$
$$= -16(t - 15)^2 + 3600$$

Now $-16(t - 15)^2 \leq 0$. Therefore, the height is at most 3600 ft. But 3600 is reached when $-16(t - 15)^2 = 0$, that is, when $t = 15$.

Answer 3600 ft, reached after 15 sec

EXAMPLE 6 A farmer wishes to build a pen with three stalls (Fig. 6) using 200 feet of fence. For what length and width will the enclosed area be greatest?

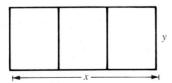

FIG. 6

SOLUTION Let x be the length and y the width. Figure 6 shows that the total amount of fence is $2x + 4y$. Therefore,
$$2x + 4y = 200 \qquad x + 2y = 100$$

158 Functions and Graphs

The area of the pen is $A = xy$. But $x = 100 - 2y$, so
$$A = (100 - 2y)y = 100y - 2y^2$$
To find the maximum value of A, complete the square:
$$\begin{aligned} A &= -2(y^2 - 50y) \\ &= -2(y^2 - 50y + 25^2) + 2 \cdot 25^2 \\ &= -2(y - 25)^2 + 1250 \end{aligned}$$
The largest value of A is 1250, which occurs when $y = 25$ and $x = 100 - 2y = 50$.

Answer length 50 ft, width 25 ft

EXERCISES

Find the linear function $f(x) = ax + b$ for which

1. $f(0) = 0$ and $f(2) = 5$
2. $f(0) = 0$ and $f(1) = -3$
3. $f(2) = 10$ and $f(6) = 8$
4. $f(4) = -1$ and $f(1) = 5$
5. $f(3) = 1.56$ and $f(4) = 1.62$
6. $f(1) = 25.4$ and $f(9) = 11.8$

If $f(x)$ is a linear function, find $f(3)$ given

7. $f(1) = 1$ and $f(5) = 13$
8. $f(4) = 0$ and $f(6) = -1$

9. A mothball of radius $\tfrac{3}{8}$ inch evaporates in 2 years. Assuming its radius is a linear function of time, find the radius after t months.

10. A property depreciates linearly from a value of $50,000 to a value of $15,000 in 20 years. Express its value after t years.

11. The population of a town increased from 12,000 in 1970 to 20,000 in 1980. Assuming the population is a linear function of time, predict the population in 1988.

12. A company had sales of $7,000,000 in 1980 and $9,500,000 in 1982. Projecting linearly, what are the predicted sales for 1990?

13. The pressure on a diver increases linearly with the depth. At the surface the pressure is 14.7 lb/in. (1 atmosphere) and it increases by 1 atmosphere with each 32.6 feet of depth. Express the pressure as a function of depth.

14. Air temperature decreases 5°F with each 1000 feet of altitude up to about 30,000 ft. If the ground temperature is 60°F, express the air temperature as a function of the altitude.

15. Suppose that the perfect age for a bride is 19 when the groom is 20, and 52 when the groom is 60. Assuming a linear relation, express the bride's age as a function of the groom's age.

16. A broker charges a $35 commission for a $1000 transaction and a $65 commission for a $3000 transaction. She uses a linear formula to compute her fees. Find that formula.

Graph

17. $y = -\tfrac{1}{2}(x - 3)^2$
18. $y = 2(x + 1)^2$

19. $y = (x + 2)^2 - 3$
20. $y = 2 - (x - 1)^2$
21. $y = x^2 - 4x + 1$
22. $y = x^2 + 2x + 4$
23. $y = -x^2 + 5x + 2$
24. $y = -x^2 - x + 1$
25. $y = 2x^2 - 8x - 5$
26. $y = 3x^2 + 12x + 13$
27. $y = x - 2x^2$
28. $y = 2x^2 + 3x$
29. $y = (x + 1)(x - 2)$
30. $y = -(x - 2)(x - 3)$
31. $y = x^2 - x + \frac{1}{4}$
32. $y = 9x^2 - 12x + 4$

33. A farmer will make a rectangular pen with 100 ft of fence, using part of a wall of his barn as one side of the pen. What is the largest area he can enclose?

34. A rectangular box has a square base. The total length of its 12 edges is 36 ft. Show that the combined area of its six faces is at most 54 ft².

35. Do Example 6 with five stalls.

36. Do Example 5 if $v_0 = 640$ ft/sec.

37. A 40-ft length of rope is cut into two pieces and each piece is folded into a square. Show that the combined areas of the two squares is at least 50 ft².

38. The average of two numbers is 10. What is the most their product can be?

39. For what value of c does the lowest point on the graph of $y = x^2 + 6x + c$ fall on the x-axis?

40. Under what conditions does the lowest point on the graph of $y = x^2 + bx + c$ fall on the y-axis?

2 POLYNOMIAL FUNCTIONS OF DEGREE GREATER THAN TWO

A function f is a **polynomial function** if

$$f(x) = a_n x^n + a_{n-1} x^{n-1} + \cdots + a_2 x^2 + a_1 x + a_0$$

Assuming $a_n \neq 0$, **the degree** of f is n. Linear and quadratic functions are polynomial functions of degree 1 and 2, respectively.

The simplest polynomial functions are the power functions $f(x) = x^n$. Let us study their graphs starting with $f(x) = x^3$. We tabulate some values for $x \geq 0$:

x	0	0.2	0.4	0.6	0.8	1	2	3	4
x^3	0	0.008	0.064	0.216	0.512	1	8	27	64

The curve $y = x^3$ starts at the origin and rises, but very slowly at first. (The cube of a small number is very small.) It reaches the height 1 at $x = 1$, then rises very fast as x increases. So the graph is very flat near $x = 0$, but zooms up once x passes 1.

The function $f(x) = x^3$ is an odd function, $(-x)^3 = -x^3$. We sketch the graph for $x \geq 0$, then obtain the rest by odd symmetry (Fig. 1a, next page).

160 Functions and Graphs

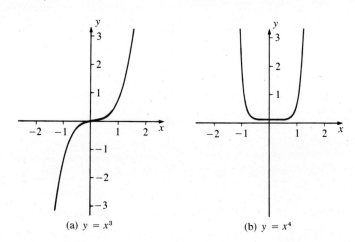

FIG. 1 (a) $y = x^3$ (b) $y = x^4$

Next, the graph of $y = x^4$. We tabulate some values using the calculator sequence $\boxed{x^2}\boxed{x^2}$ for x^4:

x	0	0.2	0.4	0.6	0.8	1	2	3	4
x^4	0	0.0016	0.0256	0.1296	0.4096	1	16	81	256

For $x \geq 0$, the graph behaves like the graph of $y = x^3$, but even more so. It rises even slower than $y = x^3$ at first. But once x passes 1, it zooms up even faster than $y = x^3$.

The function $f(x) = x^4$ is an even function, $(-x)^4 = x^4$. We sketch the graph for $x \geq 0$, then obtain the rest by reflection in the y-axis (Fig. 1b).

The graphs of odd powers

$$y = x^5 \qquad y = x^7 \qquad y = x^9 \cdots$$

look like that of $y = x^3$. The higher the exponent, the flatter the graph near $x = 0$, and the steeper for $x > 1$. The graphs of even powers

$$y = x^6 \qquad y = x^8 \qquad y = x^{10} \cdots$$

look like the graph of $y = x^4$. See Figure 2.

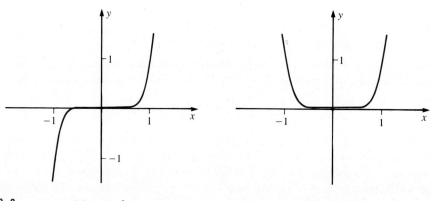

FIG. 2 (a) $y = x^5$ (b) $y = x^6$

Notice the behavior of $y = x^n$ for n odd or even. If n is odd, the graph crosses the x-axis at $x = 0$. (It slithers across, unless $n = 1$.) If n is even, the graph is tangent to the x-axis but *does not cross it*. The key is the sign of x^n. The sign changes as x passes 0 if n is odd, but does not change if n is even.

In all cases, the graph of $y = x^n$ rises beyond all bounds as x increases through positive values. We write

$$y \to \infty \quad \text{as} \quad x \to \infty$$

However, the behavior as x decreases through larger and larger negative values depends on the **parity** of n (oddness or evenness). The graph of $y = x^n$ falls below all bounds if n is odd, but rises beyond all bounds if n is even. We write

$$y \to -\infty \quad \text{as} \quad x \to -\infty \quad n \text{ odd}$$
$$y \to \infty \quad \text{as} \quad x \to -\infty \quad n \text{ even}$$

EXAMPLE 1 Graph (a) $y = (x + 2)^3$ (b) $y = 2 - (x - 1)^4$

SOLUTION (a) Shift the graph of $y = x^3$ two units to the left (Fig. 3a).

(b) Reflect the graph of $y = x^4$ in the x-axis (turn it upside down). That gives the graph of $y = -x^4$. Now shift this graph one unit to the right and two units up (Fig. 3b).

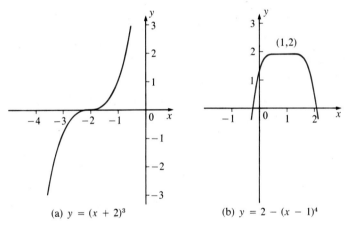

FIG. 3   (a) $y = (x + 2)^3$ (b) $y = 2 - (x - 1)^4$

Behavior for Large $|x|$

When graphing a polynomial, it helps to know the behavior as $x \to \pm\infty$. For example, take $y = 2x^3 + 7x^2 - 4x + 1$. When $|x|$ is large, the leading term $2x^3$ is much larger than the other three terms combined. (Try $x = 100$ or $x = 1000$.) Hence the graph behaves very much like the graph of $y = 2x^3$. In particular, $y \to \infty$ as $x \to \infty$ and $y \to -\infty$ as $x \to -\infty$.

Similarly, when $|x|$ is large, the graph of

$$y = a_n x^n + \cdots + a_2 x^2 + a_1 x + a_0$$

looks like that of $y = a_n x^n$. Therefore, its behavior as $x \to \pm\infty$ depends on the sign of a_n and the parity of n.

Let $y = a_n x^n + \cdots + a_2 x^2 + a_1 x + a_0$ where $a_n > 0$
Then
$$y \to \infty \quad \text{as} \quad x \to \infty$$
Also
$$y \to \infty \quad \text{as} \quad x \to -\infty \qquad n \text{ even}$$
$$y \to -\infty \quad \text{as} \quad x \to -\infty \qquad n \text{ odd}$$

Factored Polynomials

Graphing a polynomial is most convenient when the polynomial is expressed as a product of linear factors. Otherwise graphing may require plotting a large number of points or using techniques of calculus.

EXAMPLE 2 Graph $y = \tfrac{1}{3}x(x - 2)(x - 4)$

SOLUTION The factored form shows the x-intercepts: $y = 0$ at $x = 0, 2, 4$, and nowhere else. It also helps us keep track of the sign of y. See Figure 4.

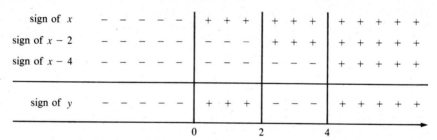

FIG. 4

The alternating signs indicate that the graph crosses the x-axis at 0, 2, and 4. Also we see that
$$y = \tfrac{1}{3}x(x - 2)(x - 4) = \tfrac{1}{3}x^3 + \text{(lower terms)}$$
so the graph zooms up as $x \to \infty$ and down as $x \to -\infty$.

This is enough information for a reasonable sketch. For greater accuracy, we plot a few points:

x	-1	1	3	5
y	-5	1	-1	5

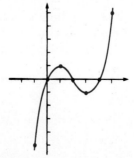

FIG. 5 $y = \tfrac{1}{3}x(x - 2)(x - 4)$

Remark In Fig. 5, the peak and valley of the curve do not occur exactly at $x = 1$ and $x = 3$. Finding their precise location requires calculus.

Fig. 4 shows that the sign of $y = \frac{1}{3}x(x - 2)(x - 4)$ alternates, changing at $x = 0$, 2, and 4. In fact, any polynomial expressed as the product of different linear factors behaves in a similar way. Thus
$$y = (x + 1)(x - 1)(x - 2)(x - 3)$$
is 0 at $x = -1, 1, 2, 3$ and changes sign at each of these values. The sign is $+$ when $x > 3$ because all factors are positive. Now we can easily trace the sign of y for all values of x. See Figure 6.

FIG. 6

EXAMPLE 3 Graph $y = (x + 1)(x - 1)(x - 2)(x - 3)$

SOLUTION Figure 6 already contains enough information for a rough sketch (Figure 7a).

The graph has valleys between -1 and 1 and between 2 and 3, and a peak between 1 and 2. To estimate their sizes, we plot a few strategic points, say, for $x = 0, 1.5,$ and 2.5. (A calculator shortens the arithmetic.)

x	0.0	1.5	2.5
y	-6.0	0.94	-1.31

Further information: $y = x^4 +$ (lower terms); no need to multiply out. Hence $y \to \infty$ as $x \to \pm\infty$. Now we have enough data for a more accurate graph (Fig. 7b).

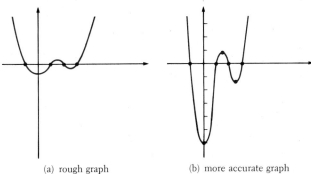

(a) rough graph (b) more accurate graph

FIG. 7 $y = (x + 1)(x - 1)(x - 2)(x - 3)$

Repeated Linear Factors

A polynomial may be the product of linear factors where some factors are repeated. For example,
$$y = (x - 1)(x - 3)^2 \qquad y = x(x - 2)^3 \qquad y = (x + 2)^2(x - 4)^5$$
Keep in mind that $(x - a)^n$ changes sign at $x = a$ if n is odd, but does not change sign if n is even.

Functions and Graphs

EXAMPLE 4 Graph (a) $y = (x - 1)(x - 3)^2$ (b) $y = (x - 1)(x - 3)^3$

SOLUTION (a) Clearly $y = 0$ at $x = 1$ and 3. At $x = 1$, the sign of y changes; hence the graph crosses the x-axis. At $x = 3$, however, the sign does not change because of the *even* power $(x - 3)^2$. Hence the graph touches the x-axis without crossing it.

Since $y = x^3 +$ (lower terms), the graph zooms up to the right and down to the left. We now have enough information for a sketch (Fig. 9a). Plotting the point $(2, 1)$ helps to estimate the peak between $x = 1$ and 3.

(b) Because of the *odd* power $(x - 3)^3$, this time y changes sign at $x = 3$. The graph crosses the x-axis at $x = 1$ and $x = 3$. Actually, it slithers across at $x = 3$. That's because near $x = 3$, the factor $x - 1$ is approximately 2, so

$$y = (x - 1)(x - 3)^3 \approx 2(x - 3)^3 \quad \text{near } x = 3$$

Since $y = x^4 +$ (lower terms), $y \to \infty$ as $x \to \pm\infty$. This is enough information for a sketch (Fig. 8b). Plotting the point $(2, -1)$ indicates the depth of the valley.

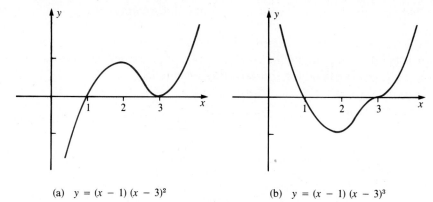

(a) $y = (x - 1)(x - 3)^2$ (b) $y = (x - 1)(x - 3)^3$

FIG. 8

EXERCISES

Graph

1. $y = \tfrac{1}{4}x^6$
2. $y = \tfrac{1}{3}x^5$
3. $y = -x^7$
4. $y = -x^8$
5. $y = \tfrac{1}{2}(x + 5)^3$
6. $y = \tfrac{1}{4}(x - 2)^4$
7. $y = (x - 3)^4 - 1$
8. $y = (x + 2)^3 + 1$
9. $y = 3 - (x + 2)^6$
10. $y = 2 - \tfrac{1}{2}(x - 1)^4$

Solve the inequality

11. $x(x - 1)(x - 2) < 0$
12. $(x + 2)(x - 1)(x - 2) > 0$
13. $x(x - 1)(x - 4)(x - 9) > 0$
14. $(x + 5)(x + 1)(x - 3)(x - 4) < 0$
15. $(x - 1)(x - 4) \geq 0$
16. $(x - 1)^3(x - 4) \leq 0$

17. $x^4 - 9x^2 > 0$

18. $x^6 + 2x^5 > 0$

19. $x^3(x + 2)^2(x + 3) < 0$

20. $(x + 2)^2(x - 4)^2(x - 5)^3 > 0$

Graph

21. $y = x(x - 1)(x - 2)$

22. $y = (x + 2)(x - 1)(x - 2)$

23. $y = \frac{1}{6}(x - 1)(x - 2)(x - 3)(x - 4)$

24. $y = \frac{1}{24}x(x - 2)(x - 3)(x - 4)$

25. $y = -x^2(x - 1)$

26. $y = -(x + 1)(x - 1)^2$

27. $y = x^3 + 5x^2 + 6x$

28. $y = x^3 - 2x^2 - 3x$

29. $y = x^2(x - 1)(x - 3)$

30. $y = x(x - 1)^2(x - 3)$

31. $y = (x^2 - 1)(x^2 - 4)$

32. $y = x^4 - 2x^2 + 1$

33. $y = \frac{1}{4}x(x - 4)^3$

34. $y = \frac{1}{2}x^3(x - 2)(x - 4)$

Graph (Suggestion First ignore the constant term.)

35. $y = x^3 - 2x + 1$

36. $y = x^3 - x^2 + 2$

37. Explain graphically why every cubic equation $ax^3 + bx^2 + cx + d = 0$ has at least one real solution.

38. Show that the graph in Figure 5 is the graph of an odd function shifted two units to the right.

39. Find a polynomial of degree 3 whose graph crosses the x-axis at $x = 1, 2,$ and 4 and has y-intercept 3.

40. Find a polynomial of degree 4 whose graph passes through $(1, 1)$ and touches the x-axis both at $x = -1$ and 0 without crossing it.

3 RATIONAL FUNCTIONS

A function f is a **rational function** if

$$f(x) = \frac{p(x)}{q(x)}$$

where $p(x)$ and $q(x)$ are polynomials. We will always assume that the quotient is in lowest terms, that is, $p(x)$ and $q(x)$ have no common polynomial factors.

Negative Power Functions

We begin with the graph of $f(x) = 1/x$. The function is odd, so we graph it for $x > 0$, and then obtain the rest by odd symmetry. Let us tabulate some values:

x	1	2	3	4	5	10	100
$1/x$	1	$\frac{1}{2}$	$\frac{1}{3}$	$\frac{1}{4}$	$\frac{1}{5}$	$\frac{1}{10}$	$\frac{1}{100}$

The values are positive and decreasing. In fact, $y \to 0$ as $x \to \infty$. Therefore, the graph approaches the y-axis as x moves to the right.

The value $x = 0$ is excluded. Still, we can see what happens when x is near 0:

x	$\frac{1}{100}$	$\frac{1}{10}$	$\frac{1}{5}$	$\frac{1}{4}$	$\frac{1}{3}$	$\frac{1}{2}$
$1/x$	100	10	5	4	3	2

As x approaches 0 *from the right*, y increases rapidly. We write
$$y \to \infty \quad \text{as} \quad x \to 0^+$$
We can now sketch the graph, using odd symmetry for $x < 0$. See Figure 1a.

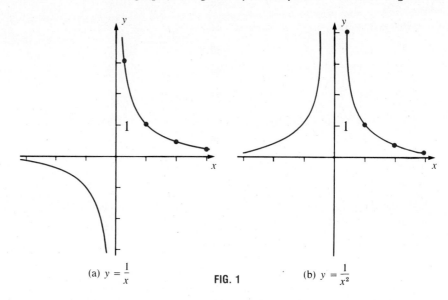

(a) $y = \dfrac{1}{x}$ FIG. 1 (b) $y = \dfrac{1}{x^2}$

Next, the graph of $f(x) = 1/x^2$. The function is even, so we graph it for $x > 0$ and then obtain the rest by reflecting in the y-axis. We tabulate some values as x increases:

x	1	2	3	4	5	10	100
$1/x^2$	1	$\frac{1}{4}$	$\frac{1}{9}$	$\frac{1}{16}$	$\frac{1}{25}$	$\frac{1}{100}$	$\frac{1}{10000}$

If $x \to \infty$, then $y \to 0$ even faster than $y = 1/x$. Now, some values as $x \to 0^+$:

x	$\frac{1}{100}$	$\frac{1}{10}$	$\frac{1}{5}$	$\frac{1}{4}$	$\frac{1}{3}$	$\frac{1}{2}$
$1/x^2$	10,000	100	25	16	9	4

As x approaches 0 from the right, the curve zooms up, even faster than $y = 1/x$. We can now sketch the graph (Fig. 1b).

The graphs of negative odd powers
$$y = \frac{1}{x^3} \quad y = \frac{1}{x^5} \quad y = \frac{1}{x^7} \cdots$$
look like the graph of $y = 1/x$. The higher the exponent, the steeper the graph near $x = 0$, and the faster it dies out as $x \to \infty$. The graphs of negative even powers
$$y = \frac{1}{x^4} \quad y = \frac{1}{x^6} \quad y = \frac{1}{x^8} \cdots$$

look like the graph of $y = 1/x^2$. Again the difference between odd and even powers is a matter of sign: if n is odd, $1/x^n$ has opposite signs for x positive and negative, but if n is even, $1/x^n$ is always positive. See Figure 2.

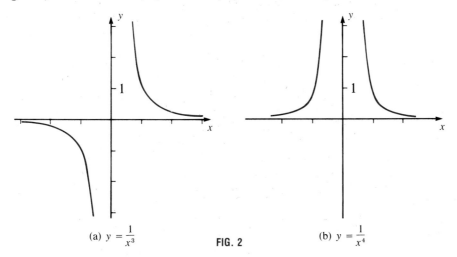

(a) $y = \dfrac{1}{x^3}$ **FIG. 2**   (b) $y = \dfrac{1}{x^4}$

Notice the special relation of the two axes to the graphs in Figs. 1 and 2. The y-axis is called a **vertical asymptote**, and the x-axis is called a **horizontal asymptote**. In general, an **asymptote** of a curve is a line that the curve approaches as it moves farther and farther from the origin.

EXAMPLE 1   Graph   (a) $y = -\dfrac{1}{(x+1)^2}$ (b) $y = \dfrac{1}{x-2} + 1$

SOLUTION (a) Reflect the graph of $y = 1/x^2$ in the x-axis. That gives the graph of $y = -1/x^2$. Then shift one unit to the left (Fig. 3a). The line $x = -1$ is a vertical asymptote, and the x-axis is a horizontal asymptote.

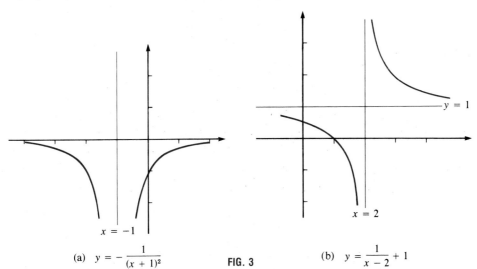

(a) $y = -\dfrac{1}{(x+1)^2}$ **FIG. 3**   (b) $y = \dfrac{1}{x-2} + 1$

(b) Shift the graph of $y = 1/x$ two units to the right and one unit up (Fig. 3b, preceding page). The line $x = 2$ is a vertical asymptote, and the line $y = 1$ is a horizontal asymptote.

Sometimes a little algebra will change the form of a function to one that is easier to graph. For example, suppose we want to graph

$$y = \frac{x-1}{x-2}$$

If the numerator were $x - 2$, we could divide the denominator into the numerator. Let us *create* an $x - 2$ in the numerator. We subtract and add 2:

$$y = \frac{x-1}{x-2} = \frac{x-2+2-1}{x-2}$$

$$= \frac{x-2+1}{x-2}$$

$$= \frac{x-2}{x-2} + \frac{1}{x-2}$$

Therefore

$$y = 1 + \frac{1}{x-2}$$

With y expressed this way, we are back to Figure 3b.

The algebra we used here is a special case of division of polynomials discussed in Chapter 9. As another example, take

$$y = \frac{3x+7}{x+1}$$

We could divide if the numerator were $3x + 3 = 3(x + 1)$. So we add and subtract 3:

$$y = \frac{3x+3-3+7}{x+1}$$

$$= \frac{3x+3}{x+1} + \frac{4}{x+1} = 3 + \frac{4}{x+1}$$

With y written this way, graphing is easier.

Further Examples

EXAMPLE 2 Graph $y = \dfrac{1}{(x+1)(x-2)}$

SOLUTION The graph does not meet the x-axis because y is never 0. It is undefined at $x = -1$ and $x = 2$. Otherwise the sign of y is the same as that of $(x + 1)(x - 2)$. See Figure 4.

FIG. 4

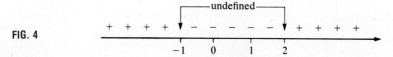

The line $x = 2$ is a vertical asymptote, for as $x \to 2^+$, the denominator approaches 0. Hence $y \to \infty$. As $x \to 2^-$ (that is, from the left), $y \to -\infty$. Similarly, the line $x = -1$ is also a vertical asymptote.

The x-axis is a horizontal asymptote, for the denominator $(x + 1)(x - 2) \to \infty$ as $x \to \pm\infty$. Hence $y \to 0$ as $x \to \pm\infty$.

We can now sketch the graph (Fig. 5). As usual, we plot a few strategic points:

x	-2	0	$\tfrac{1}{2}$	1	3
y	$\tfrac{1}{4}$	$-\tfrac{1}{2}$	$\tfrac{4}{9}$	$-\tfrac{1}{2}$	$\tfrac{1}{4}$

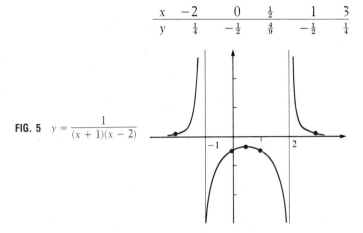

FIG. 5 $y = \dfrac{1}{(x + 1)(x - 2)}$

EXAMPLE 3 Graph $y = \dfrac{x}{(x + 1)(x - 2)}$

SOLUTION The graph is undefined at $x = -1$ and $x = 2$. Otherwise the sign of y is the same as that of $(x + 1)(x)(x - 2)$. See Figure 6.

FIG. 6

We see that the graph crosses the y-axis at $x = 0$. Also, exactly as in Example 2, the lines $x = -1$ and $x = 2$ are vertical asymptotes.

One further observation: The x-axis is a horizontal asymptote. As $x \to \pm\infty$, the denominator gets much larger than the numerator, so $y \to 0$. To be more precise, write

$$y = \frac{x}{(x + 1)(x - 2)} = \frac{x}{x^2 - x + 2}$$

Divide numerator and denominator by x^2, the largest power in the denominator:

$$y = \frac{\dfrac{1}{x}}{1 - \dfrac{1}{x} + \dfrac{2}{x^2}}$$

As $x \to \pm\infty$, the numerator $\to 0$ and the denominator $\to 1 - 0 + 0 = 1$. Hence $y \to 0/1 = 0$.

We now have enough information to sketch the graph (Fig. 7, next page).

FIG. 7.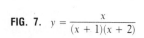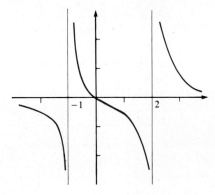

EXAMPLE 4 Graph $y = \dfrac{4}{x^2 + 1}$

SOLUTION We can get a lot of free information about the graph by just looking at the equation.

(1) y is defined for all values of x; there are no vertical asymptotes.

(2) The function $4/(x^2 + 1)$ is even; the graph is symmetric in the y-axis.

(3) $y > 0$ for all x; the graph is entirely above the x-axis.

(4) y decreases as x increases. By symmetry, the largest value of y occurs for $x = 0$.

(5) $y \to 0$ as $x \to \pm\infty$; the x-axis is a horizontal asymptote.

This is enough to give the shape of the graph (Fig. 8). For greater accuracy, we compute a few points:

x	0	0.5	1	1.5	2	3	4
y	4	3.2	2.0	1.23	0.80	0.40	0.24

FIG. 8. $y = \dfrac{4}{x^2 + 1}$

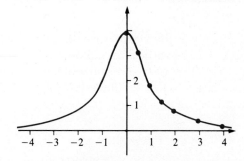

EXERCISES

Graph

1. $y = \dfrac{1}{x - 5}$

2. $y = \dfrac{1}{x + 3}$

3. $y = -\dfrac{1}{x + 1}$

4. $y = -\dfrac{1}{2(x + 6)}$

5. $y = \dfrac{1}{x+2} - 1$ 6. $y = 2 - \dfrac{1}{x}$

7. $y = \dfrac{1}{(x-4)^2}$ 8. $y = -\dfrac{1}{(x+1)^2}$

9. $y = \dfrac{x+4}{x+5}$ 10. $y = \dfrac{x+1}{x-2}$

11. $y = \dfrac{2x-1}{x-1}$ 12. $y = \dfrac{3x+4}{x+2}$

13. $y = \dfrac{1}{x^2-9}$ 14. $y = \dfrac{1}{x^2+3x}$

15. $y = \dfrac{x}{x^2+1}$ 16. $y = \dfrac{1}{x^4+1}$

17. $y = \dfrac{x}{(x-2)^2}$ 18. $y = \dfrac{x-1}{x^2}$

19. $y = \dfrac{x}{x^2-4x+3}$ 20. $y = \dfrac{-x}{x^2+2x-3}$

21. $y = \dfrac{x}{(x-2)(x+1)}$ 22. $y = \dfrac{x+1}{x^2-4}$

23. $y = \dfrac{1}{32x^2(x-1)^2}$ 24. $y = \dfrac{1}{x(x-1)(x-2)}$

25. Graph $y = \dfrac{x^2}{x^2+3}$ (Hint Write the numerator as $x^2 + 3 - 3$.)

26. Show that the graph of $y = \dfrac{1}{(x+1)(x+5)}$ is symmetric about the line $x = -3$.

27. Find a quadratic polynomial $q(x)$ such that the graph of $y = 1/q(x)$ has vertical asymptotes $x = -1$ and $x = 1$, and y-intercept -5.

28. Find a polynomial $q(x)$ of degree 4 such that the graph of $y = 1/q(x)$ has y-intercept 1 and exactly two vertical asymptotes, $x = -3$ and $x = 1$.

4 VARIATION

In this section we study a common type of functional relation between quantities called **variation**.

Direct Variation

We say that y **varies directly with** x, or y **is proportional to** x if
$$y = kx$$

for some constant k. Similarly, we say that y **varies directly with** x^n or y **is proportional to** x^n if

$$y = kx^n$$

The constant k in each formula is called the **constant of proportionality**.

Examples

(1) An employee's weekly pay p varies directly with the number h of hours worked:

$$p = kh$$

Here the constant of proportionality k is the hourly rate of pay.

(2) Near the surface of a planet, the distance d an object falls in t sec varies directly with t^2:

$$d = kt^2$$

This type of formula holds for any planet, but the constant k depends on the mass of the planet.

(3) The volume V of a sphere is proportional at the cube of its radius r.

$$V = \tfrac{4}{3}\pi r^3$$

Here $k = \tfrac{4}{3}\pi$.

In each of the examples, the *ratio* of the two quantities is constant:

$$\frac{p}{h} = k \qquad \frac{d}{t^2} = k \qquad \frac{V}{r^3} = \frac{4}{3}\pi$$

Thus, if two spheres have radii r_1 and r_2, and volumes V_1 and V_2, then

$$\frac{V_1}{r_1^3} = \frac{V_2}{r_2^3} \qquad \text{or equivalently} \qquad \frac{V_1}{V_2} = \frac{r_1^3}{r_2^3}$$

EXAMPLE 1 The amount of newsprint (paper) used by a daily newspaper is proportional to its circulation. With 100,000 subscribers, it uses 150 tons of newsprint per week. If the circulation increases to 120,000, how much newsprint will it need?

SOLUTION 1 Let p be the number of tons of newsprint used weekly, and s the number of subscribers. Since p is proportional to s,

$$p = ks$$

First find k using the data $p = 150$ when $s = 100{,}000$:

$$150 = k(100{,}000)$$

$$k = \frac{150}{100{,}000} = 0.0015$$

Therefore

$$p = 0.0015s$$

When $s = 120{,}000$,

$$p = (0.0015)(120{,}000) = \underline{180 \text{ tons}}$$

SOLUTION 2 The ratio p/s is constant, so

$$\frac{p_2}{s_2} = \frac{p_1}{s_1} \quad \text{or equivalently} \quad \frac{p_2}{p_1} = \frac{s_2}{s_1}$$

Set $p_1 = 150$, $s_1 = 100{,}000$, $s_2 = 120{,}000$ and solve for p_2:

$$\frac{p_2}{150} = \frac{120{,}000}{100{,}000} = 1.2$$

$$p_2 = (150)(1.2) = \underline{180 \text{ tons}}$$

EXAMPLE 2 The braking distance needed to stop a car varies with the square of the speed. For a certain car the braking distance at 30 mph (44 ft/sec) is 57 ft.
(a) Find a formula for the braking distance in general.
(b) What is the braking distance at 60 mph?

SOLUTION (a) Let d be the braking distance (in feet) at speed s (in ft/sec). Then

$$d = ks^2$$

To find k, substitute $d = 57$ and $s = 44$:

$$57 = k \cdot 44^2$$

$$k = \frac{57}{44^2}$$

Hence the desired formula is

$$d = \frac{57}{44^2} s^2 = 57\left(\frac{s}{44}\right)^2$$

(b) Substitute $s = 88$ in the formula:

$$d = 57\left(\tfrac{88}{44}\right)^2 = 57(2)^2 = \underline{228 \text{ ft}}$$

Remark Note that doubling the speed multiplies the distance by 4. Similarly, tripling the speed multiplies the distance by 9. That's because the distance varies with the *square* of the speed. So drive carefully.

Inverse Variation

We say that y **varies inversely with** x, or y **is inversely proportional to** x if

$$y = \frac{k}{x}$$

Similarly, we say that y **varies inversely with** x^n, or y **is inversely proportional to** x^n if

$$y = \frac{k}{x^n}$$

The point of inverse variation is that y gets *smaller* as x gets larger, and vice versa.

Examples

(1) Driving time t between two points at constant speed s varies inversely with the speed:

$$t = \frac{k}{s}$$

(2) The intensity of illumination I on a screen from a source of light is inversely proportional to the square of the distance d between them:

$$I = \frac{k}{d^2}$$

EXAMPLE 3 At what distance is the illumination on a screen 10 times as great as it is at 7 ft?

SOLUTION We have $I = k/d^2$ from the preceding example. The constant k is not given, but that does not matter. Whatever k is, the illumination at 7 ft is $k/7^2$. We want d so that the illumination is 10 times as much. So d must satisfy

$$\frac{k}{d^2} = 10 \left(\frac{k}{7^2}\right)$$

Solve for d (the k drops out):

$$d^2 = \frac{7^2}{10}$$

$$d = \frac{7}{\sqrt{10}} \approx \underline{2.2 \text{ ft}}$$

Joint Variation

A quantity may depend on several quantities, not just one. We say that z **varies jointly with** x and y or **varies directly with** x and y if

$$z = kxy$$

We say that z **varies directly with** x **and inversely with** y if

$$z = \frac{kx}{y}$$

Many other combinations of direct and inverse variation are possible.

Examples
(1) The volume V of a cylinder varies directly with its height h and the square of its radius r:

$$V = \pi r^2 h \quad \text{that is,} \quad V = kr^2 h \quad \text{where} \quad k = \pi$$

(2) (Newton's Law of Gravitation) The gravitational attraction F of two bodies is directly proportional to the product of their masses, m_1 and m_2, and inversely proportional to the square of the distance d between them:

$$F = \frac{km_1 m_2}{d^2}$$

EXAMPLE 4 The volume V of a gas varies directly with the temperature T and inversely with the pressure P. The volume is 100 in³ when the temperature is 400° F and the pressure is 50 lb/in². Find the volume when the temperature is 375° F and the pressure is 75 lb/in².

SOLUTION We have
$$V = \frac{kT}{P}$$
To find k, set $V = 100$, $T = 400$, and $P = 50$:
$$100 = \frac{k \cdot 400}{50} = 8k$$
$$k = \frac{100}{8} = 12.5$$
Hence
$$V = \frac{12.5\,T}{P}$$
Now set $T = 375$ and $P = 75$:
$$V = \frac{(12.5)(375)}{75} = (12.5)(5) = \underline{62.5 \text{ in}^3}$$

EXERCISES

Express the statement as an equation with the correct constant of proportionality

1. y varies directly with x, and $y = 5$ when $x = 3$.
2. u varies directly with v, and $u = \frac{1}{2}$ when $v = 10$.
3. A is proportional to r^2, and $A = 4\pi$ when $r = 2$.
4. w is proportional to s^3, and $w = 8$ when $s = 4$.
5. The circumference C of a circle is proportional to its diameter d.
6. The diagonal d of a square is proportional to its side s.
7. y varies inversely with x, and $y = 0.35$ when $x = 10$.
8. y varies inversely with the square of x, and $y = 0.16$ when $x = 25$.
9. z varies directly with x and the square of y, and $z = 5$ when $x = 10$ and $y = 2$.
10. r varies directly with the cube of s and inversely with the square of t, and $r = 1$ when $s = 2$ and $t = 4$.
11. If y is inversely proportional to x^2, what happens to y when x is doubled? tripled?
12. If y is proportional to x^3, what happens to y when x is doubled? tripled?
13. The resistance R to blood flow in an artery is directly proportional to the length L of the artery and inversely proportional to the 4-th power of its diameter D. What happens to R if L is tripled and D is doubled?
14. The pitch (frequency) of a violin string is proportional to the tension in the string and inversely proportional to the length. What happens to the pitch if both the tension and the length are decreased by 10%?

15. At constant speed the gas consumed by a car is proportional to the distance driven. At 55 mph, a car uses 9 gallons of gas in 200 miles. How much will it use in 450 miles?

16. The amount of litter along a stretch of highway is approximately proportional to the length of the stretch. A cleanup crew collected one truckload of litter in 7.5 miles. How many truckloads do they expect along a 50-mile stretch?

17. (Hooke's Law) The force needed to stretch a spring a short distance x varies directly with x. If a force of 5 pounds will stretch a certain spring 6 inches, what force will stretch it 10 inches?

18. (Ohm's Law) The current I in an electrical circuit varies directly with the voltage V. If I is 6 amperes when V is 15 volts, what is I when V is 40 volts?

19. In warm weather, the attendance at math lectures varies inversely with the square root of the number of degrees above 70° F. At 10 AM the temperature was 71° and 160 out of 200 attended. How many students out of 200 attended the 3 PM lecture when the temperature was 79°?

20. At the INV Company, the lowest paid employees do the most work. In fact, work done varies inversely with salary. If a secretary at $10,000 does 8 hours of work a day, how much work does a vice-president at $45,000 do?

21. Weight varies inversely with the square of the distance from the center of the earth. An astronaut weighs 180 lb at the surface of the earth. What will he weigh 1000 miles up? (The radius of the earth is 4000 miles.)

22. The thickness of a $\frac{1}{4}$-lb hamburger is inversely proportional to the square of its radius. The thickness is $\frac{1}{2}$ inch if the radius is 2 inches. What is the thickness if the radius is $1\frac{3}{4}$ inches?

23. A diver standing at the end of a diving board causes the tip of the board to deflect downward a distance proportional to the cube of its length. If the deflection is 3 inches for an 8-ft board, how much is it for a 10-ft board?

24. (Kepler's Third Law) The time required for a planet to make one circuit around the sun varies with the $\frac{3}{2}$ power of its average distance from the sun. For the earth, this period is one year. How long is it for Jupiter, whose average distance from the sun is 5.2 that of the earth from the sun?

25. The cost of insulating an attic varies jointly with the area of the attic floor and the thickness of the insulation. For an attic with floor area 1000 ft² and insulation 6 inches thick, the cost is $350. What is the cost for an attic of area 1400 ft² and insulation 8 inches thick?

26. The volume of a pyramid with square base varies directly with the height and the square of the side of the base. When the height is 20 cm and the side of the base is 9 cm, the volume is 540 cm³. Find the volume when the height is 18 cm and the side of the base is 10 cm.

27. A lamp gives satisfactory illumination for reading at 3 ft. At what distance will a lamp twice as strong give the same illumination? Illumination varies directly with the strength of the lamp and inversely with the square of the distance.

28. The kinetic energy of a moving body varies directly with its mass and the square of its velocity. A body of mass 5g moving at 10 cm/sec has kinetic energy 250 erg. Find the kinetic energy of a body of mass 8g moving at 20 cm/sec.

29. Show that the area of a square inscribed in a circle is proportional to the square of the radius.

30. Show that the area of a regular hexagon (6-sided figure) is proportional to the square of the side length.

True or false? Why?

31. If x varies directly with y, then y varies inversely with x.

32. If u varies inversely with the square of v, then v varies inversely with the square root of u.

5 CONIC SECTIONS

This section deals with equations whose graphs are important curves: circles, parabolas, ellipses, and hyperbolas.

Circles

We start with the equation

$$x^2 + y^2 = r^2 \qquad r > 0$$

Now $x^2 + y^2$ is the distance squared of a point (x, y) from $(0, 0)$. So the equation says

$$[\text{distance from } (0, 0)]^2 = r^2$$

that is,

$$[\text{distance from } (0, 0)] = r$$

Hence the graph is a circle of radius r with center at the origin (Fig. 1a).

By the same reasoning, the graph of

$$(x - h)^2 + (y - k)^2 = r^2$$

is a circle of radius r with center at (h, k). See Figure 1b.

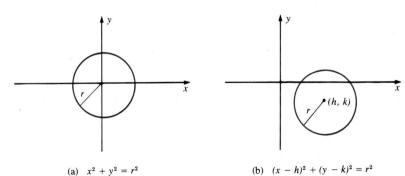

(a) $x^2 + y^2 = r^2$ (b) $(x - h)^2 + (y - k)^2 = r^2$

FIG. 1

> **Circles** The graph of
> $$(x - h)^2 + (y - k)^2 = r^2$$
> is the circle of radius r and center (h, k).

EXAMPLE 1 Show that the graph is a circle:
$$x^2 + y^2 + 2x - 6y + 6 = 0$$

SOLUTION Write the equation as
$$(x^2 + 2x) + (y^2 - 6y) = -6$$
Now complete the square in x and in y:
$$(x^2 + 2x + 1) + (y^2 - 6y + 9) = -6 + 1 + 9$$
$$(x + 1)^2 + (y - 3)^2 = 4$$
This equation has the form $(x - h)^2 + (y - k)^2 = r^2$, with $h = -1$, $k = 3$, and $r = 2$. Its graph is the circle of radius 2 and center $(-1, 3)$.

EXAMPLE 2 The points $(-2, 1)$ and $(3, 4)$ are ends of a diameter of a circle. Find the equation of that circle.

SOLUTION By the distance formula, the length of the diameter is
$$\sqrt{(3 + 2)^2 + (4 - 1)^2} = \sqrt{5^2 + 3^2} = \sqrt{34}$$
Hence the radius is $\frac{1}{2}\sqrt{34}$.

The center of the circle is the midpoint of the diameter. By the midpoint formula, that is the point
$$\left(\frac{-2 + 3}{2}, \frac{1 + 4}{2}\right) = \left(\frac{1}{2}, \frac{5}{2}\right)$$
The equation of the circle is $(x - h)^2 + (y - k)^2 = r^2$, where $h = \frac{1}{2}$, $k = \frac{5}{2}$, and $r = \frac{1}{2}\sqrt{34}$.
Answer $(x - \frac{1}{2})^2 + (y - \frac{5}{2})^2 = \frac{17}{2}$

Remark The graph of $x^2 + y^2 = r^2$ is not the graph of a function because each x between $-r$ and r gives *two* values, $y = \pm\sqrt{r^2 - x^2}$. However, the *upper half* of the circle is the graph of a function, $y = +\sqrt{r^2 - x^2}$, and the *lower half* is also the graph of a function, $y = -\sqrt{r^2 - x^2}$.

Parabolas

As pointed out in Section 1, the curve
$$y = ax^2$$
is called a parabola (Fig. 2a,b). Its **axis** is the y-axis, and its **vertex** is the origin. In general, the graph of quadratic function,
$$y = ax^2 + bx + c$$

is also a parabola. By completing the square, we can write this equation in the form

$$y = a(x - h)^2 + k \quad \text{or} \quad y - k = a(x - h)^2$$

The graph is a shifted version of the parabola $y = ax^2$. It is still vertical, but its vertex is moved to the point (h, k). See Figure 2c.

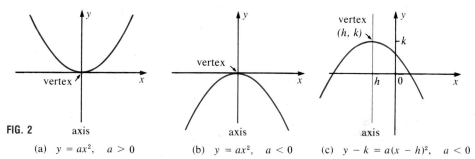

FIG. 2
(a) $y = ax^2, \quad a > 0$
(b) $y = ax^2, \quad a < 0$
(c) $y - k = a(x - h)^2, \quad a < 0$

Suppose we interchange x and y in $y = ax^2$. We get $x = ay^2$, whose graph is the same parabola but horizontal, having axis on the x-axis and vertex at the origin (Fig. 3a,b). Similarly $x - h = a(y - k)^2$ is a horizontal parabola with vertex at (h, k). See Figure 3c.

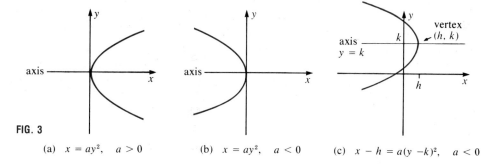

FIG. 3
(a) $x = ay^2, \quad a > 0$
(b) $x = ay^2, \quad a < 0$
(c) $x - h = a(y - k)^2, \quad a < 0$

Parabolas The graphs of

(1) $\quad y - k = a(x - h)^2 \quad\quad$ (2) $\quad x - h = a(y - k)^2$

are parabolas with vertices at (h, k).

Parabola (1) is vertical with axis $x = h$. It opens upward if $a > 0$, downward if $a < 0$.

Parabola (2) is horizontal with axis $y = k$. It opens to the right if $a > 0$, to the left if $a < 0$.

EXAMPLE 3 Graph (a) $x = 1 - \tfrac{1}{4}y^2$ (b) $x = y^2 - 2y$

SOLUTION (a) Write the equation as

$$x - 1 = -\tfrac{1}{4}(y - 0)^2$$

The graph is a parabola with vertex (1, 0). The parabola is horizontal and opens to the left. To get an idea of its width, plot a few points. For example, if $x = 0$, then $y = \pm 2$. Thus the y-intercepts are $(0, \pm 2)$. Two other points are $(-\frac{5}{4}, \pm 3)$. See Figure 4a.

(b) Complete the square:
$$x + 1 = y^2 - 2y + 1$$
$$x + 1 = (y - 1)^2$$

The graph is a horizontal parabola with vertex $(-1, 1)$ and opening to the right. A few easily computed points are $(0, 0)$, $(0, 2)$, $(3, 3)$, and $(3, -1)$. See Figure 4b.

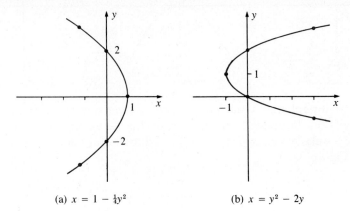

FIG. 4 (a) $x = 1 - \frac{1}{4}y^2$ (b) $x = y^2 - 2y$

Ellipses

Let us graph the equation
$$\frac{x^2}{a^2} + \frac{y^2}{b^2} = 1$$

We solve for y and find *two* values for each x:
$$y = \frac{b}{a}\sqrt{a^2 - x^2} \quad \text{and} \quad y = -\frac{b}{a}\sqrt{a^2 - x^2}$$

If the factor b/a were not there, we would have
$$y = \sqrt{a^2 - x^2} \quad \text{and} \quad y = -\sqrt{a^2 - x^2}$$
functions whose graphs are the upper and lower halves of the circle $x^2 + y^2 = a^2$.

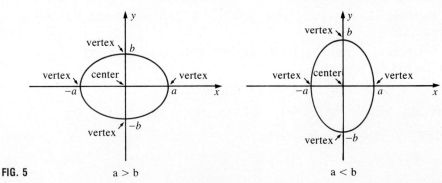

FIG. 5 $a > b$ $a < b$

The factor b/a stretches these semicircles in the y-direction if $b/a > 1$, shrinks them if $b/a < 1$. Hence our graph is an oval-shaped curve called an **ellipse**. See Figure 5.

The points $(\pm a, 0)$ and $(0, \pm b)$ are **vertices** of the ellipse. The origin is its **center**. The larger of the two numbers a and b is the **semi-major axis**; the smaller is the **semi-minor axis**.

> **Ellipses** The graph of
> $$\frac{x^2}{a^2} + \frac{y^2}{b^2} = 1 \qquad a, b > 0$$
> is an ellipse with center $(0, 0)$ and vertices $(\pm a, 0)$ and $(0, \pm b)$.

Note that if $a = b$, the graph is the circle $x^2 + y^2 = b^2$. Thus, circles are special cases of ellipses.

EXAMPLE 4 Graph (a) $4x^2 + 9y^2 = 36$ (b) $4x^2 + y^2 = 4$

SOLUTION (a) Divide both sides by 36:
$$\frac{x^2}{9} + \frac{y^2}{4} = 1 \quad \text{or} \quad \frac{x^2}{3^2} + \frac{y^2}{2^2} = 1$$
The graph is an ellipse with vertices $(\pm 3, 0)$ and $(0, \pm 2)$. See Figure 6a.

(b) Divide both sides by 4:
$$x^2 + \frac{y^2}{4} = 1 \quad \text{or} \quad \frac{x^2}{1^2} + \frac{y^2}{2^2} = 1$$
The graph is an ellipse with vertices $(\pm 1, 0)$ and $(0, \pm 2)$. See Figure 6b.

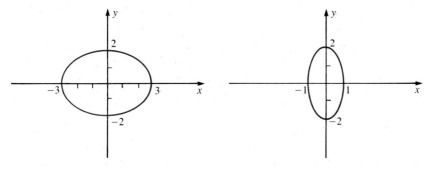

FIG. 6 (a) $4x^2 + 9y^2 = 36$ (b) $4x^2 + y^2 = 4$

Hyperbolas

For our last curve of this section, we consider the equation
$$\frac{x^2}{a^2} - \frac{y^2}{b^2} = 1 \qquad a, b > 0$$
Because of the minus sign, both x^2/a^2 and y^2/b^2 can be large, as long as their *difference* is 1. Therefore, the graph is not bounded as it is for the ellipse. The graph

is symmetric in both the x- and the y-axes. For if (x, y) satisfies the equation, so do $(x, -y)$ and $(-x, y)$. Therefore, we can sketch the part in the first quadrant, and then get the rest of the graph by reflecting in both axes.

Solve for y in the first quadrant:

$$y = \frac{b}{a}\sqrt{x^2 - a^2}$$

The square root requires that $x \geq a$. So let x start at a and increase. Then y starts at 0 and increases. When x is very large, $\sqrt{x^2 - a^2}$ is slightly less than $\sqrt{x^2} = x$. Hence y is slightly less than $(b/a)x$. Therefore as $x \to \infty$, the graph approaches the line $y = (b/a)x$ from below.

With this information, we can sketch the graph in the first quadrant (Fig. 7a). Then we complete the graph by symmetry (Fig. 7b).

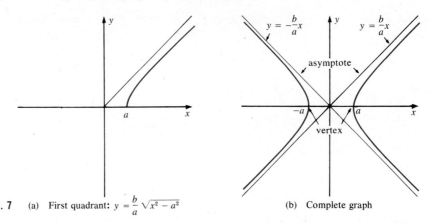

FIG. 7 (a) First quadrant: $y = \frac{b}{a}\sqrt{x^2 - a^2}$ (b) Complete graph

The curve is called a **hyperbola**. The points $(\pm a, 0)$ are the **vertices** of the hyperbola. The x- and y-axes are its **axes**, the origin is its **center**, and the lines $y = \pm(b/a)x$ are its **asymptotes**.

By similar reasoning, we find that the graph of

$$\frac{y^2}{b^2} - \frac{x^2}{a^2} = 1$$

is a hyperbola whose branches open upward and downward, whose vertices are $(0, \pm b)$, and whose asymptotes are again the lines $y = \pm(b/a)x$.

Hyperbolas   The graphs of

$$(1) \quad \frac{x^2}{a^2} - \frac{y^2}{b^2} = 1 \qquad (2) \quad \frac{y^2}{b^2} - \frac{x^2}{a^2} = 1$$

are hyperbolas with asymptotes $y = \pm \frac{b}{a} x$.

Hyperbola (1) has vertices $(\pm a, 0)$ and opens horizontally.
Hyperbola (2) has vertices $(0, \pm b)$ and opens vertically.

EXAMPLE 5 Graph (a) $4x^2 - 9y^2 = 36$ (b) $16y^2 - 9x^2 = 144$

SOLUTION (a) Divide both sides by 36:

$$\frac{x^2}{9} - \frac{y^2}{4} = 1 \qquad \frac{x^2}{3^2} - \frac{y^2}{2^2} = 1$$

The graph is a hyperbola with vertices $(\pm 3, 0)$ and asymptotes $y = \pm \frac{2}{3}x$. See Figure 8a.

(b) Divide by 144:

$$\frac{y^2}{9} - \frac{x^2}{16} = 1 \qquad \frac{y^2}{3^2} - \frac{x^2}{4^2} = 1$$

The graph is a hyperbola with vertices $(0, \pm 3)$ and asymptotes $y = \pm \frac{3}{4}x$. See Figure 8b.

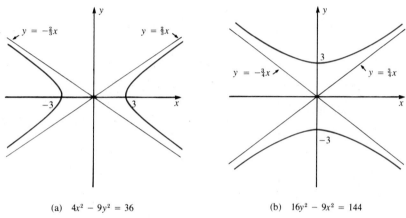

(a) $4x^2 - 9y^2 = 36$ (b) $16y^2 - 9x^2 = 144$

FIG. 8

Suggestion Instead of memorizing the formula for the asymptotes, reason this way. If x and y are very large, then x^2/a^2 and y^2/b^2 are large numbers that are nearly equal. Hence

$$\frac{y^2}{b^2} \approx \frac{x^2}{a^2} \qquad y^2 \approx \frac{b^2}{a^2} x^2 \qquad y \approx \pm \frac{b}{a} x$$

Conic Sections

Circles, parabolas, ellipses, and hyperbolas are called **conic sections**. Ancient geometers obtained them by cutting a cone with a plane. (*Section* means cutting, as in *cross-section*.) See Figure 9, next page.

These curves have many fascinating and useful properties. Perhaps the most famous is that the orbits of the planets around the sun are ellipses. (The sun is not at the center of the ellipse, however!) For further information, we refer you to books on analytic geometry and calculus.

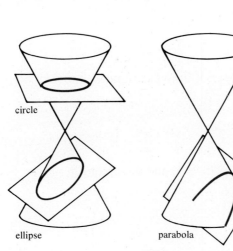
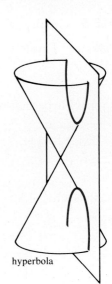

FIG. 9 Sections of a cone

EXERCISES

Find the center and radius of the circle

1. $x^2 + y^2 - 4x = 0$
2. $x^2 + y^2 + 2y = 3$
3. $x^2 + y^2 + 6x + 2y + 6 = 0$
4. $x^2 + y^2 - 10x - 4y + 28 = 0$
5. $x^2 - 4x + 1 = 2y - y^2$
6. $x^2 + y^2 = x + y + 2$
7. $2x^2 + 2y^2 = x - y$
8. $3x^2 + 3y^2 - 4x - 9y = 1$

Show that no points satisfy the equation even though it looks like the equation of a circle. (Complete the square.)

9. $x^2 + y^2 - 4x - 4y + 9 = 0$
10. $x^2 + y^2 = 10x - 27$

Find an equation for the circle

11. center $(0, 0)$, passing through $(4, -3)$
12. center $(0, 2)$, passing through $(-5, 14)$
13. tangent to both axes, radius 5, center in the first quadrant
14. opposite ends of a diameter are $(0, 0)$, $(-8, 6)$

Graph

15. $x = 2y^2$
16. $x = -3y^2$
17. $x = 1 - 2y^2$
18. $x = 3y^2 - 3$
19. $x = 2(y + 1)^2$
20. $x - 4 = (y - 1)^2$
21. $x^2 + 4x - y = 0$
22. $y^2 - 4x + y = 0$
23. $x = -2y^2 + 4y - 2$
24. $x = y^2 - y + 1$

25. $\dfrac{x^2}{25} + \dfrac{y^2}{9} = 1$ 26. $\dfrac{x^2}{4} + \dfrac{y^2}{16} = 1$

27. $x^2 + 9y^2 = 9$ 28. $16x^2 + y^2 = 4$

29. $x^2 + 4y^2 = 1$ 30. $9x^2 + y^2 = 1$

31. $4x^2 + 25y^2 = 100$ 32. $49x^2 + 4y^2 = 196$

33. $\dfrac{x^2}{4} - \dfrac{y^2}{9} = 1$ 34. $\dfrac{x^2}{9} - \dfrac{y^2}{4} = 1$

35. $\dfrac{y^2}{4} - \dfrac{x^2}{9} = 1$ 36. $\dfrac{y^2}{9} - \dfrac{x^2}{4} = 1$

37. $y^2 - 4x^2 = 4$ 38. $4x^2 - y^2 = 4$

39. $x^2 - y^2 = 2$ 40. $y^2 - x^2 = 3$

Find an equation for the curve

41. parabola, horizontal, vertex (2, 0), and passing through (0, 6)

42. parabola, axis the line $x = 1$, and passing through (0, 0) and (1, 4)

43. ellipse, vertices $(\pm 3, 0)$, $(0, \pm 5)$

44. hyperbola, vertices $(\pm 4, 0)$, asymptotes $y = \pm \frac{1}{2}x$

Verify the statement

45. Each point on the parabola $4x = y^2$ is equidistant from the point (1, 0) and the line $x = -1$.

46. The graph of $Ax^2 + By^2 = C$, where $A, B, C > 0$, is an ellipse.

47. The circles $x^2 + y^2 = 9$ and $x^2 + y^2 - 16x - 2y + 59 = 0$ do not intersect.

48. The circle $x^2 + y^2 = 9$ lies inside the circle $x^2 + y^2 - 2x + 2y - 23 = 0$.

REVIEW EXERCISES FOR CHAPTER 6

Find a linear function $f(x) = ax + b$ such that

1. $f(1) = 4$ and $f(9) = 16$ 2. $f(0) = 5{,}000{,}000$ and $f(10) = 1{,}200{,}000$

3. It cost $400 to clean up a football stadium after a game attended by 20,000, and $800 after a game attended by 50,000. Assuming a linear relation, express clean-up cost as a function of attendance.

4. A property has depreciated in 8 years from $40,000 to $35,000. Projecting linearly, what will be its value after another 12 years?

Graph

5. $y = 1 - \frac{1}{4}(x - 5)^2$ 6. $y = \frac{1}{2}(x + 1)^3$

7. $y = x^2 + 6x + 1$ 8. $y = x^3(x - 1)$

9. $y = x^3 - 4x^2 + 4x$

10. $x^2 + 2x + y = 1$

11. $x^2 + y^2 + 16y = 0$

12. $\dfrac{x^2}{16} - \dfrac{y^2}{25} = 1$

13. $\dfrac{x^2}{4} + y^2 = 1$

14. $y = \dfrac{1}{x - 2}$

15. $y = \dfrac{1}{(x + 1)(x + 2)}$

16. $y = \dfrac{x}{x^2 - 9}$

17. Find an equation for the circle tangent to the y-axis, with center (2, 5).

18. A farmer will make a rectangular pen inside his barn with 40 feet of chicken wire, using a corner of the barn for two sides of the pen. Find the maximum area he can enclose.

19. The average amount of popcorn sold at a movie theater is proportional to the attendance. On Thursday, 300 people came to the theater and 36 pounds of popcorn were sold. How much popcorn should be sold on weekend nights if 1000 people come?

20. Suppose that w varies directly with x and y, and inversely with z. If $x = 3$, $y = 5$, and $z = 10$, then $w = 16$. Find w when $x = 2$, $y = 6$, and $z = 4$.

7

Exponential and Logarithm Functions

1 EXPONENTIAL FUNCTIONS

Introduction

In this chapter, we study exponential functions and their inverses, logarithm functions. These functions are of great importance for theoretical reasons, and also in practice because they arise in so many applications. For example, here are a few problems that will be solved in this chapter:

(1) What are the monthly payments on a 36-month auto loan of $2500 at 15% annual interest compounded monthly?

(2) The great San Francisco earthquake measured 8.3 on the Richter scale. How many times stronger was it than a quake measuring 5.5, the level at which serious damage begins?

(3) I deposit $6000 in a savings account at 8% annual interest compounded quarterly. What is the value of the account after 7 years?

(4) Molten lava at 900°C cools in air at 25°C. How long until the lava cools to 40°?

(5) (Brain teaser) What is the most valuable object in the world *by weight*? (It is not made of gold, which is valuable but very heavy.) For the answer, see Example 6, page 202.

Solutions to these and other problems are expressed in terms of exponentials and logarithms. Until recently, computations that convert these solutions to numerical answers required use of logarithm tables. However, that method is now giving way to use of scientific calculators which make the computations easier, faster, and more accurate.

Let us emphasize that calculators replace only the *computational* features of logarithms. Nothing can replace logarithm *functions*; they are here to stay.

In Sections 3–5, we will set up solutions, and then compute numerical answers using a calculator. These computations can also be done using log tables. For readers who prefer that method, there is a discussion of computation by logarithms in the Appendix.

Exponential Functions

In Chapter 2, we defined b^r, where $b > 0$ and r is an integer or a rational number. Now we introduce exponential functions $f(x) = b^x$, where x takes *all real values*.

A strict definition of b^x is technical. For example, if $x = \sqrt{2} = 1.4142 \cdots$, then $b^{\sqrt{2}}$ can be defined through a sequence of closer and closer approximations:

$$b,\ b^{1.4},\ b^{1.41},\ b^{1.414},\ b^{1.4142},\ \ldots$$

We will simply assert that a function b^x exists, that all rules of exponents hold, and that the function has a smooth graph.

Let us graph $f(x) = 2^x$. We tabulate some values:

x	0	1	2	3	4	5	6	7	8	9	10
2^x	1	2	4	8	16	32	64	128	256	512	1024

Notice how fast the values increase. The graph zooms up sharply to the right. Now some negative values of x:

x	-10	-9	-8	-7	-6	-5	-4	-3	-2	-1	0
2^x	0.00	0.00	0.00	0.01	0.02	0.03	0.06	0.12	0.25	0.50	1.00

Notice how fast the values decrease toward 0 as x moves to the left. The graph approaches the x-axis as a horizontal asymptote (Fig. 1).

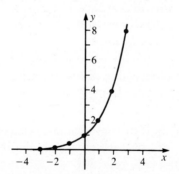

FIG. 1 $y = 2^x$

Because $y = 2^x$ increases so rapidly, we can plot only a small part of the graph using the same scales on both axes. (If one unit is $\frac{1}{8}$ inch, then at $x = 10$, the graph would be about 14 feet off the top of the page. At $x = 20$, it would be about 2.76 *miles* up, at $x = 30$ about 2824 miles up, and at $x = 45$ about the distance of the sun.)

For any number $b > 1$, the graph of $y = b^x$ is similar. The larger b is, the faster the graph rises to the right, and the faster it approaches the x-axis to the left (Fig. 2).

FIG. 2

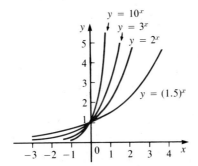

The graph of $y = b^{-x}$ is the reflection of the graph of $y = b^x$ in the y-axis (Fig. 3). It *decreases* as $x \to \infty$ and *increases* as $x \to -\infty$. For the graph of $y = (\frac{1}{2})^x$ we simply plot $y = 2^{-x}$.

FIG. 3 $y = b^{-x}$, $b > 1$

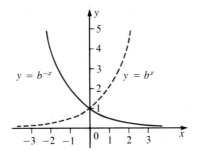

Exponential Function For each $b > 0$, there is a function $f(x) = b^x$, called the **exponential function to the base** b, with the following properties:

(1) b^x is defined for all x, and $b^x > 0$

(2) $b^{x_1} b^{x_2} = b^{x_1 + x_2}$ $\dfrac{b^{x_1}}{b^{x_2}} = b^{x_1 - x_2}$ $(b^{x_1})^{x_2} = b^{x_1 x_2}$

(3) If $b > 1$, then b^x is an increasing function taking each positive value once. Also
$$b^x \to \infty \quad \text{as} \quad x \to \infty \qquad b^x \to 0 \quad \text{as} \quad x \to -\infty$$

EXAMPLE 1 Without a calculator show that
  (a) $9 < 3^{2.47} < 27$ (b) $2 < 16^{0.29} < 4$

SOLUTION (a) Write the inequality as
$$3^2 < 3^{2.47} < 3^3$$
Since $f(x) = 3^x$ is an increasing function, the inequality is correct.

(b) Note that $2 = 16^{1/4}$ and $4 = 16^{1/2}$. Therefore, the assertion is that
$$16^{0.25} < 16^{0.29} < 16^{0.50}$$
Since $f(x) = 16^x$ is an increasing function, this is correct.

EXAMPLE 2 A bacteria colony has 10,000 members at time $t = 0$ and doubles every 3 hours. Verify that its growth is described by the exponential formula $N(t) = 10,000 \cdot 2^{t/3}$. Sketch the graph of $N(t)$.

SOLUTION There are two things to check: (a) that $N(0) = 10,000$ and (b) that $N(t)$ doubles every 3 hours, $N(t + 3) = 2N(t)$.

(a) $N(0) = 1000 \cdot 2^0 = 10,000 \cdot 1 = 10,000$

(b) $N(t + 3) = 10,000 \cdot 2^{(t+3)/3}$
$= 10,000 \cdot 2^{(t/3)+1} = 10,000 \cdot 2^{t/3} \cdot 2^1$
$= 2(10,000 \cdot 2^{t/3}) = 2N(t)$

For the graph (Fig. 4), use a scale of 10,000 on the vertical axis, and the data

t	0	3	6	9
$2^{t/3}$	1	2	4	8

(With a calculator you can easily compute more points for greater accuracy.)

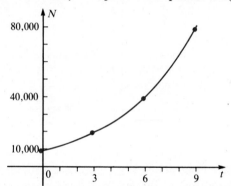

FIG. 4

EXAMPLE 3 A 3-gm sample of a radioactive substance decays, losing half its mass each day. Verify that the mass after t days is given by $M(t) = 3 \cdot 2^{-t}$. Sketch the graph of $M(t)$.

SOLUTION We must check that (a) $M(0) = 3$, and (b) that $M(t + 1) = \frac{1}{2}M(t)$.

(a) $M(0) = 3 \cdot 2^{-0} = 3 \cdot 1 = 3$

(b) $M(t + 1) = 3 \cdot 2^{-(t+1)} = 3 \cdot 2^{-t} \cdot 2^{-1}$
$= \frac{1}{2}(3 \cdot 2^{-t}) = \frac{1}{2}M(t)$

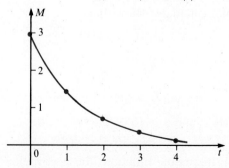

FIG. 5

For the graph (Fig. 5), we plot a few points:

t	0	1	2	3	4
$3 \cdot 2^{-t}$	3	$\frac{3}{2}$	$\frac{3}{4}$	$\frac{3}{8}$	$\frac{3}{16}$

EXAMPLE 4 A cold object at temperature 0°C is plunged into warm water at 40°C. After t minutes, its temperature is $T(t) = 40(1 - 2^{-t})$. Graph $T(t)$.

SOLUTION Compute some values of $1 - 2^{-t}$ and multiply by 40:

t	0	1	2	3	4	5
$1 - 2^{-t}$	0	$\frac{1}{2}$	$\frac{3}{4}$	$\frac{7}{8}$	$\frac{15}{16}$	$\frac{31}{32}$
$40(1 - 2^{-t})$	0	20	30	35	37.5	38.75

As t increases, the temperature approaches 40°. Hence the line $T = 40$ is a horizontal asymptote of the graph (Fig. 6)

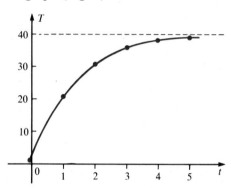

FIG. 6

Computing Exponential Functions

To compute (approximate) numerical values such as $2^{4.37}$ or $74^{-13.5}$ or $6^{5/12}$ use the $\boxed{y^x}$ key on your calculator.

$2^{4.37}$ $2\ \boxed{y^x}\ 4\ .\ 3\ 7\ \boxed{=}$ 20.677645

$74^{-13.5}$ $7\ 4\ \boxed{y^x}\ 1\ 3\ .\ 5\ \boxed{+/-}\ \boxed{=}$ 5.826017×10^{-26}

$6^{5/12}$ $6\ \boxed{y^x}\ \boxed{(}\ 5\ \boxed{\div}\ 1\ 2\ \boxed{)}\ \boxed{=}$ 2.109744

These can also be computed using logarithms (see the Appendix).

Rate of Growth

We have seen that 2^x increases very fast, and 3^x, 5^x, 10^x faster yet. But even a weak-looking exponential such as $(1.01)^x$ eventually increases very fast. It starts out slowly:

x	1	2	3	5	10	20	30	50
$(1.01)^x$	1.01	1.02	1.03	1.05	1.10	1.22	1.35	1.64

Apparently $(1.01)^x$ is much smaller than x. But hold on:

x	100	200	300	400	500	600	700
$(1.01)^x$	2.7	7.3	19.8	53.5	144.8	391.6	1059

Thus, $(1.01)^x$ passes x between 600 and 700. From then on, it's no contest:

x	800	1000	2000	5000	10000
$(1.01)^x$	2865	20,959	439,286,205	4.0×10^{21}	1.6×10^{43}

After a slow start, $(1.01)^x$ overwhelms x. Each time x increases by 1 unit, $(1.01)^x$ increases by 1%. At first that doesn't amount to much, but as $(1.01)^x$ grows, a 1% increase becomes more important. When $x = 700$, it amounts to an increase of about 10.6; when $x = 2000$, it amounts to an increase of about 4,392,862. Thus, starting with $1, it is much better to receive 1% interest a day than to receive $1 a day with no interest.

Similarly, any exponential function b^x overwhelms any linear function $cx + d$, provided $b > 1$. In fact, b^x overwhelms any polynomial function of any degree. This mathematical fact led the British economist Thomas Malthus to a gloomy conclusion. In 1798, he predicted great human suffering because the world population was growing exponentially, while the food supply was increasing only linearly. Fortunately, we have survived for a number of reasons, especially improved agricultural methods.

EXERCISES

Graph for $-3 \leq x \leq 3$

1. $y = 3^x$
2. $y = 3^{-x}$
3. $y = 4^{-x}$
4. $y = 4^x$
5. $y = (\frac{3}{2})^x$
6. $y = (\frac{3}{4})^x$
7. $y = \frac{1}{10} \cdot 5^x$
8. $y = \frac{1}{5} \cdot 2^x$
9. $y = 2^{x-1}$
10. $y = 3^{x-2}$
11. $y = 2^{|x|}$
12. $y = 3^{-|x|}$
13. $y = 3^{-x/2}$
14. $y = 2^{x/2}$
15. $y = 1 - 2^{-x}$
16. $y = 2^x + 1$

Verify the estimate (without a calculator)

17. $32 < 2^{5.41} < 64$
18. $27 < 3^{3.62} < 81$
19. $27 < 9^{1.66} < 81$
20. $64 < 4^{3.47} < 128$
21. $0.04 < 5^{-1.3} < 0.20$
22. $0.125 < 2^{-2.34} < 0.250$

23. Radioactive carbon-14 decays to half its mass in 5568 years. Verify that the mass of a 10-gm sample after t years is $M(t) = 10 \cdot 2^{-t/5568}$.

24. A bacteria colony starts with 20,000 members and doubles every 8 hours. Show that its population after t hours is $N(t) = 20000 \cdot 2^{t/8}$.

25. (See Exercise 23) How long until the carbon-14 sample decays to 2.5 gm?

26. (See Exercise 24) How long until the population reaches 320,000?

27. A hot pie is removed from a 350°F oven and set in 70° air to cool. After t minutes, its temperature is $T(t) = 70 + 280 \cdot 2^{-t/5}$. Graph $T(t)$.

28. A piece of hot steel at 425°C is plunged into an oil bath at 25°C. After t minutes, its temperature is $T(t) = 25 + 400 \cdot 2^{-t}$. Graph $T(t)$.

29. Show graphically that $2^x > x$ for all real x.

30. Show graphically that the graphs of $y = 10^{-x}$ and $y = -x - 5$ intersect at exactly one point, very close to $(5, 0)$.

31. Verify that $2^{10} = 1024$. Use this to make the rough estimate (without a calculator) that $2^{60} \approx 10^{18}$.

32. Verify that $\pi^2 \approx 10$. Use this to make a rough estimate (without a calculator) of π^{100}.

The remaining exercises require a calculator.

33. Make a table comparing the values of 2^x, 3^x, and 5^x for $x = 5, 10, 20, 30, 50, 100$.

34. Make a table comparing the values of x^2, x^3 and $(1.01)^x$ for $x = 100, 1000, 2000, 5000, 10000$.

35. Approximately how many miles is 2^{45} inches?

36. If the number 5^{68} is written out in decimal form, how many digits does it have?

By experimenting on your calculator, find the smallest *integer* x beyond which

37. $(1.01)^x > x$ **38.** $(1.001)^x > x$

39. $(1.0001)^x > 5$ **40.** $(1.01)^x > 100x$

2 LOGARITHMS AND LOGARITHM FUNCTIONS

If $b > 1$, the exponential function b^x is increasing and takes each positive value once. Thus to each $s > 0$ corresponds a real unique number r such that $b^r = s$.

Logarithm to the Base b For each positive real number s there is a unique real number r such that $b^r = s$. The number r is called the **logarithm of s to the base b**, and is denoted $\log_b s$. The statements

$$b^r = s \quad \text{and} \quad \log_b s = r$$

are equivalent.

Examples

Exponential Statement	Equivalent Logarithmic Statement
$3^1 = 3$	$\log_3 3 = 1$
$4^3 = 64$	$\log_4 64 = 3$
$5^6 = 15{,}625$	$\log_5 15{,}625 = 6$
$7^{1/2} = \sqrt{7}$	$\log_7 \sqrt{7} = \tfrac{1}{2}$
$2^{-5} = \tfrac{1}{32}$	$\log_2 \tfrac{1}{32} = -5$
$10^{-4} = 0.0001$	$\log_{10} 0.0001 = -4$

Finding $\log_b s$ means answering the question $b^? = s$.

Exponential and Logarithm Functions

EXAMPLE 1 Find

(a) $\log_2 16$ (b) $\log_3 \frac{1}{9}$ (c) $\log_4 8$ (d) $\log_5 1$ (e) $\log_6 (-2)$

SOLUTION (a) $\log_2 16$ is the number r such that $2^r = 16$. But $2^4 = 16$, so $r = \underline{4}$.

(b) $\log_3 \frac{1}{9}$ is the number r such that $3^r = \frac{1}{9}$. But $3^{-2} = \frac{1}{9}$, so $r = \underline{-2}$.

(c) $\log_4 8$ is the number r such that $4^r = 8$. But $4^{3/2} = 8$, so $r = \underline{\frac{3}{2}}$.

(d) $\log_5 1$ is the number r such that $5^r = 1$. But $5^0 = 1$, so $r = \underline{0}$.

(e) $b^r > 0$ for every real number r. Hence $b^r = -2$ is impossible. No solution.

EXAMPLE 2 Find x if $\log_2 (3x + 1) = 6$.

SOLUTION The equation is equivalent to
$$3x + 1 = 2^6 = 64$$
Solving for x gives $x = \underline{21}$.

Logarithms inherit some important algebraic properties from exponentials.

Rules for Logarithms

(1) $\log_b rs = \log_b r + \log_b s$

(2) $\log_b \dfrac{r}{s} = \log_b r - \log_b s$

(3) $\log_b r^s = s \log_b r$

These rules are practically restatements of properties of b^x. Take Rule (1), for instance. Suppose that $\log_b r = x$ and $\log_b s = y$. Then $r = b^x$ and $s = b^y$, so
$$rs = b^x \cdot b^y = b^{x+y}$$
which means that
$$\log_b rs = x + y = \log_b r + \log_b s$$
Verifications of Rules (2) and (3) are similar; we leave them as exercises.

EXAMPLE 3 Given $\log_7 2 \approx 0.3562$, $\log_7 3 \approx 0.5646$, $\log_7 5 \approx 0.8271$ compute (a) $\log_7 10$ (b) $\log_7 \dfrac{3}{25}$ (c) $\log_7 \sqrt[3]{16}$

SOLUTION (a) By Rule (1),
$$\log_7 10 = \log_7 (2 \cdot 5) = \log_7 2 + \log_7 5$$
$$\approx 0.3562 + 0.8271 = \underline{1.1833}$$

(b) By Rules (2) and (3),
$$\log_7 \frac{3}{25} = \log_7 3 - \log_7 25 = \log_7 3 - \log_7 5^2$$
$$= \log_7 3 - 2 \log_7 5$$
$$\approx 0.5646 - 2(0.8271) = \underline{-1.0896}$$

(c) By Rule (3),
$$\log_7 \sqrt[3]{16} = \log_7 (2^4)^{1/3} = \log_7 2^{4/3}$$
$$= \frac{4}{3} \log_7 2 \approx \frac{4}{3}(0.3562) \approx \underline{0.4749}$$

EXAMPLE 4 Express as a single logarithm
(a) $-3(\log_b 4 + \log_b 5)$
(b) $2 \log_b 7 - \log_b 12$
(c) $\log_b 2 + \frac{1}{2} \log_b 6 - 3 \log_b 5$

SOLUTION
(a) $-3(\log_b 4 + \log_b 5) = -3 \log_b (4 \cdot 5)$
$$= -3 \log_b 20 = \log_b 20^{-3} = \log_b \underline{\frac{1}{8000}}$$

(b) $2 \log_b 7 - \log_b 12 = \log_b 7^2 - \log_b 12 = \log_b \underline{\frac{49}{12}}$

(c) $\log_b 2 + \frac{1}{2} \log_b 6 - 3 \log_b 5 = \log_b 2 + \log_b 6^{1/2} - \log_b 5^3$
$$= \log_b \frac{2 \cdot 6^{1/2}}{5^3} = \log_b \underline{\frac{2\sqrt{6}}{125}}$$

Warning Rules (1), (2), and (3) apply to logs of products, quotients, and powers, *not* to sums and differences. There is no nice formula for $\log_b (r + s)$. Note also, there is no special formula for $(\log_b r) \cdot (\log_b s)$ or for $(\log_b r)/(\log_b s)$.

EXAMPLE 5 Solve for x: $\log_2 x + \log_2 (x - 6) = 4$

SOLUTION By Rule (1), the equation is equivalent to
$$\log_2 x(x - 6) = 4$$
Express it in exponential form:
$$x(x - 6) = 2^4$$
This is a quadratic equation for x:
$$x^2 - 6x = 16$$
$$x^2 - 6x - 16 = 0$$
$$(x - 8)(x + 2) = 0$$
Hence $x = 8$ and $x = -2$ are the only possible solutions. But we reject $x = -2$ because $\log_2 x$ is not defined for the negative value $x = -2$. That leaves only

$x = 8$. Now if $x = 8$, then

$$\log_2 x + \log_2 (x - 6) = \log_2 8 + \log_2 2 = 3 + 1 = 4 \quad \text{(Correct!)}$$

Hence the equation has one solution: $x = \underline{8}$.

Logarithm Functions

We saw in Chapter 5, Section 5, that an increasing function has an inverse function. For $b > 1$, the exponential function $y = b^x$ is increasing. Therefore, it has an inverse function $x = g(y)$, which we write as $x = \log_b y$.

Let us reverse x and y and study the inverse function in the form $y = \log_b x$. Its graph is the reflection of $y = b^x$ in the line $y = x$. The graph is steadily increasing, and y takes every real value once (Fig. 1).

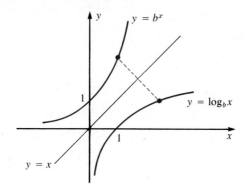

FIG. 1 $y = \log_b x$

Logarithm Function For each $b > 1$, there is a function $y = \log_b x$, called the **logarithm function to the base b**, with the following properties:

(1) $y = \log_b x$ is equivalent to $x = b^y$

(2) $\log_b x$ is defined for $x > 0$

(3) $\log_b x$ is an increasing function that takes each real value once

(4) $\log_b x \to \infty$ as $x \to \infty$ $\quad \log_b x \to -\infty$ as $x \to 0+$

The following properties are clear from Figure 1.

$$\text{Let } b > 1. \text{ Then } \begin{cases} \log_b x < 0 & \text{for } 0 < x < 1 \\ \log_b 1 = 0 & \\ \log_b x > 0 & \text{for } x > 1 \end{cases}$$

EXAMPLE 6 Graph $y = \log_2 x$

SOLUTION The graph has the general shape of the lower curve in Fig. 1. For more precision we plot a few points (Fig. 2).

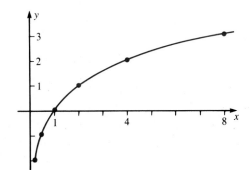

x	$\frac{1}{4}$	$\frac{1}{2}$	1	2	4	8
$\log_2 x$	-2	-1	0	1	2	3

FIG. 2 $y = \log_2 x$

EXAMPLE 7 Solve the inequality $\log_3 x > 4$

SOLUTION Note that $4 = \log_3 81$, so the inequality can be written as
$$\log_3 x > \log_3 81$$
Since $\log_3 x$ is an increasing function, $\underline{x > 81}$.

Recall that if f and g are a pair of inverse functions, then
$$g[f(x)] = x \quad \text{and} \quad f[g(y)] = y$$
Applied to $f(x) = b^x$ and $g(y) = \log_b y$, these say

$$\boxed{\log_b b^x = x \qquad b^{\log_b y} = y}$$

Examples

$$\log_5 5^{3.8} = 3.8 \qquad 4^{\log_4 7} = 7 \qquad 10^{4 \log_{10} 2} = (10^{\log 2})^4 = 2^4 = 16$$

EXERCISES

Write an equivalent logarithmic statement
1. $2^{10} = 1024$
2. $5^4 = 625$
3. $10^{-2} = 0.01$
4. $3^{-5} = \frac{1}{243}$
5. $8^{-1/3} = \frac{1}{2}$
6. $25^{-3/2} = \frac{1}{125}$

Write an equivalent exponential statement
7. $\log_{10} 10{,}000 = 4$
8. $\log_2 128 = 7$
9. $\log_4 32 = 2.5$
10. $\log_{16} 8 = 0.75$
11. $\log_3 \dfrac{1}{81\sqrt{3}} = -4.5$
12. $\log_2 \dfrac{1}{4096} = -12$

Without a calculator find

13. $\log_2 256$
14. $\log_3 243$
15. $\log_9 9$
16. $\log_8 1$
17. $\log_{10} \frac{1}{1000}$
18. $\log_2 \frac{1}{64}$
19. $\log_3 9\sqrt[4]{3}$
20. $\log_4 128$
21. $\log_8 \frac{1}{\sqrt[3]{2}}$
22. $\log_{100} \frac{1}{\sqrt[4]{10}}$
23. $\log_5 5^{12}$
24. $\log_6 6^{-29}$
25. $\log_8 64^{15}$
26. $\log_2 8^{17/3}$
27. $10^{\log_{10} 9}$
28. $6^{\log_6 31}$
29. $2^{3 \log_2 5}$
30. $2^{-\log_2 5}$
31. $(\log_3 \sqrt{3})^2$
32. $\frac{\log_3 27}{\log_3 9}$

Given $\log_{11} 2 \approx 0.2891$, $\log_{11} 3 \approx 0.4582$, and $\log_{11} 5 \approx 0.6712$, compute

33. $\log_{11} 33$
34. $\log_{11} \frac{121}{5}$
35. $\log_{11} 45$
36. $\log_{11} \frac{9}{16}$
37. $\log_{11} \left(\frac{1}{27}\sqrt{5}\right)$
38. $\log_{11} \frac{1}{\sqrt[3]{80}}$

Express as a single logarithm

39. $\log_b 6 + 3 \log_b 2$
40. $3 \log_b 5 - \log_b 7$
41. $\log_b 3 + \frac{1}{2} \log_b 5 - \frac{1}{2} \log_b 7$
42. $\frac{2}{3} \log_b 8 + \frac{1}{2} \log_b 9$
43. $\log_b rs - \log_b st + \log_b tu - \log_b uv$
44. $\log_b r + 2 \log_b r^2 + 3 \log_b r^3$

Solve for x

45. $\log_{12} x = 0$
46. $\log_5 x = 2$
47. $\log_8 x = -\frac{2}{3}$
48. $\log_{36} x = -\frac{1}{4}$
49. $\log_3 (x + 5) = 4$
50. $\log_3 (x + 7) = -1$
51. $\log_4 (2x + 3) = -\frac{1}{2}$
52. $\log_{10} (7x - 8) = 3$
53. $\log_9 (x^2 - 10) = 1$
54. $\log_5 (x^2 + 10) = 1$
55. $\log_x 8 = 3$
56. $\log_x 49 = 2$

Solve the inequality

57. $\log_5 x > 3$
58. $\log_2 x < 4$
59. $-2 \leq \log_{10} x \leq -1$
60. $3 \leq \log_4 x \leq 3.5$

Solve for x

61. $\log_{10} (4x - 3) = \log_{10} (2x + 9)$
62. $\log_3 x^2 = \log_3 5x$
63. $\log_6 x + \log_6 (x - 1) = 1$
64. $\log_6 x - \log_6 (x - 1) = 1$
65. $\log_3 (2x + 1) - \log_3 (2x - 1) = 3$
66. $\log_3 x + \log_3 (x - 10) = 2$
67. $\log_2 \sqrt{x} + \log_2 \frac{1}{x} = 3$
68. $2 \log_2 x - \log_2 (3x + 8) = 1$

Graph

69. $y = \log_3 x$
70. $y = \log_4 x$

71. $y = \log_{10} x$
72. $y = \log_2 |x|$
73. $y = \log_2 x^2$
74. $y = \log_2 \sqrt{x}$

The graph of $y = \log_{10} x$ rises very slowly.

75. Where does it reach the level $y = 1$? 2? 10?
76. How much does it rise as x increases from 1,000,000 to 10,000,000?
77. Show graphically that $x > \log_{10} x$ for all $x > 0$.
78. Show that $\log_b (r + s) \neq \log_b r + \log_b s$ by taking $s = 1$.
79. Justify Rule (2) for logarithms.
80. Justify Rule (3) for logarithms.

3 COMMON LOGARITHMS, EXPONENTIAL EQUATIONS

For numerical computations, the most useful base of logarithms is 10. The logarithm of x to the base 10 is called the **common logarithm** of x and is written log x. When you see log x without a base, then base 10 is understood. For example,

$$\log 1000 = 3 \qquad \log 0.01 = -2$$

Most scientific calculators have a key $\boxed{\log}$. To find the common logarithm of a number, just enter your number, then press $\boxed{\log}$. For example,

$$2\ 5\ .\ 4\ \boxed{\log} \quad \text{gives} \quad \log 25.4 \approx 1.404834$$
$$.\ 1\ 8\ 5\ \boxed{\log} \quad \text{gives} \quad \log 0.185 \approx -0.732828$$

Remember that the function log x is the inverse of the exponential function 10^x. Therefore, on a calculator with a key $\boxed{10^x}$, you can find common logs from the sequence $\boxed{\text{INV}}\ \boxed{10^x}$. For example,

$$8\ 3\ 1\ 9\ \boxed{\text{INV}}\ \boxed{10^x} \quad \text{gives} \quad \log 8319 \approx 3.920071$$

Common logs can also be found from tables. Readers who prefer that method may want to consult the Appendix before proceeding with the rest of this chapter.

Exponential Equations

An **exponential equation** is an equation in which the unknown occurs in an exponent. The usual strategy for solving such an equation is to take logarithms.

EXAMPLE 1 Solve (a) $7^x = 55$ (b) $2^{-x} = 0.0094$

SOLUTION (a) Take common logs on both sides:

$$x \log 7 = \log 55$$
$$x = \frac{\log 55}{\log 7}$$

For some purposes, this form of the answer may be good enough. If you need a numerical answer, use the calculator sequence:

$$5\ 5\ \boxed{\log}\ \boxed{\div}\ 7\ \boxed{\log}\ \boxed{=} \quad \underline{2.059}$$

(b) Take common logs on both sides:
$$2^{-x} = 0.0094$$
$$-x \log 2 = \log 0.0094$$
$$x = -\frac{\log 0.0094}{\log 2} \approx \underline{6.734}$$

EXAMPLE 2 Solve $3^{4x-5} = 11^x$

SOLUTION Take logs on both sides:
$$(4x - 5)\log 3 = x \log 11$$
Both log 3 and log 11 are constants, so this is a linear equation in x. Solve:
$$(4 \log 3 - \log 11)x = 5 \log 3$$
$$x = \frac{5 \log 3}{4 \log 3 - \log 11}$$
To evaluate x, one possible calculator sequence is
$$\boxed{3}\;\boxed{\log}\;\boxed{\times}\;\boxed{5}\;\boxed{\div}\;\boxed{(}\;\boxed{3}\;\boxed{\log}\;\boxed{\times}\;\boxed{4}\;\boxed{-}\;\boxed{1}\;\boxed{1}\;\boxed{\log}\;\boxed{)}\;\boxed{=}\;\underline{2.751}$$

EXAMPLE 3 Solve $2^x - 2^{-x} = 8$

SOLUTION This is really a disguised quadratic equation. For if $2^x = y$, then $2^{-x} = 1/y$, and the equation is
$$y - \frac{1}{y} = 8$$
$$y^2 - 8y - 1 = 0$$
By the quadratic formula,
$$2^x = y = \frac{8 \pm \sqrt{68}}{2} = 4 \pm \sqrt{17}$$
But, $4 - \sqrt{17} < 0$ and 2^x cannot be negative. That leaves only $2^x = 4 + \sqrt{17}$. Taking logs,
$$x \log 2 = \log(4 + \sqrt{17})$$
$$x = \frac{\log(4 + \sqrt{17})}{\log 2} \approx \underline{3.022}$$

Compound Interest

Exponents occur naturally in questions of compound interest. For example, suppose that $1000 is deposited in a savings account paying 6% annual interest. At the end of a year, the interest is $1000(0.06). This is added to the account, bringing it to
$$1000 + 1000(0.06) = 1000(1.06)$$
The next year starts with this amount. At the end of the year, the balance is again

multiplied by 1.06, so the account grows to
$$[1000(1.06)](1.06) = 1000(1.06)^2$$
Each year the current balance is multiplied by 1.06, so the powers of 1.06 build up.

Interest can be computed in various ways. For example, 6% annual interest compounded semi-annually means that 3% interest is added every 6 months; 6% compounded quarterly means that 1.5% interest is added every quarter (3 months), etc.

In general, suppose an account is opened with amount P (principal) and draws interest i per period. After each period the balance is multiplied by $(1 + i)$. Hence, in n periods the principal grows by a factor of $(1 + i)^n$.

Compound Interest Suppose that a sum P is invested at an interest rate i per period. Then its value A after n periods is
$$A = P(1 + i)^n$$

EXAMPLE 4 Suppose that $3000 is deposited into an account paying 8% annual interest, compounded quarterly. Find (a) the value after 5 years, and (b) when the account first exceeds $5000.

SOLUTION (a) With 4 payments per year, $i = \frac{1}{4}(0.08) = 0.02$ and $n = 5 \cdot 4 = 20$. Hence
$$A = 3000(1.02)^{20} \approx \$4457.84$$

(b) We want the smallest positive *integer* n such that
$$3000(1.02)^n > 5000$$
$$(1.02)^n > \tfrac{5}{3} \approx 1.67$$
Taking logs, we have
$$n \log 1.02 > \log 1.67$$
$$n > \frac{\log 1.67}{\log 1.02} \approx 25.9$$
Hence, $n = 26$. The account first exceeds $5000 after 26 quarterly periods, that is, $6\tfrac{1}{2}$ years.

EXAMPLE 5 At what annual interest rate will money double in 10 years if compounded yearly?

SOLUTION The desired rate i satisfies
$$P(1 + i)^{10} = 2P$$
Therefore
$$(1 + i)^{10} = 2$$
$$1 + i = 2^{1/10}$$
$$i = 2^{1/10} - 1 \approx 0.0718$$
Thus, the interest rate is 7.18%.

EXAMPLE 6 The world's most valuable postage stamp is a magenta-colored, 1856 British Guiana one-penny stamp. As a hedge against inflation, a group of investors bought it in 1970 for $280,000. They kept it in a vault until 1980, when they sold it for $850,000. Find the average return on their investment, that is, an interest rate that compounded yearly would have produced the same 10-year growth.

SOLUTION The equivalent interest rate i satisfies

$$280{,}000(1 + i)^{10} = 850{,}000$$

$$(1 + i)^{10} = \frac{85}{28}$$

$$i = \left(\frac{85}{28}\right)^{1/10} - 1 \approx 0.1174$$

Their average return was <u>11.74%</u>. (They did well because the inflation rate averaged about 7% yearly for the decade.)

Significant Figures

When a number is written with a decimal point, the number of **significant figures** is the number of digits from the left-most non-zero digit to the right-most digit. For example,

$$0.1174 \qquad 0.002983 \qquad 5.169 \qquad 12.85 \qquad 300.7$$

all have 4 significant figures.

If a number is expressed in scientific notation as $c \times 10^n$, where $1 \leq c < n$, the number of significant figures is the number of significant figures in c. For example,

$$4.15 \times 10^7 = 41{,}500{,}000$$

has 3 significant figures.

In the exercises, you will be asked to estimate certain answers to 3 significant figures. That means, for instance, that you should estimate

$$\begin{array}{lll} 41{,}538{,}667 & \text{by} & 41{,}500{,}000 \\ 36.8217 & \text{by} & 36.8 \\ 0.0618952 & \text{by} & 0.0619, \quad \text{etc.} \end{array}$$

EXERCISES

Estimate x to 3 significant figures

1. $2^x = 10$
2. $3^x = 77$
3. $3^{-x} = 0.002$
4. $5^{-x} = 0.043$
5. $(3.5)^x = 4$
6. $(9.1)^x = 5.8$
7. $(0.99)^x = 0.41$
8. $(0.34)^x = 0.000072$
9. $(1.05)^x = 2.4$
10. $(1.12)^x = 3.15$
11. $6^{0.47x} = 2070$
12. $2^{0.08x} = 7.51$
13. $2^{3x-1} = 497$
14. $5^{2x+3} = 804$

15. $10^x = 174 \cdot 2^x$
16. $7^{2x} = 5 \cdot 4^x$
17. $3^{4x-1} = 5^{2x}$
18. $8^{2x-1} = 11^x$
19. $8^{x+3} = 9^{-x}$
20. $10^{-x} = (\frac{1}{2})^{4x+5}$
21. $12^{3x} = 10^{x^2}$
22. $5^{-x^2} = 8^{-3x}$
23. $8^x = 9^x$
24. $7^x = 8^{-x}$
25. $3^x - 3^{-x} = 2$
26. $5^x - 5^{-x} = 1$
27. $2^x + 2^{-x} = 1$
28. $10^x + 10^{-x} = 2$
29. $7^x - 6 \cdot 7^{-x} = 1$
30. $3^x - 5 \cdot 3^{-x} = 4$
31. $9^x - 3^x = 2$
32. $4^x - 2^x = 6$

33. To what amount does $1000 grow in 5 years at 10% annual interest compounded (a) semi-annually, and (b) quarterly?

34. To what amount does $1000 grow in 3 years at 12% annual interest compounded (a) quarterly, and (b) monthly?

35. Ben Franklin said that a penny saved is a penny earned. Suppose Ben had invested a penny on his 21st birthday in 1727 in an account that compounded interest semi-annually. Compute how much that penny would have earned by 1982 at a yearly interest rate of (a) 6% (b) 8% (c) 10%

36. At 6% compounded monthly, how much should be deposited into a trust fund to be worth $25,000 in 21 years?

37. According to a rule of thumb, money at p% interest compounded yearly doubles in about $72/p$ years. Check this for $p = 8, 9,$ and 12.

38. What annual interest rate compounded monthly is equivalent to 18% compounded yearly?

39. At 6% compounded monthly, how long will it take $10,000 to grow to $15,000?

40. At 12% compounded monthly, how long will it take money to triple?

41. (See Example 6) Suppose the investors had bought the stamp in 1972 for $310,000. What average yearly return would they have had on their 8-year investment?

42. I invest $1000 in an account at 5% annual interest compounded semi-annually. After 4 years, the bank raises the interest rate to 6%. Find the value of the account at the end of 10 years.

What is the effect of the key sequence on a given number?

43. $\boxed{10^x}\ \boxed{\log}\ \boxed{\log}\ \boxed{10^x}$

44. $\boxed{\sqrt{\ }}\ \boxed{\log}\ \boxed{\times}\ 2\ \boxed{=}\ \boxed{10^x}$

45. Find the largest number x on your calculator so that $x\ \boxed{10^x}\ \boxed{10^x}$ does not result in **Error**.

46. Find the smallest number x on your calculator so that $x\ \boxed{\log}\ \boxed{\log}\ \boxed{\log}$ does not result in **Error**.

4 LOGARITHMS TO OTHER BASES

Natural Logarithms

Next to the base 10, the most important base for logarithms is an irrational number called e, which arises in many applications. We leave the precise definition of e to calculus. To 10 decimal places,

$$e \approx 2.71828\ 18285$$

The function $\log_e x$ is called the **natural logarithm function**, and is written $\ln x$. It is the inverse of the exponential function $y = e^x$. Thus,

$$\ln e^x = x \quad \text{and} \quad e^{\ln x} = x$$

EXAMPLE 1 Solve $y = ae^{kt}$ for t.

SOLUTION Divide by a, then take natural logs on both sides:

$$\frac{y}{a} = e^{kt}$$

$$\ln \frac{y}{a} = kt$$

$$\underline{t = \frac{1}{k} \ln \frac{y}{a}}$$

Most scientific calculators have a key $\boxed{e^x}$ or a key $\boxed{\ln}$ or both:

$$2\,.\,9\,8\,4\ \boxed{e^x} \quad \text{gives} \quad e^{2.984} \approx 19.766725$$
$$2\,.\,9\,8\,4\ \boxed{\ln} \quad \text{gives} \quad \ln 2.984 \approx 1.093265$$

Since e^x and $\ln x$ are inverse functions,

$$\boxed{\text{INV}}\ \boxed{e^x} \quad \text{is equivalent to} \quad \boxed{\ln}$$
$$\boxed{\text{INV}}\ \boxed{\ln} \quad \text{is equivalent to} \quad \boxed{e^x}$$

Therefore you can calculate both e^x and $\ln x$ if you have 2 of the 3 keys: $\boxed{\text{INV}}$, $\boxed{e^x}$, $\boxed{\ln}$.

EXAMPLE 2 Molten lava at 900°C cools in air at 25°C. After h hours, its temperature is $T = 25 + 875e^{-0.08h}$. Find the time in which the lava cools to 40°C.

SOLUTION Solve for h in the equation

$$25 + 875e^{-0.08h} = 40$$
$$e^{-0.08h} = \tfrac{15}{875}$$

Take natural logs:

$$-0.08h = \ln \tfrac{15}{875}$$
$$h = -\frac{1}{0.08} \ln \frac{15}{875}$$

For a numerical answer, use the calculator sequence

$$15 \boxed{\div} 875 \boxed{=} \boxed{\ln} \boxed{\div} .08 \boxed{=} \boxed{+/-} \qquad \underline{50.8 \text{ hours}}$$

The number e arises in continuous compounding of interest. A yearly interest rate i can be compounded quarterly, monthly, daily, hourly, etc. If it is compounded n times a year, then the periodic rate is i/n. After t years, the principal P grows to

$$A = P\left(1 + \frac{i}{n}\right)^{nt}$$

As n grows larger and larger, the compounding theoretically approaches "continuous compounding." It is shown in calculus that

$$\left(1 + \frac{i}{n}\right)^n \to e^i$$

Therefore, the principal P grows to Pe^{it}.

> **Continuous Compounding** Suppose that a sum P is invested at an annual interest rate i compounded continuously. Then its value A after t years is
> $$A = Pe^{it}$$

For example, in one year a deposit P at 6% compounded continuously grows to $Pe^{0.06} \approx P \times 1.0618$. Hence 6% compounded continuously is equivalent to about 6.18% compounded annually.

EXAMPLE 3 At what annual rate of interest compounded continuously will money double in 10 years?

SOLUTION The desired rate i satisfies

$$Pe^{10i} = 2P$$
$$e^{10i} = 2$$

Take natural logs:

$$10i = \ln 2$$
$$i = \tfrac{1}{10} \ln 2 \approx 0.0693$$

Hence the desired rate is about $\underline{6.93\%}$. (Compare to Example 5, page 201, where we found that 7.18% interest is needed if the compounding is yearly.)

Other Bases

Common logarithms and natural logarithms are available on calculators and in tables, but logarithms to other bases are not. So how can we compute, say, $\log_7 55$? Well, if $\log_7 55 = r$, then

$$7^r = 55$$

We take common logs on both sides:
$$r \log 7 = \log 55$$
$$r = \frac{\log 55}{\log 7}$$

In the same way, we can express $\log_b x$ in terms of common logs, for any base b. Suppose $\log_b x = r$. Then
$$b^r = x$$
We take common logs on both sides:
$$r \log b = \log x$$
$$r = \frac{\log x}{\log b}$$
But $r = \log_b x$. Therefore,

$$\boxed{\log_b x = \left(\frac{1}{\log b}\right) \log x}$$

Think of this formula as
$$\log_b x = k \log x \qquad \text{where} \qquad k = \frac{1}{\log b}$$
It shows that logs to base b are proportional to logs to base 10. For example,
$$\log_2 x = \frac{1}{\log 2} \log x \approx \frac{1}{0.3010} \log x \approx 3.322 \log x$$
$$\ln x = \frac{1}{\log e} \log x \approx \frac{1}{0.4334} \log x \approx 2.302 \log x$$

EXAMPLE 4 Compute (a) $\log_4 13$ (b) $\log_3 0.00217$

SOLUTION (a) $\log_4 13 = \dfrac{\log 13}{\log 4} \approx \underline{1.8502}$

(b) $\log_3 0.00217 = \dfrac{\log 0.00217}{\log 3} \approx \underline{-5.5825}$

Relation Between $\log_a x$ and $\log_b x$

Given any two bases $a, b > 1$, both $\log_a x$ and $\log_b x$ are proportional to $\log x$. Hence they must be proportional to each other, for
$$\log_a x = \frac{\log x}{\log a} \qquad \log_b x = \frac{\log x}{\log b}$$
We divide $\log_a x$ by $\log_b x$:
$$\frac{\log_a x}{\log_b x} = \frac{\log b}{\log a} = \log_a b$$

Solving for $\log_b x$, we have

$$\boxed{\log_b x = \left(\frac{1}{\log_a b}\right) \log_a x}$$

Examples

$$\log_8 x = \frac{\log_2 x}{\log_2 8} = \frac{1}{3} \log_2 x \qquad \log_8 x = \frac{\log_5 x}{\log_5 8} = \frac{\log_5 x}{\frac{\log 8}{\log 5}} \approx 0.7740 \log_5 x$$

EXERCISES

Find x to three significant figures

1. $e^x = 22$
2. $e^{-x} = 0.045$
3. $e^{3x-1} = 119$
4. $e^{2x+1} = 60.7$
5. $\ln x = -1.44$
6. $\ln x = 5.63$
7. $\ln e^x = 65$
8. $e^{\ln x} = 7.28$
9. $10^x = e$
10. $e^x = 10$
11. $\log_x 10 = 3.1$
12. $\log_x 0.02 = -4.7$

Express x in terms of y

13. $y = Ae^{-4x}$
14. $y = Be^{2x+1}$
15. $y = e^{-x^2}$
16. $y = e^{\sqrt{x}}$
17. $y = \ln(3x + 5)$
18. $y = 4 + \ln x$
19. $y = \dfrac{e^x + e^{-x}}{2}$
20. $y = \dfrac{e^x - e^{-x}}{2}$
21. $y = \dfrac{a}{b + ce^{-kx}}$
22. $10^y = e^x$

Compute to three significant figures

23. $\log_7 5$
24. $\log_{11} 8$
25. $\log_2 314$
26. $\log_6 88.3$
27. $\log_5 0.00047$
28. $\log_7 0.0000192$

If $x > 1$, which is bigger?

29. $\ln x$ or $\log x$
30. $\ln x$ or $\log x^2$
31. $\log_6 x$ or $\log_7 x$
32. e^{1000} or 10^{434}

Estimate the base b to three significant figures

33. $\log_b x = \tfrac{3}{4} \log x$ for all $x > 0$

34. $\log_b x = 1.68 \log x$ for all $x > 0$

35. Show that $(\ln 10)(\log e) = 1$

36. Find a relation between $\log_4 x$ and $\log_8 x$.

37. Hot tea at 200°F cools in a room at 68°F. After t minutes its temperature is $T = 68 + 132e^{-0.4t}$. How long until it cools to 130°F, the ideal drinking temperature?

38. A cold piece of metal at 0°C is placed in warm air at 25°C. After t minutes its temperature rises to $T = 25(1 - e^{-0.096t})$. When will the temperature reach 25°?

39. According to one projection, U.S. population (in millions) is increasing by the formula $P = 225e^{0.0078t}$ where t is in years and $t = 0$ in 1980. If so, when will the population reach 300 million?

40. At h feet above sea level, atmospheric pressure in lb/in² is approximately $p = 14.7e^{-0.000039h}$. Find the altitude (to the nearest 100 ft) where the pressure is 6 lb/in².

41. Which is better for the investor, 8% compounded quarterly or $7\tfrac{3}{4}$% compounded continuously?

42. How much must be invested at 7% compounded continuously to be worth $5000 in 10 years?

43. At what annual interest rate compounded continuously will money triple in 10 years?

44. How long will it take money at 8% compounded continuously to triple?

45. In calculus it is shown that for large values of n,

$$1 + \frac{1}{2} + \frac{1}{3} + \frac{1}{4} + \cdots + \frac{1}{n} \approx \ln n + \gamma$$

where $\gamma \approx 0.5772$. Estimate how many terms are needed until the sum $1 + \frac{1}{2} + \frac{1}{3} + \cdots$ exceeds 10.

46. Suppose you were lucky enough to have $1000 and lucky enough to deposit it in a bank that gives 100% yearly interest. Compute the value of the account after one year if the interest is compounded (a) every day (b) every hour (c) every minute. Compare with continuous compounding.

5 APPLICATIONS

Exponential Growth and Decay

Suppose a city had population 75,000 in 1980 and is growing at a steady 3% per year. Then each year the population is multiplied by 1.03, so after t years it is given by the formula

$$P(t) = 75000 \cdot (1.03)^t$$

This formula holds not only for integer values of t but for all real $t \geq 0$. It is a continuously increasing function that increases by 3% in any one-year period. For, given $P(t)$, then one year later the population is

$$P(t+1) = 75000 \cdot (1.03)^{t+1} = 75000 \cdot (1.03)^t \cdot (1.03)$$
$$= (1.03)[75000 \cdot (1.03)^t] = (1.03)P(t)$$

a 3% increase.

This example is a typical case of exponential growth. We say that a quantity **increases exponentially** if it obeys a **growth law** of the type $P(t) = P_0 \cdot b^t$, where $b > 1$. The quantity starts with the initial value P_0 at $t = 0$ and increases. In any time period of a given length it increases by a fixed same factor. For example, from time t to time $t + t_0$, the increase is from $P(t)$ to

$$P(t + t_0) = P_0 \cdot b^{t+t_0} = P_0 \cdot b^t \cdot b^{t_0}$$
$$= b^{t_0}(P_0 \cdot b^t) = b^{t_0} \cdot P(t)$$

Thus, $P(t)$ is multiplied by the same factor b^{t_0} no matter what t is.

Exponential Growth A quantity **grows exponentially** if it satisfies a growth law of the form

$$P(t) = P_0 \cdot b^t \qquad b > 1$$

The initial value is $P_0 = P(0)$. The function $P(t)$ increases by a fixed percentage in a fixed period of time.

Note that

$$P(t) = P_0 \cdot b^{kt} \qquad b > 1, \, k > 0$$

is also an exponential growth law, since it can be written as

$$P(t) = P_0 \cdot (b^k)^t = P_0 \cdot b_1^t \qquad b_1 = b^k > 1$$

In applications of exponential growth, the base b must be found from the data. Sometimes that requires a computation, but often it can be done by inspection.

EXAMPLE 1 A bacteria colony starts with 5000 members and increases exponentially. Find the growth law if the population (a) increases by 15% every hour (b) doubles every hour (c) doubles every 3 hours (d) triples every 10 hours

SOLUTION In each case, the growth law is of the form $P(t) = 5000 \cdot b^t$.

(a) Take $b = 1.15$. Hence $\underline{P(t) = 5000 \cdot (1.15)^t}$.

(b) Take $b = 2$. Hence $\underline{P(t) = 5000 \cdot 2^t}$.

(c) Since $2^{t/3}$ doubles every 3 hours, take $b = 2^{1/3}$. Hence $\underline{P(t) = 5000 \cdot 2^{t/3}}$.

(d) Since $3^{t/10}$ triples every 10 hours, take $b = 3^{1/10}$. Hence $\underline{P(t) = 5000 \cdot 3^{t/10}}$.

EXAMPLE 2 The population of a city increased from 160,000 in 1975 to 190,000 in 1980. Assuming exponential growth, find its growth law.

SOLUTION Let $t = 0$ in 1975. We are given that

$$P(0) = 160{,}000 \qquad \text{and} \qquad P(5) = 190{,}000$$

A more common unit of loudness is the **decibel**. Ten decibels equal one bel. To a sound of intensity I is assigned the decibel reading

$$d(I) = 10 \log \frac{I}{I_0}$$

Thus, freeway noise has 80 decibels.

The Richter scale for measuring intensities of earthquakes works on the same principle. To a quake of intensity I is assigned the number

$$r(I) = \log \frac{I}{I_0}$$

where I_0 is a standard very low intensity used for comparison purposes.

EXAMPLE 6 The great San Francisco earthquake of 1906 measured 8.3 on the Richter scale. How many times stronger was it than a quake measuring 5.5, the level at which serious damage begins?

SOLUTION Let I_1 denote the intensity of the San Francisco quake and I_2 the intensity of the other quake. Then,

$$r(I_1) = \log \frac{I_1}{I_0} = 8.3 \quad \text{and} \quad r(I_2) = \log \frac{I_2}{I_0} = 5.5$$

$$I_1 = 10^{8.3} I_0 \quad \text{and} \quad I_2 = 10^{5.5} I_0$$

Divide I_1 by I_2, and I_0 drops out:

$$\frac{I_2}{I_1} = \frac{10^{8.3}}{10^{5.5}} = 10^{8.3-5.5} = 10^{2.8} \approx 631$$

Answer about 631 times stronger

Another application of $\log(I/I_0)$ is to the magnitude of stars. See Exercises 21–24.

Annuities and Payment of Debts

We have seen applications of exponentials to compound interest. Now we give two more business applications, both derived from the formula for compound interest.

An **annuity** means a periodic payment of a fixed amount. Suppose you make n periodic deposits of amount P into a savings account paying interest i per period. Let A_n be the value of the account just after the n-th deposit. We would like a formula for A_n.

Let us illustrate with A_4. Just after the 4-th deposit, the value A_4 consists of 4 amounts:

P	just deposited
$P(1 + i)$	value of the previous deposit
$P(1 + i)^2$	value of the deposit before that
$P(1 + i)^3$	value of the first deposit

Hence

$$A_4 = P[1 + (1 + i) + (1 + i)^2 + (1 + i)^3]$$

By the same reasoning,
$$A_n = P[1 + (1 + i) + (1 + i)^2 + \cdots + (1 + i)^{n-1}]$$
Now, the expression in brackets has sum
$$\frac{(1 + i)^n - 1}{i}$$
as we will show in Chapter 10. Hence we obtain a formula for A_n.

Annuity Suppose that periodic payments P are made into an account at interest rate i per period. Then the value A_n after n payments is
$$A_n = P\frac{(1 + i)^n - 1}{i}$$

EXAMPLE 7 At the birth of their child, a couple plans to save for its college education. If they deposit $50 a month in a savings account paying 6% annual interest compounded monthly, find the value of the account on their child's 18th birthday.

SOLUTION Use the annuity formula with $P = 50$, $i = \frac{1}{12}(0.06) = 0.005$, and $n = 18 \cdot 12 = 216$. Then,
$$A_n = 50\frac{(1.005)^{216} - 1}{0.005} \approx \underline{\$19{,}367.66}$$

Closely related to the annuity formula is one for payment of debts.

Payment of Debts Suppose that a debt of amount A is repaid in n periods at interest rate i per period. Then the periodic payment is
$$P = A\frac{i}{1 - (1 + i)^{-n}}$$

EXAMPLE 8 Find the monthly payments on a 36-month auto loan of $2500, at 15% annual interest compounded monthly.

SOLUTION Use the formula with $A = 2500$, $n = 36$, and $i = \frac{1}{12}(0.15) = 0.0125$:
$$P = 2500\frac{0.0125}{1 - (1.0125)^{-36}} \approx \underline{\$86.66}$$

EXERCISES

In exercises concerning growth or decay, assume an exponential law.

In Exs. 1–4, find the growth (or decay) law for

1. a bacteria colony that starts with 10,000 members and triples in 12 hours
2. a yeast culture that starts with 5 gm and increases by 30% in 2 hours

3. a bad investment of $5000 that loses 2% a year
4. a 10-gm sample of a radioactive substance that decays to 0.5 gm in 100 days
5. Radioactive radium-F has a half-life of 138.3 days. How long will it take a sample to lose 40% of its mass?
6. Thorium-X has a half-life of 3.64 days. How long will it take a sample to decay to 1% of its original mass?
7. The 1970 population of San Antonio was 707,500. At what yearly rate must the city grow to reach 1,000,000 by the year 2000?
8. A bacteria colony doubles in 5 hours. After 9 hours its population is 10^6. What was its original population?
9. City A has a population of 400,000 and is increasing at the rate of 2% a year. City B has a population of 360,000 and is increasing at 3% a year. How long until B catches up to A?
10. At what rate of increase would city B in Exercise 9 catch up to city A in 10 years?
11. A 5-lb sample of radioactive material contains 2 lb of radium-F, half-life 138.3 days, and 3 lb of thorium-X, half-life 3.64 days. When will the sample contain equal amounts of the two elements?
12. Wood from the Lascaux caves in France was found to contain 14.5% of the normal concentration of ^{14}C. Estimate the age of the famous Lascaux cave paintings. (See Example 5.)
13. Express the growth law $M(t) = M_0 \cdot 2^{t/5}$ in the form $M(t) = M_0 \cdot e^{kt}$.
14. Express the decay law $M(t) = M_0 \cdot 10^{-t}$ in the form $M(t) = M_0 \cdot e^{-kt}$.
15. A hot object at temperature T_0 is plunged into a cooling bath at temperature B. Let $T(t)$ denote its temperature after t minutes. It is known that the *difference* $T(t) - B$ decreases exponentially. Show that
$$T(t) - B = (T_0 - B)a^{-t} \qquad a > 1$$
What is its "ultimate" temperature?
16. A piece of hot steel at 425°C is plunged into oil at 25°C. In one minute it cools to 225°. When will its temperature reach 30°? (Use Exercise 15.)
17. How many times stronger is an earthquake measuring 5.5 on the Richter scale than one measuring 5.1?
18. How many times louder is a jet landing at 118 decibels than normal speech at 60 decibels?
19. By how many decibels do two sounds differ if one is twice as loud as the other?
20. Compare the intensities of two sounds that differ by one decibel.

The **magnitude** of a star of brightness B is defined to be
$$m(B) = -2.5 \log \frac{B}{B_0}$$

where B_0 is a standard brightness used for comparison. (The minus sign causes $m(B)$ to increase as B decreases.) Apply this formula in Exs. 21–24.

21. Find the magnitude of a star of brightness (a) B_0 (b) $\frac{1}{100} B_0$

22. The brightest star in the sky is Sirius, with magnitude -1.5. The faintest star visible to the naked eye has magnitude 6.0. How many times brighter is Sirius than such a star?

23. How many times brighter is the Sun, magnitude -26.8 than Sirius, magnitude -1.5?

24. The second brightest star in the sky is Canopus with magnitude -0.7. Find the magnitude of a star $\frac{1}{50}$ as bright.

25. Monthly deposits of $30 are made into an account paying 9% annual interest compounded monthly. Find the value of the account after 10 years.

26. Quarterly deposits of $250 are made into an account paying 8% interest compounded quarterly. Find the value of the account after 15 years.

27. After how many monthly deposits of $50 into an account paying 6% annual interest compounded monthly will the savings first exceed $10,000?

28. After how many yearly deposits of $500 into an account paying 10% annual interest will the savings first exceed $7000?

29. How much must be deposited monthly to save $25,000 in 21 years at $\frac{1}{2}$% interest per month?

30. Suppose that in Example 7, the interest is compounded quarterly, not monthly. What is the value of the account on the child's 18th birthday?

31. Find the monthly payments on a 4-year auto loan of $3000 at 18% annual interest compounded monthly.

32. Find the monthly payments on a 20-year mortgage of $40,000 at 12% annual interest compounded monthly.

33. The Andersons can afford monthly mortgage payments of $400. How large a 25-year mortgage can they afford at 12% annual interest compounded monthly?

34. If I can pay $60 a month for two years, how much can I afford to borrow at 18% annual interest, compounded monthly?

35. A 20-year savings plan calls for equal monthly deposits into an account paying 0.7% monthly interest. How long does it take to save half the 20-year total?

36. You deposit $40 a month into an account paying 6% annual interest compounded monthly. After 5 years the bank increases the interest rate to 8%. What is the value of your account after 10 years of deposits?

REVIEW EXERCISES FOR CHAPTER 7

1. Without a calculator find
   (a) $\log_2 32\sqrt[3]{2}$
   (b) $\log_5 \frac{1}{125}$
   (c) $\log_7 7^8$
   (d) $\log_{12} 12$

Exponential and Logarithm Functions

2. Express the statement in terms of exponentials
   (a) $\log_3 1 = 0$ (b) $\log_4 128 = 3.5$ (c) $\log_{32} \frac{1}{2} = -\frac{1}{5}$ (d) $\log_2 65536 = 16$

Graph

3. $y = 1 + (\frac{1}{2})^x$

4. $y = 9^{x/2}$

5. $y = \log_2 \left(\frac{x}{4}\right)$

6. $y = \log_{10}(x - 2)$

7. Find a function $f(x)$ for which $f(x + 1) = \frac{3}{2}f(x)$.

8. Express as a single logarithm: $\frac{1}{3}\log_7 1000 - 2\log_7 5 - 3\log_7 2$

Solve for x

9. $\log_9 (2x - 1) = -\frac{1}{2}$

10. $\log_2 x - \log_2 (4x + 5) = 3$

Find x to 3 significant figures

11. $5^{3x+1} = 14$

12. $(1.06)^x = 8$

Express x in terms of y

13. $y = \frac{1}{2} \ln (3x - 2)$

14. $y = \dfrac{e^{2x} - 1}{e^{2x} + 1}$

15. To what amount does $5000 grow in 10 years at 7% annual interest compound (a) quarterly (b) monthly?

16. A deposit of $1000 grew to $1346.11 in 4 years at interest compounded quarterly. What was the annual interest rate?

17. An investor placed $10,000 in account A paying 6% compounded monthly and, at the same time, $10,000 in account B paying 9% compounded monthly. After 4 years, she closed out A and put it all into B. What was the value of the investment 10 years after the initial deposits?

18. A TV commercial warns consumers that their heating bills will double in 10 years. Assuming continued inflation of at least 8% per year, show that this is not so bad.

19. A radioactive material loses 20% of its mass in 20 days. Estimate its half-life.

20. A $50,000 mortgage is paid off in 20 years in equal monthly payments. If the interest rate is 1% per month, what is the total amount of money paid back?

Systems of Equations and Inequalities

1 SYSTEMS OF EQUATIONS IN TWO VARIABLES

A 2×2 **linear system** in x and y is a pair of linear equations
$$\begin{cases} a_1 x + b_1 y = c_1 \\ a_2 x + b_2 y = c_2 \end{cases}$$
A **solution** of the system is a pair of real numbers (x, y) that makes both equations true statements at the same time (simultaneously).

The strategy for solving a linear system is to eliminate one of the variables. This reduces the system to one equation in the other variable.

EXAMPLE 1 Solve the system $\begin{cases} 2x + y = 11 \\ x - y = 4 \end{cases}$

SOLUTION 1 Suppose (x, y) is a solution. Then both equations are true statements. Adding them will produce another true statement *and* will eliminate y:

$$\begin{array}{r} 2x + y = 11 \\ x - y = 4 \\ \hline \end{array}$$
(add) $\quad 3x = 15$

The y drops out, leaving $3x = 15$. Hence, if (x, y) is a solution, then $x = 5$. Substituting $x = 5$ into the equation $x - y = 4$ shows that $y = 1$. Therefore, the only possible solution is (5, 1). But (5, 1) *is* a solution since

$$2 \cdot 5 + 1 = 11 \quad \text{and} \quad 5 - 1 = 4$$

SOLUTION 2 Let us eliminate x this time. Multiply the second equation by 2 and subtract it from the first equation:

$$\begin{array}{r} 2x + y = 11 \\ 2x - 2y = 8 \\ \hline \end{array}$$
(subtract) $\quad 3y = 3$

Hence, if (x, y) is a solution, then $y = 1$. From $x - y = 4$, it follows that $x = 5$.

EXAMPLE 2 Solve $\begin{cases} 5x + 2y = -1 \\ 4x + 3y = 1 \end{cases}$

217

SOLUTION To eliminate y, multiply the first equation by 3, the second by -2, and add. (Or, multiply the second by 2 and subtract.)

$$\begin{array}{r|rr} 3 & 5x + 2y = -1 & \quad 15x + 6y = -3 \\ -2 & 4x + 3y = 1 & \quad \underline{-8x - 6y = -2} \\ & & \quad 7x = -5 \end{array}$$

Hence $x = -\frac{5}{7}$. Now find y by substituting $x = -\frac{5}{7}$ into either of the given equations, say the first:

$$5(-\tfrac{5}{7}) + 2y = -1$$
$$2y = -1 + \tfrac{25}{7} = \tfrac{18}{7}$$
$$y = \tfrac{9}{7}$$

Answer $\left(-\dfrac{5}{7}, \dfrac{9}{7}\right)$

Check $5x + 2y = 5\left(-\dfrac{5}{7}\right) + 2\left(\dfrac{9}{7}\right) = \dfrac{-25 + 18}{7} = \dfrac{-7}{7} = -1$

$4x + 3y = 4\left(-\dfrac{5}{7}\right) + 3\left(\dfrac{9}{7}\right) = \dfrac{-20 + 27}{7} = \dfrac{7}{7} = 1$

Examples 1 and 2 illustrate a way to eliminate one of the variables: by adding or subtracting multiples of the equations. Another way is by substitution. Solve one equation for x in terms of y, or for y in terms of x. Then substitute into the other equation. For instance, take the system in Example 1:

$$\begin{cases} 2x + y = 11 \\ x - y = 4 \end{cases}$$

From the first equation, $y = 11 - 2x$. Substitute this expression into the second equation:

$$x - (11 - 2x) = 4$$
$$3x = 15$$
$$x = 5$$

Now substitute $x = 5$ into either of the original equations, say the second:

$$5 - y = 4$$
$$y = 1$$

You may also solve the system by obtaining either $x = y + 4$ or $y = x - 4$ from the second equation and substituting into the first.

EXAMPLE 3 A treasure hunter found a bar made of gold and silver. The bar weighs 15,693 gm and has volume 932 cc. The density of gold is 18.8 gm/cc, and that of silver is 10.6 gm/cc. What volume of each metal does the bar contain?

SOLUTION Let G be the volume of gold (in cc) and S the volume of silver in the bar. Then $G + S = 932$. Each cc of gold weighs 18.8 gm, so the gold weighs 18.8G. Similarly, the silver weighs 10.6S. Therefore G and S satisfy the system

$$\begin{cases} G + S = 932 \\ 18.8G + 10.6S = 15{,}693 \end{cases}$$

From the first equation, $S = 932 - G$. Substitute this value into the second equation:

$$18.8G + (10.6)(932 - G) = 15{,}693$$
$$(18.8 - 10.6)G = 15{,}693 - (10.6)(932)$$
$$G = \frac{15{,}693 - (10.6)(932)}{8.2} = 709$$

Since $S = 932 - G$,
$$S = 932 - 709 = 223$$

Answer 709 cc of gold and 223 cc of silver

Inconsistent and Dependent Systems

Not every system has a solution. For example, if the system
$$\begin{cases} x - 2y = 1 \\ x - 2y = -4 \end{cases}$$
had a solution, then $1 = -4$, which is false. Therefore no solution exists. Such a system is called **inconsistent**.

Another inconsistent system is
$$\begin{cases} 3x - y = 1 \\ 15x - 5y = 7 \end{cases}$$
Look at the left sides: $15x - 5y = 5(3x - y)$. Therefore, if $3x - y = 1$, then $15x - 5y = 5$, not 7. If you do not notice this, it doesn't matter; the algebra will take care of itself. When you try to eliminate y, you will end up with a contradiction, $0 = 2$ or $5 = 7$.

At the opposite extreme are systems that have infinitely many solutions, for example,
$$\begin{cases} x + 2y = 4 \\ 3x + 6y = 12 \end{cases}$$
The second equation is 3 times the first; it is really the same equation. Hence each pair (x, y) that satisfies the first equation also satisfies the second. There are infinitely many such pairs, for we can assign to y any real value t. Then, as long as $x = 4 - 2t$, the equation is satisfied. Hence, all pairs of the form
$$(4 - 2t, t) \qquad t \text{ any real number}$$
are solutions.

If you try to eliminate one variable in this example, *both* variables will drop out and you will end up with $0=0$. Such systems are called **dependent**; they consist of two equivalent equations.

Graphical Interpretation

We have seen 2×2 linear systems with (a) exactly one solution, (b) no solutions, and (c) infinitely many solutions. A look at a graph shows why these are the only possibilities.

The graph of a linear system consists of two straight lines. A solution of the system is a pair (x, y) that satisfies both equations. On the graph, (x, y) is a point that lies on both lines. Thus, solutions show up as points of intersection.

There are three possibilities (Fig. 1). Either the two lines (a) intersect in one point (one solution), or (b) are parallel (no solutions), or (c) coincide (all points on the one line are solutions).

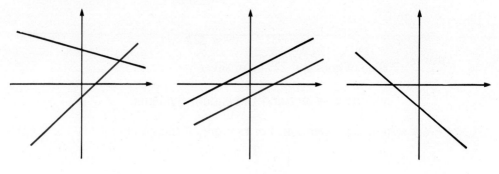

(a) Intersecting lines: one intersection

(b) Parallel lines: no intersections

(c) Coinciding lines: all points are intersections

FIG. 1

Non-linear Systems

Sometimes we deal with systems of two equations in x and y, where not both are linear, that is, **non-linear systems.** Examples are

$$\begin{cases} xy = 3 \\ x + 2y = 7 \end{cases} \qquad \begin{cases} x^2 + xy + y^2 = 3 \\ x + y = 1 \end{cases} \qquad \begin{cases} x^2 + y^2 = 25 \\ x^2 - y^2 = 1 \end{cases}$$

EXAMPLE 4 Solve the system $\begin{cases} xy = 3 \\ x + 2y = 7 \end{cases}$

SOLUTION From the second equation, $x = 7 - 2y$. Substitute this expression into the first equation:

$$(7 - 2y)y = 3$$
$$0 = 2y^2 - 7y + 3$$
$$0 = (2y - 1)(y - 3)$$

Therefore $y = \tfrac{1}{2}$ or $y = 3$. If $y = \tfrac{1}{2}$, then $x = 7 - 2y = 6$. If $y = 3$, then $x = 7 - 2y = 1$.

Answer $(6, \tfrac{1}{2})$, $(1, 3)$

EXAMPLE 5 Find all points of intersection of the parabola $y = 4 - x^2$ and the line $y = x + 1$.

SOLUTION A sketch (Fig. 2) indicates two points of intersection: one in the first quadrant and one in the third. Roughly, these points are near $(-2, -1)$ and $(1, 2)$. To find their exact position, solve the system

$$\begin{cases} y = 4 - x^2 \\ y = x + 1 \end{cases}$$

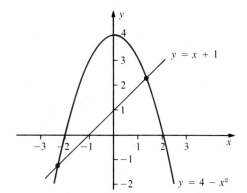

FIG. 2

Substitute $y = x + 1$ into the first equation and solve for x:

$$x + 1 = 4 - x^2$$
$$x^2 + x - 3 = 0$$

Hence

$$x = \frac{-1 + \sqrt{13}}{2} \quad \text{or} \quad x = \frac{-1 - \sqrt{13}}{2}$$

$$y = x + 1 = \frac{1 + \sqrt{13}}{2} \qquad y = x + 1 = \frac{1 - \sqrt{13}}{2}$$

Answer

$$\left(\frac{-1 + \sqrt{13}}{2}, \frac{1 + \sqrt{13}}{2}\right) \approx (1.3, 2.3)$$

$$\left(\frac{-1 - \sqrt{13}}{2}, \frac{1 - \sqrt{13}}{2}\right) \approx (-2.3, -1.3)$$

EXERCISES

Solve the system

1. $\begin{cases} x - y = 3 \\ x + 2y = 6 \end{cases}$
2. $\begin{cases} 2x + y = 2 \\ -3x + y = -3 \end{cases}$
3. $\begin{cases} x - 2y = 1 \\ 2x + y = 9 \end{cases}$

4. $\begin{cases} x + 3y = 0 \\ 2x - y = 7 \end{cases}$
5. $\begin{cases} 2x - 7y = -1 \\ 3x + 4y = 2 \end{cases}$
6. $\begin{cases} 6x + 5y = -2 \\ 5x - 3y = 1 \end{cases}$

7. $\begin{cases} \frac{1}{3}x + \frac{1}{2}y = 0 \\ 2x + y = 6 \end{cases}$
8. $\begin{cases} 5x - \frac{1}{2}y = 12 \\ 3x - y = 2 \end{cases}$
9. $\begin{cases} 1.3x + 4.6y = 10.7 \\ 8.1x - 3.8y = 40.7 \end{cases}$

10. $\begin{cases} 4.4x + 5.6y = 33.52 \\ 11.6x + 4.9y = 123.88 \end{cases}$
11. $\begin{cases} 9x - 7y = 0 \\ 4x + 3y = 0 \end{cases}$
12. $\begin{cases} 8x + 5y = 0 \\ 2x - 9y = 0 \end{cases}$

13. $\begin{cases} x + 4y = 7 \\ 2x + 8y = 10 \end{cases}$
14. $\begin{cases} 2x - 3y = -4 \\ 6x - 9y = 1 \end{cases}$
15. $\begin{cases} \frac{1}{2}x - \frac{3}{4}y = 1 \\ 4x - 6y = 8 \end{cases}$

16. $\begin{cases} 9x + 12y = 21 \\ x + \frac{4}{3}y = \frac{7}{3} \end{cases}$
17. $\begin{cases} \dfrac{1}{x} + \dfrac{1}{y} = 1 \\ \dfrac{1}{x} - \dfrac{1}{y} = -3 \end{cases}$
18. $\begin{cases} 2^x + 2^y = 17 \\ 2^x - 2^y = 15 \end{cases}$

19. $\begin{cases} \sqrt{x} + \sqrt{y} = 7 \\ 3\sqrt{x} - 2\sqrt{y} = 11 \end{cases}$

20. $\begin{cases} x - \sqrt{y} = 1 \\ 3x + 7\sqrt{y} = 8 \end{cases}$

Sketch the curves and compute all points of intersection

21. $\begin{cases} y = x^2 \\ y = 2x - 3 \end{cases}$

22. $\begin{cases} y = x^2 + 2x + 2 \\ y = x + 4 \end{cases}$

23. $\begin{cases} \dfrac{x^2}{4} + y^2 = 1 \\ y = 3x \end{cases}$

24. $\begin{cases} xy = 1 \\ x + 3y = 6 \end{cases}$

25. $\begin{cases} y = x^2 + 2x \\ y = 4 - x^2 \end{cases}$

26. $\begin{cases} y = x^3 \\ y = 4x \end{cases}$

27. $\begin{cases} x^2 + y^2 = 4 \\ (x-1)^2 + y^2 = 9 \end{cases}$

28. $\begin{cases} x^2 + y^2 = 9 \\ x^2 - y^2 = 1 \end{cases}$

29. $\begin{cases} \dfrac{x^2}{9} - \dfrac{y^2}{25} = 1 \\ \dfrac{1}{3}x - \dfrac{2}{5}y = 1 \end{cases}$

30. $\begin{cases} \dfrac{x^2}{4} - \dfrac{y^2}{9} = 1 \\ 2x - y = -6 \end{cases}$

31. $\begin{cases} x^2 + y^2 = 25 \\ x + y = 1 \end{cases}$

32. $\begin{cases} x^2 + y^2 = 16 \\ x - y = 3 \end{cases}$

33. For what values of c does the line $y = x + c$ intersect the parabola $y = x^2 - 4x + 3$?

34. For what values of c is the system $\begin{cases} 2x - 3y = 1 \\ 4x + cy = c \end{cases}$ inconsistent?

35. Jogging 4 miles and walking 4 miles takes me 104 minutes. Jogging 6 miles and walking 2 miles takes me 84 minutes. How long does it take me to jog a mile? Walk a mile?

36. In a certain isosceles triangle, each of the two equal base angles exceeds the third angle by 24°. Find all three angles.

37. With the current, a river boat goes 18 miles in 2 hours. Against the current, it goes 21 miles in 3 hours. Find its speed in still water and the speed of the current.

38. A broker's commission is a fixed amount plus a percentage of the transaction. On a $1000 sale, the fee is $25, and on a $3000 sale it is $55. Find a formula for the commission.

39. Five triple burgers and four bags of french fries cost $8.48. Six burgers and three fries cost $9.24. Find the prices of one burger and one bag of fries.

40. One year a university sold 400 A-stickers for preferred parking and 900 B-stickers for regular parking. The income was $55,800. The next year they sold 500 A-stickers and 950 B-stickers for $64,500. Find the prices of an A-sticker and a B-sticker.

41. If a team wins 3 out of their next 4 games, they will raise their winning percentage to 50%. If they win 8 straight, they will be up to 62.5%. How many wins and losses do they have now?

42. Find a and b so that the graph of $y = ax + b/x$ passes through $(2, \tfrac{1}{2})$ and $(3, 7)$. Sketch the graph.

43. A bar containing 100 cc of copper and 150 cc of zinc weighs 1925 gm. Another bar con-

taining 500 cc of copper and 80 cc of zinc weighs 5002 gm. Find the densities (in gm/cc) of copper and zinc.

44. An investor bought 10 ounces of gold and 200 ounces of silver for $9400. On the same day, another investor bought 30 oz of gold and 1000 oz of silver for $34,200. What were the prices of gold and silver that day?

2 SYSTEMS OF EQUATIONS IN THREE VARIABLES

A 3×3 **linear system** is a set of three linear equations

$$\begin{cases} a_1 x + b_1 y + c_1 z = d_1 \\ a_2 x + b_2 y + c_2 z = d_2 \\ a_3 x + b_3 y + c_3 z = d_3 \end{cases}$$

A **solution** is a triple of real numbers (x, y, z) that makes all three equations true statements.

The general strategy for solving such a system is to eliminate one variable at a time until you reach one equation in one unknown. This can be done in several ways, but we will concentrate on a practical method called **Gaussian elimination.** The idea is to transform a given system into one of the form

$$\begin{cases} a_1' x + b_1' y + c_1' z = d_1' \\ b_2' y + c_2' z = d_2' \\ c_3' z = d_3' \end{cases}$$

This is called **echelon** or **triangular form.** A system in echelon form is especially easy to solve. The third equation gives z immediately. Knowing z, we find y from the second equation. Knowing y and z, we then find x from the first equation.

In reducing a given system to echelon form, we must be careful to end up with an **equivalent** system, one with exactly the same set of solutions. Therefore we must use operations that do not change solutions. There are three such operations. (We have already used two of these in Section 1.)

Operations That Produce Equivalent Systems

(1) Interchanging two equations

(2) Multiplying both sides of an equation by the same non-zero number

(3) Adding a multiple of one equation to another equation

EXAMPLE 1 Solve the system $\begin{cases} x + y + 2z = 1 \\ -2x + y + 5z = 4 \\ 3x + 7y + z = 0 \end{cases}$

SOLUTION First we eliminate x from the second and third equations without changing the first equation. Step 1: Add 2 times the first equation to the second. Step 2: Add -3 times the first equation to the third.

Step 1 $\begin{cases} x + y + 2z = 1 \\ 3y + 9z = 6 \\ 3x + 7y + z = 0 \end{cases}$ Step 2 $\begin{cases} x + y + 2z = 1 \\ 3y + 9z = 6 \\ 4y - 5z = -3 \end{cases}$

Next, let us knock out the $4y$ in the third equation. Step 3: Multiply the second equation by $\frac{1}{3}$, so the coefficient of y becomes 1. Step 4: Add -4 times the second equation to the third.

Step 3 $\begin{cases} x + y + 2z = 1 \\ y + 3z = 2 \\ 4y - 5z = -3 \end{cases}$ Step 4 $\begin{cases} x + y + 2z = 1 \\ y + 3z = 2 \\ -17z = -11 \end{cases}$

After Step 4 we have an equivalent system in echelon form. From the third equation, $z = \frac{11}{17}$. Then, from the second equation,

$$y = 2 - 3z = 2 - 3(\tfrac{11}{17}) = \tfrac{1}{17}$$

Finally, from the first equation,

$$x = 1 - y - 2z = 1 - \tfrac{1}{17} - 2(\tfrac{11}{17}) = -\tfrac{6}{17}$$

Answer $(-\tfrac{6}{17}, \tfrac{1}{17}, \tfrac{11}{17})$

Remark 1 *Theoretically*, there is no reason to check the answer. The system in Step 4 is equivalent to the given system; hence its solutions are guaranteed to be the same. In practice, however, it is always good to check against mistakes in arithmetic.

Remark 2 A system of the form

$\begin{cases} a_1 x + b_1 y + c_1 z = d_1 \\ a_2 x + b_2 y = d_2 \\ a_3 x = d_3 \end{cases}$ can be written as $\begin{cases} c_1 z + b_1 y + a_1 x = d_1 \\ b_2 y + a_2 x = d_2 \\ a_3 x = d_3 \end{cases}$

so it is actually in echelon form.

EXAMPLE 2 Find a quadratic polynomial $y = ax^2 + bx + c$ whose graph passes through $(1, -2)$, $(3, 1)$, and $(4, -1)$.

SOLUTION Since the graph passes through $(1, -2)$, we may substitute $x = 1$ and $y = -2$ in $y = ax^2 + bx + c$:

$$-2 = a + b + c \quad \text{that is,} \quad a + b + c = -2$$

Similarly, we substitute $x = 3$, $y = 1$ and $x = 4$, $y = -1$:

$$9a + 3b + c = 1 \quad \text{and} \quad 16a + 4b + c = -1$$

Thus, we have a system for (a, b, c):

$$\begin{cases} a + b + c = -2 \\ 9a + 3b + c = 1 \\ 16a + 4b + c = -1 \end{cases}$$

It is easiest to eliminate c from the second and third equations. We subtract the first equation from each:

$$\begin{cases} a + b + c = -2 \\ 8a + 2b = 3 \\ 15a + 3b = 1 \end{cases}$$

Next, let us eliminate the $3b$ in the third equation. If you are good at fractions, add $-\frac{3}{2}$ times the second equation to the third. We're not, so we multiply the third by 2, then subtract 3 times the second:

$$\begin{cases} a + b + c = -2 \\ 8a + 2b = 3 \\ 30a + 6b = 2 \end{cases} \qquad \begin{cases} a + b + c = -2 \\ 8a + 2b = 3 \\ 6a = -7 \end{cases}$$

Hence $a = -\frac{7}{6}$. Then, from the second equation,

$$b = \tfrac{1}{2}(3 - 8a) = \tfrac{1}{2}[3 - 8(-\tfrac{7}{6})] = \tfrac{1}{2}(\tfrac{74}{6}) = \tfrac{37}{6}$$

Finally, from the first equation,

$$c = -2 - a - b = -2 - (-\tfrac{7}{6}) - \tfrac{37}{6} = -2 - \tfrac{30}{6} = -7$$

Answer $y = -\tfrac{7}{6}x^2 + \tfrac{37}{6}x - 7$. See Figure 1.

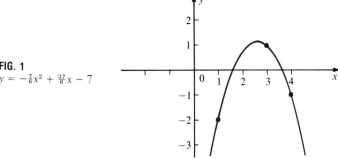

FIG. 1
$y = -\tfrac{7}{6}x^2 + \tfrac{37}{6}x - 7$

Two Equations in Three Unknowns

A linear system of 2 equations in 3 unknowns is a pair of linear equations such as

$$\begin{cases} x + y - z = 3 \\ x - y + 3z = 5 \end{cases} \qquad \begin{cases} x + y + z = 0 \\ 2x + 2y + 2z = 7 \end{cases}$$

The system may have no solutions like the one on the right above. Otherwise it has infinitely many solutions.

EXAMPLE 3 Solve $\begin{cases} x + y - z = 3 \\ x - y + 3z = 5 \end{cases}$

SOLUTION Let us write the system as

$$\begin{cases} x + y = 3 + z \\ x - y = 5 - 3z \end{cases}$$

For each real value of z, this is an ordinary 2×2 system. So we think of z as a constant and solve for x and y:

$$x = 4 - z \qquad y = 2z - 1$$

Hence, if z is assigned any real value t, then the triple $(4 - t, 2t - 1, t)$ is a solution.

Answer $(4 - t, 2t - 1, t)$ t any real number

Inconsistent and Dependent Systems

A 3 × 3 system may be **inconsistent,** that is, not have any solutions. Here are two examples:

$$\text{(A)} \quad \begin{cases} x + 2y + 3z = 1 \\ x + 2y + 3z = 2 \\ 4x - 3y - z = 6 \end{cases} \qquad \text{(B)} \quad \begin{cases} x + 2y + 3z = 1 \\ x + y - z = 2 \\ 5x + 6y - z = 7 \end{cases}$$

In Example A, the first two equations are contradictory. In Example B, there is a relation among the left-hand sides,

$$5x + 6y - z = (x + 2y + 3z) + 4(x + y - z)$$

but the same relation does *not* hold among the right-hand sides:

$$7 \neq 1 + 4 \cdot 2$$

If you try to eliminate variables, you will end up with a contradiction such as $0 = -2$.

At the opposite extreme, a 3 × 3 system may have infinitely many solutions. This happens when the system has solutions but is **dependent,** that is, equivalent to a system with fewer equations.

EXAMPLE 4 Solve $\begin{cases} x + y + z = 1 \\ x - 2y + 2z = 4 \\ 2x - y + 3z = 5 \end{cases}$

SOLUTION Subtract the first equation from the second and twice the first from the third:

$$\begin{cases} x + y + z = 1 \\ -3y + z = 3 \\ -3y + z = 3 \end{cases}$$

The second and third equations are identical. Hence the system is equivalent to

$$\begin{cases} x + y + z = 1 \\ -3y + z = 3 \end{cases} \quad \text{which we write as} \quad \begin{cases} x + y = 1 - z \\ -3y = 3 - z \end{cases}$$

Solving for x and y gives

$$x = 2 - \tfrac{4}{3}z \qquad y = \tfrac{1}{3}z - 1$$

There are infinitely many solutions: if z is assigned any real value t, then the triple $(2 - \tfrac{4}{3}t, \tfrac{1}{3}t - 1, t)$ is a solution.

Answer

$$(2 - \tfrac{4}{3}t, \tfrac{1}{3}t - 1, t) \qquad t \text{ any real number}$$

Our discussion in this section applies to 4 × 4 systems and higher. The ideas are the same, but the computations are longer. In the next section, we discuss a way of shortening the work.

EXERCISES

Solve the system

1. $\begin{cases} x + y + z = 5 \\ -x + 2y + z = 4 \\ 2x - y - 3z = -8 \end{cases}$

2. $\begin{cases} x + y + 2z = 3 \\ 2x - 3y - z = -9 \\ 3x + 2y + 2z = 0 \end{cases}$

3. $\begin{cases} 3x + y + z = -1 \\ x - 4y - 2z = 9 \\ 2x + 5y = 2 \end{cases}$

4. $\begin{cases} 4x + y - z = 2 \\ -x + 2y + 2z = -2 \\ 5y + z = 1 \end{cases}$

5. $\begin{cases} 2x - y - 3z = 1 \\ x + 4y + 2z = -1 \\ 3x - y - z = 1 \end{cases}$

6. $\begin{cases} 2x + y + 3z = 1 \\ -x + 4y + 2z = 0 \\ 3x + y + z = -1 \end{cases}$

7. $\begin{cases} x + y - z = 0 \\ x - y + z = 0 \\ -x + y + z = 0 \end{cases}$

8. $\begin{cases} 4x + 3y - z = 0 \\ 2x + y + z = 0 \\ y - 3z = 0 \end{cases}$

9. $\begin{cases} x + y = -4 \\ y + z = 3 \\ x + z = 2 \end{cases}$

10. $\begin{cases} x + 2y = 3 \\ y + 2z = 4 \\ 2x + z = 5 \end{cases}$

11. $\begin{cases} 7x + y + z = -1 \\ 9x + 2y + 2z = 3 \\ 12x + 5y - z = -6 \end{cases}$

12. $\begin{cases} 6x + y - 7z = 2 \\ -4x + y + 9z = 4 \\ 10x - y - 15z = 2 \end{cases}$

13. $\begin{cases} \frac{1}{2}x + \frac{1}{3}y - \frac{5}{6}z = 0 \\ \frac{1}{2}x - \frac{1}{10}y + \frac{2}{5}z = \frac{4}{5} \\ 3x - \frac{1}{3}y - \frac{5}{12}z = \frac{9}{4} \end{cases}$

14. $\begin{cases} x + y + 3z = \frac{1}{4} \\ \frac{1}{3}x + \frac{1}{2}y + \frac{4}{3}z = 0 \\ 5x + 3y + 5z = \frac{3}{4} \end{cases}$

15. $\begin{cases} x + y + z = -0.1 \\ 3x - 2y + 2z = -7.6 \\ 5x + 3y + 4z = -1.5 \end{cases}$

16. $\begin{cases} x + 2y + 3z = -1.9 \\ 3x - 2y - 4z = 1.4 \\ 4x + 5y + 3z = -0.4 \end{cases}$

17. $\begin{cases} x + y - z - 4w = 6 \\ 2x - 3y + 2z + 5w = 1 \\ 3x - 5y + 4z + 9w = -2 \\ -x + 6y - 3z - 7w = 3 \end{cases}$

18. $\begin{cases} 4x + 2y + 3z + 5w = 0 \\ 7x + y + 3z + 5w = 4 \\ 6x + 3y + 4z + 7w = 0 \\ 3x + y + 2z + 3w = 1 \end{cases}$

Find the quadratic function $y = ax^2 + bx + c$ whose graph passes through the points

19. $(-1, 1), (0, 0), (1, 2)$

20. $(1, 1), (2, 2), (3, 7)$

21. $(-2, -5), (0, 3), (2, 0)$

22. $(-3, 2), (1, 1), (2, -1)$

Find all solutions

23. $\begin{cases} x + 2y - z = 1 \\ x - y + 3z = 6 \end{cases}$

24. $\begin{cases} 2x + y + z = 10 \\ 3x + y - z = 8 \end{cases}$

25. $\begin{cases} 2x - y + z = 1 \\ 3x + y + z = 0 \\ 7x - y + 3z = 2 \end{cases}$

26. $\begin{cases} 11x + 10y + 9z = 5 \\ x + 2y + 3z = 1 \\ 3x + 2y + z = 1 \end{cases}$

27. $\begin{cases} x + y + 2z = 1 \\ 3x + 5y + 7z = 2 \\ 4x + 6y + 9z = 5 \end{cases}$

28. $\begin{cases} x - y - 4z = 1 \\ 2x + 2y - z = 0 \\ 5x + 3y - 6z = 4 \end{cases}$

29. The average age of a father, mother, and child is 23. The father is 2 years older than the mother and 22 years older than the child. Find all three ages.

30. A collection of 52 coins, pennies, nickels, and dimes, totals $2.75. If the dimes were pennies and the pennies dimes, the total would be $2.57. How many of each type of coin are there?

31. For a rock concert all 1000 seats are sold at $5, $7, and $10, bringing total receipts of $7600. For a less popular group the seats are reduced to $3, $6, and $9, bringing receipts of $6450. Find the number of cheap, medium-priced, and expensive seats.

32. Amy Smith has $10,000 in three accounts, one paying 5% yearly, one 6%, and one 10%. The combined yearly interest is $830. It increases to $900 when the first two accounts raise their interest to 6% and 8%. How much does she have in each account?

33. Describe all ways of designing a 60-minute exercise program of calisthenics, running, and swimming that will burn 500 calories. Assume calisthenics burns 6 cal/min, running 8 cal/min, and swimming 10 cal/min.

34. Describe all ways of investing $10,000 in conservative bonds paying 6% yearly, moderate bonds paying 8%, and risky bonds paying 12%, so that the combined yearly income is $900.

3 SOLUTION OF LINEAR SYSTEMS IN MATRIX NOTATION

In this section, we discuss a convenient shorthand method for solving systems of linear equations.

Suppose we want to solve the system

$$\begin{cases} x + y + 2z = 1 \\ -2x + y + 5z = 4 \\ 3x + 7y + z = 0 \end{cases}$$

There is no need to write the x's, y's, and z's over and over again. We abbreviate the system by the notation

(1) $$\begin{bmatrix} 1 & 1 & 2 & 1 \\ -2 & 1 & 5 & 4 \\ 3 & 7 & 1 & 0 \end{bmatrix}$$

Such a rectangular array of numbers is called a **matrix**. The matrix shown here has 3 **rows** and 4 **columns**. In our scheme, each row of the matrix represents an equation. Thus, the second row

$$-2 \quad 1 \quad 5 \quad 4 \qquad \text{represents} \qquad -2x + y + 5z = 4$$

To solve the given system, we reduce it to echelon form. The reduced system will be represented by a matrix of the form

(2) $$\begin{bmatrix} a_1 & b_1 & c_1 & d_1 \\ 0 & b_2 & c_2 & d_2 \\ 0 & 0 & c_3 & d_3 \end{bmatrix}$$

So the problem becomes: given a matrix (1), reduce it to a matrix (2) that represents an equivalent system. How? By imitating the three operations on equations that produce equivalent systems. The corresponding operations *on matrices* are called **elementary row operations.**

Operations on Equations	Elementary Row Operations
(1) Interchanging two equations	Interchanging two rows
(2) Multiplying both sides of an equation by k, where $k \neq 0$	Multiplying each entry in a row by k, where $k \neq 0$
(3) Adding k times one equation to another equation	Adding k times the entries in one row to the corresponding entries of another row

We will abbreviate the elementary row operations by the notation

$R_i \leftrightarrow R_j$ Interchange Row i and Row j

$R_i \rightarrow kR_i$ Multiply Row i by k

$R_i \rightarrow R_i + kR_j$ Add k times Row j to Row i

EXAMPLE 1 Use matrix notation to solve the system
$$\begin{cases} 3x + 7y + z = 0 \\ -2x + y + 5z = 4 \\ x + y + 2z = 1 \end{cases}$$

SOLUTION We will reduce the corresponding matrix to echelon form using a sequence of elementary row operations. So you can see how it works, we will display the matrices and the corresponding linear systems side by side.

$$\begin{cases} 3x + 7y + z = 0 \\ -2x + y + 5z = 4 \\ x + y + 2z = 1 \end{cases} \qquad \begin{bmatrix} 3 & 7 & 1 & 0 \\ -2 & 1 & 5 & 4 \\ 1 & 1 & 2 & 1 \end{bmatrix}$$

↓ Interchange Eq. 1 and Eq. 3   ↓ $R_1 \leftrightarrow R_3$

$$\begin{cases} x + y + 2z = 1 \\ -2x + y + 5z = 4 \\ 3x + 7y + z = 0 \end{cases} \qquad \begin{bmatrix} 1 & 1 & 2 & 1 \\ -2 & 1 & 5 & 4 \\ 3 & 7 & 1 & 0 \end{bmatrix}$$

| Add 2 times Eq. 1 to Eq. 2 | $R_2 \rightarrow R_2 + 2R_1$
↓ Add -3 times Eq. 1 to Eq. 3   ↓ $R_3 \rightarrow R_3 - 3R_1$

$$\begin{cases} x + y + 2z = 1 \\ 3y + 9z = 6 \\ 4y - 5z = -3 \end{cases} \qquad \begin{bmatrix} 1 & 1 & 2 & 1 \\ 0 & 3 & 9 & 6 \\ 0 & 4 & -5 & -3 \end{bmatrix}$$

↓ Multiply Eq. 2 by $\tfrac{1}{3}$ ↓ $R_2 \rightarrow \tfrac{1}{3}R_2$

$$\begin{cases} x + y + 2z = 1 \\ y + 3z = 2 \\ 4y - 5z = -3 \end{cases} \qquad \begin{bmatrix} 1 & 1 & 2 & 1 \\ 0 & 1 & 3 & 2 \\ 0 & 4 & -5 & -3 \end{bmatrix}$$

↓ Add −4 times Eq. 2 to Eq. 3 ↓ $R_3 \to R_3 - 4R_2$

$$\begin{cases} x + y + 2z = 1 \\ y + 3z = 2 \\ -17z = -11 \end{cases} \qquad \begin{bmatrix} 1 & 1 & 2 & 1 \\ 0 & 1 & 3 & 2 \\ 0 & 0 & -17 & -11 \end{bmatrix}$$

The system is now in triangular form, from which we find the solution in the usual way.

Answer $\left(-\frac{6}{17}, \frac{1}{17}, \frac{11}{17}\right)$

EXAMPLE 2 Use matrix notation to solve the system

$$\begin{cases} x + y + z - w = -2 \\ x + 2y - z - w = 4 \\ 3x + y + 2z - 4w = -4 \\ 3y + 4z + 5w = -7 \end{cases}$$

SOLUTION We represent the system by a matrix, and then reduce that matrix to echelon form by a sequence of elementary row operations:

$$\begin{bmatrix} 1 & 1 & 1 & -1 & -2 \\ 1 & 2 & -1 & -1 & 4 \\ 3 & 1 & 2 & -4 & -4 \\ 0 & 3 & 4 & 5 & -7 \end{bmatrix} \to \begin{bmatrix} 1 & 1 & 1 & -1 & -2 \\ 0 & 1 & -2 & 0 & 6 \\ 0 & -2 & -1 & -1 & 2 \\ 0 & 3 & 4 & 5 & -7 \end{bmatrix} \quad \begin{matrix} R_2 \to R_2 - R_1 \\ R_3 \to R_3 - 3R_1 \end{matrix}$$

$$\to \begin{bmatrix} 1 & 1 & 1 & -1 & -2 \\ 0 & 1 & -2 & 0 & 6 \\ 0 & 0 & -5 & -1 & 14 \\ 0 & 0 & 10 & 5 & -25 \end{bmatrix} \quad \begin{matrix} R_3 \to R_3 + 2R_2 \\ R_4 \to R_4 - 3R_2 \end{matrix}$$

$$\to \begin{bmatrix} 1 & 1 & 1 & -1 & -2 \\ 0 & 1 & -2 & 0 & 6 \\ 0 & 0 & -5 & -1 & 14 \\ 0 & 0 & 0 & 3 & 3 \end{bmatrix} \quad R_4 \to R_4 + 2R_3$$

The last row represents the equation $3w = 3$, hence $w = 1$. We work backward to find the other variables:

Row 3: $-5z - w = 14$ $\qquad z = -\frac{1}{5}(14 + w) = -\frac{1}{5}(14 + 1) = -3$
Row 2: $y - 2z = 6$ $\qquad y = 2z + 6 = 2(-3) + 6 = 0$
Row 1: $x + y + z - w = -2$ $\qquad x = -2 - y - z + w = -2 - 0 + 3 + 1 = 2$

Answer $(2, 0, -3, 1)$

Remark With some extra work, the final triangular matrix in Example 2 can be further reduced to

$$\begin{bmatrix} 1 & 0 & 0 & 0 & 2 \\ 0 & 1 & 0 & 0 & 0 \\ 0 & 0 & 1 & 0 & -3 \\ 0 & 0 & 0 & 1 & 1 \end{bmatrix}$$

from which you can read off $x = 2$, $y = 0$, $z = -3$, and $w = 1$.

EXAMPLE 3 Solve $\begin{cases} x - 4y + 3z = 1 \\ 2x + y - 4z = 5 \\ 4x + 11y - 18z = 8 \end{cases}$

SOLUTION

$$\begin{bmatrix} 1 & -4 & 3 & 1 \\ 2 & 1 & -4 & 5 \\ 4 & 11 & -18 & 8 \end{bmatrix} \rightarrow \begin{bmatrix} 1 & -4 & 3 & 1 \\ 0 & 9 & -10 & 3 \\ 0 & 27 & -30 & 4 \end{bmatrix} \quad \begin{aligned} R_2 &\rightarrow R_2 - 2R_1 \\ R_3 &\rightarrow R_3 - 4R_1 \end{aligned}$$

$$\rightarrow \begin{bmatrix} 1 & -4 & 3 & 1 \\ 0 & 9 & -10 & 3 \\ 0 & 0 & 0 & -5 \end{bmatrix} \quad R_3 \rightarrow R_3 - 3R_2$$

Hold it! Row 3 represents the equation

$$0 \cdot x + 0 \cdot y + 0 \cdot z = -5$$

a contradiction. Hence the system is inconsistent; it has <u>no solutions</u>.

EXAMPLE 4 Solve $\begin{cases} x + 2y - 2z = 5 \\ 2x - y + 3z = 7 \\ -x + 8y - 12z = 1 \end{cases}$

SOLUTION

$$\begin{bmatrix} 1 & 2 & -2 & 5 \\ 2 & -1 & 3 & 7 \\ -1 & 8 & -12 & 1 \end{bmatrix} \rightarrow \begin{bmatrix} 1 & 2 & -2 & 5 \\ 0 & -5 & 7 & -3 \\ 0 & 10 & -14 & 6 \end{bmatrix} \quad \begin{aligned} R_2 &\rightarrow R_2 - 2R_1 \\ R_3 &\rightarrow R_3 + R_1 \end{aligned}$$

$$\rightarrow \begin{bmatrix} 1 & 2 & -2 & 5 \\ 0 & -5 & 7 & -3 \\ 0 & 0 & 0 & 0 \end{bmatrix} \quad R_3 \rightarrow R_3 + 2R_2$$

The row of zeros indicates that the system is dependent. It is equivalent to the system of *two* equations:

$$\begin{cases} x + 2y - 2z = 5 \\ -5y + 7z = -3 \end{cases} \text{ or } \begin{cases} x + 2y = 5 + 2z \\ -5y = -3 - 7z \end{cases}$$

Solving, we find

$$x = \tfrac{19}{5} - \tfrac{4}{5}z \qquad y = \tfrac{3}{5} + \tfrac{7}{5}z$$

Now z can take any real value t, so there are infinitely many solutions.

Answer $(\tfrac{19}{5} - \tfrac{4}{5}t, \tfrac{3}{5} + \tfrac{7}{5}t, t)$ t any real number

EXERCISES

1–18. Do Exercises 1–18 of Section 2 using matrices.

19–22. Do Exercises 25–28 of Section 2 using matrices.

Solve using matrices

23. $\begin{cases} x - 2y + z = 0 \\ 3x + 4y - z = 0 \\ x - 12y + 5z = 0 \end{cases}$
24. $\begin{cases} 5x + 2y - 3z = 0 \\ 2x - y + 4z = 0 \\ 3x + 3y - 7z = 0 \end{cases}$

25. $\begin{cases} x + y - 2z + 2w = 0 \\ 2x - y - z + 3w = 1 \\ -3x + 2y + 3z + w = 4 \\ x + 3y - 2z + 8w = 8 \end{cases}$

26. $\begin{cases} x + 3z - w = 1 \\ -2x + y + z + 2w = 0 \\ x + y + 10z - w = 4 \\ 2y - z + 3w = 2 \end{cases}$

27. $\begin{cases} 2a - b + c - d - e = 0 \\ 3a + 2b - c + d - 3e = -2 \\ a + 3b + 2c - 2d = 6 \\ -2a + b - 4c + 3d + e = 0 \\ 2b - 2c + d + 2e = 7 \end{cases}$

28. $\begin{cases} a + b + c = 1 \\ 2a - 2b - 4c - d - 3e = 0 \\ 3b + d + e = 3 \\ 3a + 4b + c - 2d = 0 \\ 4a - 9d + 4e = -8 \end{cases}$

Find a cubic function $y = ax^3 + bx^2 + cx + d$ whose graph passes through the points

29. (1, 0), (2, −1), (3, 1), (4, 9)   **30.** (−1, 0), (1, 1), (3, 10), (5, 3)

4 DETERMINANTS

In this section, we deal with square matrices of sizes 2×2 and 3×3. Then, in Section 5, we go on to general $n \times n$ matrices. It is convenient to denote matrices by the notation

$$\begin{bmatrix} a_{11} & a_{12} \\ a_{21} & a_{22} \end{bmatrix}, \quad \begin{bmatrix} a_{11} & a_{12} & a_{13} \\ a_{21} & a_{22} & a_{23} \\ a_{31} & a_{32} & a_{33} \end{bmatrix}, \quad \ldots \quad \begin{bmatrix} a_{11} & a_{12} & \cdots & a_{1n} \\ a_{21} & a_{22} & \cdots & a_{2n} \\ \vdots & \vdots & & \vdots \\ a_{n1} & a_{n2} & \cdots & a_{nn} \end{bmatrix}$$

Thus, a_{ij} is the entry in the i-th row and j-th column.

To each square matrix A is assigned a real *number* called its **determinant** and denoted

$$\det A \quad \text{or} \quad |A| \quad \text{or} \quad \begin{vmatrix} a_{11} & \cdots & a_{1n} \\ \vdots & & \vdots \\ a_{n1} & \cdots & a_{nn} \end{vmatrix}$$

Let us start with a 2×2 matrix A. Its determinant is defined as follows:

Determinant of a 2×2 Matrix

$$|A| = \begin{vmatrix} a_{11} & a_{12} \\ a_{21} & a_{22} \end{vmatrix} = a_{11}a_{22} - a_{12}a_{21}$$

For example,

$$\begin{vmatrix} 5 & 4 \\ 6 & 2 \end{vmatrix} = 5 \cdot 2 - 4 \cdot 6 = -14 \qquad \begin{vmatrix} -1 & 3 \\ -3 & 7 \end{vmatrix} = (-1)7 - 3(-3) = 2$$

Why make such an unusual definition? One reason is that expressions such as $a_{11}a_{22} - a_{12}a_{21}$ arise naturally in solutions of 2×2 linear systems. To see how, let us solve the system

$$\begin{cases} a_{11}x_1 + a_{12}x_2 = b_1 \\ a_{21}x_1 + a_{22}x_2 = b_2 \end{cases}$$

(From now on, we'll write the unknowns as x_1, x_2, x_3, ... instead of x, y, z, ...). For convenience, we assume that the coefficients a_{ij} are all non-zero.

To eliminate x_2 we multiply the first equation by a_{22}, the second by $-a_{12}$, and then add:

$$\begin{aligned} a_{11}a_{22}x_1 + a_{12}a_{22}x_2 &= b_1 a_{22} \\ -a_{12}a_{21}x_1 - a_{12}a_{22}x_2 &= -a_{12}b_2 \end{aligned}$$

(add) $\qquad (a_{11}a_{22} - a_{12}a_{21})x_1 = b_1 a_{22} - a_{12} b_2$

Assuming $a_{11}a_{22} - a_{12}a_{21} \neq 0$, we divide both sides by this number to find x_1. A similar elimination gives x_2. The results are

$$x_1 = \frac{b_1 a_{22} - a_{12} b_2}{a_{11} a_{22} - a_{12} a_{21}} \qquad x_2 = \frac{a_{11} b_2 - b_1 a_{21}}{a_{11} a_{22} - a_{12} a_{21}}$$

Both numerators and both denominators are determinants! Therefore, we can write these formulas as

$$x_1 = \frac{\begin{vmatrix} b_1 & a_{12} \\ b_2 & a_{22} \end{vmatrix}}{\begin{vmatrix} a_{11} & a_{12} \\ a_{21} & a_{22} \end{vmatrix}} \qquad x_2 = \frac{\begin{vmatrix} a_{11} & b_1 \\ a_{21} & b_2 \end{vmatrix}}{\begin{vmatrix} a_{11} & a_{12} \\ a_{21} & a_{22} \end{vmatrix}}$$

These formulas are the simplest case of a general statement called **Cramer's rule**. (It is not hard to check that they hold even if some coefficients are zero, provided the denominator is not zero.)

Cramer's Rule for Two Equations in Two Unknowns The linear system

$$\begin{cases} a_{11}x_1 + a_{12}x_2 = b_1 \\ a_{21}x_1 + a_{22}x_2 = b_2 \end{cases} \quad \text{where} \quad |A| = \begin{vmatrix} a_{11} & a_{12} \\ a_{21} & a_{22} \end{vmatrix} \neq 0$$

has a unique solution (x_1, x_2) given by

$$x_1 = \frac{\begin{vmatrix} b_1 & a_{12} \\ b_2 & a_{22} \end{vmatrix}}{|A|} \qquad x_2 = \frac{\begin{vmatrix} a_{11} & b_1 \\ a_{21} & b_2 \end{vmatrix}}{|A|}$$

Note that A is the matrix of the coefficients on the left side of the given system. To compute x_1, replace the x_1-column of A by the column of constants on the right side of the system. Then take the determinant and divide by $|A|$. For x_2, replace the x_2-column of A by the column of constants.

Cramer's rule applies only if $|A| \neq 0$. When $|A| = 0$, the system is either inconsistent or dependent.

EXAMPLE 1 Use Cramer's rule to solve $\begin{cases} 3x_1 - 4x_2 = 7 \\ 5x_1 + 9x_2 = -1 \end{cases}$

SOLUTION The determinant of the system is
$$|A| = \begin{vmatrix} 3 & -4 \\ 5 & 9 \end{vmatrix} = 3 \cdot 9 - (-4) \cdot 5 = 47$$
Hence, by Cramer's rule,
$$x_1 = \tfrac{1}{47} \begin{vmatrix} 7 & -4 \\ -1 & 9 \end{vmatrix} = \underline{\tfrac{59}{47}} \qquad x_2 = \tfrac{1}{47} \begin{vmatrix} 3 & 7 \\ 5 & -1 \end{vmatrix} = \underline{-\tfrac{38}{47}}$$

Now we define the determinant of a 3×3 matrix:

Determinant of a 3×3 Matrix

$$\begin{vmatrix} a_{11} & a_{12} & a_{13} \\ a_{21} & a_{22} & a_{23} \\ a_{31} & a_{32} & a_{33} \end{vmatrix} = a_{11}a_{22}a_{33} - a_{11}a_{23}a_{32} + a_{12}a_{23}a_{31} \\ - a_{12}a_{21}a_{33} + a_{13}a_{21}a_{32} - a_{13}a_{22}a_{31}$$

A 3×3 determinant can be evaluated in several ways by grouping the six terms in pairs and removing a common factor from each pair, for example,
$$a_{11}(a_{22}a_{33} - a_{23}a_{32}) - a_{12}(a_{21}a_{33} - a_{23}a_{31}) + a_{13}(a_{21}a_{32} - a_{22}a_{31})$$
Each quantity in parentheses is a 2×2 determinant, so we can write
$$\begin{vmatrix} a_{11} & a_{12} & a_{13} \\ a_{21} & a_{22} & a_{23} \\ a_{31} & a_{32} & a_{33} \end{vmatrix} = a_{11}\begin{vmatrix} a_{22} & a_{23} \\ a_{32} & a_{33} \end{vmatrix} - a_{12}\begin{vmatrix} a_{21} & a_{23} \\ a_{31} & a_{33} \end{vmatrix} + a_{13}\begin{vmatrix} a_{21} & a_{22} \\ a_{31} & a_{32} \end{vmatrix}$$
The coefficients of the 2×2 determinants are elements of the first row of A, with suitable signs. There are similar formulas based on the second and third rows, and also on each of the three columns.

To describe these six formulas, let us take a typical entry, a_{ij}. We cross out the i-th row and j-th column. What remains is a 2×2 matrix, whose determinant M_{ij} is called the **minor** of a_{ij}. For instance,

$$M_{31} = \begin{vmatrix} \cancel{a_{11}} & a_{12} & a_{13} \\ \cancel{a_{21}} & a_{22} & a_{23} \\ \cancel{a_{31}} & \cancel{a_{32}} & \cancel{a_{33}} \end{vmatrix} = \begin{vmatrix} a_{12} & a_{13} \\ a_{22} & a_{23} \end{vmatrix} \qquad M_{23} = \begin{vmatrix} a_{11} & a_{12} & \cancel{a_{13}} \\ \cancel{a_{21}} & \cancel{a_{22}} & \cancel{a_{23}} \\ a_{31} & a_{32} & \cancel{a_{33}} \end{vmatrix} = \begin{vmatrix} a_{11} & a_{12} \\ a_{31} & a_{32} \end{vmatrix}$$

Now a 3×3 determinant can be expressed in terms of minors.

Expansion by Minors of Rows

$$\begin{vmatrix} a_{11} & a_{12} & a_{13} \\ a_{21} & a_{22} & a_{23} \\ a_{31} & a_{32} & a_{33} \end{vmatrix} = \begin{cases} a_{11}M_{11} - a_{12}M_{12} + a_{13}M_{13} & \text{(by first row)} \\ -a_{21}M_{21} + a_{22}M_{22} - a_{23}M_{23} & \text{(by second row)} \\ a_{31}M_{31} - a_{32}M_{32} + a_{33}M_{33} & \text{(by third row)} \end{cases}$$

Expansion by Minors of Columns

$$\begin{vmatrix} a_{11} & a_{12} & a_{13} \\ a_{21} & a_{22} & a_{23} \\ a_{31} & a_{32} & a_{33} \end{vmatrix} = \begin{cases} a_{11}M_{11} - a_{21}M_{21} + a_{31}M_{31} & \text{(by first column)} \\ -a_{12}M_{12} + a_{22}M_{22} - a_{32}M_{32} & \text{(by second column)} \\ a_{13}M_{13} - a_{23}M_{23} + a_{33}M_{33} & \text{(by third column)} \end{cases}$$

The sign in front of the term $a_{ij}M_{ij}$ is $(-1)^{i+j}$. It is the sign in the i-th row and j-th column of

$$\begin{bmatrix} + & - & + \\ - & + & - \\ + & - & + \end{bmatrix}$$

If we attach this sign to the minor M_{ij}, we get the **cofactor** C_{ij}. Thus,

$$C_{ij} = (-1)^{i+j} M_{ij}$$

Written with cofactors instead of minors, the expansion formulas have all plus signs. For example,

$$|A| = a_{21}C_{21} + a_{22}C_{22} + a_{23}C_{23} \quad \text{(by second row)}$$

EXAMPLE 2 Evaluate $|A| = \begin{vmatrix} 2 & -3 & 1 \\ 2 & 4 & -5 \\ 3 & 0 & 2 \end{vmatrix}$

(a) by the first row (b) by the second column

SOLUTION

(a) $|A| = 2M_{11} - (-3)M_{12} + (1)M_{13}$

$$= 2 \begin{vmatrix} 4 & -5 \\ 0 & 2 \end{vmatrix} + 3 \begin{vmatrix} 2 & -5 \\ 3 & 2 \end{vmatrix} + \begin{vmatrix} 2 & 4 \\ 3 & 0 \end{vmatrix}$$

$$= 2(8) + 3(19) + (-12) = \underline{61}$$

(b) $|A| = -(-3)M_{12} + 4M_{22} - 0 \cdot M_{32}$

$$= 3 \begin{vmatrix} 2 & -5 \\ 3 & 2 \end{vmatrix} + 4 \begin{vmatrix} 2 & 1 \\ 3 & 2 \end{vmatrix} = 3(19) + 4(1) = \underline{61}$$

One use for 3×3 determinants is in solutions of 3×3 linear systems. Let us state Cramer's rule for such systems. We omit the proof, which is not hard but tedious.

Cramer's Rule for Three Equations in Three Unknowns

The linear system

$$\begin{cases} a_{11}x_1 + a_{12}x_2 + a_{13}x_3 = b_1 \\ a_{21}x_1 + a_{22}x_2 + a_{23}x_3 = b_2 \\ a_{31}x_1 + a_{32}x_2 + a_{33}x_3 = b_3 \end{cases} \quad \text{where} \quad |A| = \begin{vmatrix} a_{11} & a_{12} & a_{13} \\ a_{21} & a_{22} & a_{23} \\ a_{31} & a_{32} & a_{33} \end{vmatrix} \neq 0$$

has a unique solution (x_1, x_2, x_3) given by

$$x_1 = \frac{\begin{vmatrix} b_1 & a_{12} & a_{13} \\ b_2 & a_{22} & a_{23} \\ b_3 & a_{32} & a_{33} \end{vmatrix}}{|A|} \quad x_2 = \frac{\begin{vmatrix} a_{11} & b_1 & a_{13} \\ a_{21} & b_2 & a_{23} \\ a_{31} & b_3 & a_{33} \end{vmatrix}}{|A|} \quad x_3 = \frac{\begin{vmatrix} a_{11} & a_{12} & b_1 \\ a_{21} & a_{22} & b_2 \\ a_{31} & a_{32} & b_3 \end{vmatrix}}{|A|}$$

EXAMPLE 3 Solve by Cramer's rule $\begin{cases} 2x_1 - 3x_2 + x_3 = 1 \\ 2x_1 + 4x_2 - 5x_3 = 2 \\ 3x_1 + 2x_3 = 0 \end{cases}$

SOLUTION The determinant in all three denominators is

$$|A| = \begin{vmatrix} 2 & -3 & 1 \\ 2 & 4 & -5 \\ 3 & 0 & 2 \end{vmatrix}$$

According to Example 2, we have $|A| = 61$. Hence, by Cramer's rule,

$$x_1 = \frac{1}{61} \begin{vmatrix} 1 & -3 & 1 \\ 2 & 4 & -5 \\ 0 & 0 & 2 \end{vmatrix} = \frac{|A_1|}{61} \qquad x_2 = \frac{1}{61} \begin{vmatrix} 2 & 1 & 1 \\ 2 & 2 & -5 \\ 3 & 0 & 2 \end{vmatrix} = \frac{|A_2|}{61}$$

$$x_3 = \frac{1}{61} \begin{vmatrix} 2 & -3 & 1 \\ 2 & 4 & 2 \\ 3 & 0 & 0 \end{vmatrix} = \frac{|A_3|}{61}$$

To compute $|A_1|$, we expand by its third row, taking advantage of the two zeros there:

$$|A_1| = \begin{vmatrix} 1 & -3 & 1 \\ 2 & 4 & -5 \\ 0 & 0 & 2 \end{vmatrix} = 0 - 0 + 2 \begin{vmatrix} 1 & -3 \\ 2 & 4 \end{vmatrix} = 2(10) = 20$$

Next, we expand $|A_2|$ by its second column:

$$|A_2| = \begin{vmatrix} 2 & 1 & 1 \\ 2 & 2 & -5 \\ 3 & 0 & 2 \end{vmatrix} = -1 \cdot \begin{vmatrix} 2 & -5 \\ 3 & 2 \end{vmatrix} + 2 \begin{vmatrix} 2 & 1 \\ 3 & 2 \end{vmatrix} - 0 = -(19) + 2(1) = -17$$

Finally, we expand $|A_3|$ by its third row:

$$|A_3| = \begin{vmatrix} 2 & -3 & 1 \\ 2 & 4 & 2 \\ 3 & 0 & 0 \end{vmatrix} = 3 \begin{vmatrix} -3 & 1 \\ 4 & 2 \end{vmatrix} - 0 + 0 = 3(-10) = -30$$

Therefore

$$(x_1, x_2, x_3) = \left(\frac{|A_1|}{61}, \frac{|A_2|}{61}, \frac{|A_3|}{61} \right) = \left(\frac{20}{61}, -\frac{17}{61}, -\frac{30}{61} \right)$$

Remark Doing Example 3 required evaluating four determinants. Because these contained some 0 entries, the computations were not as long as they could have been. In the next section we discuss ways of reducing the work in computing determinants, even when they contain no zeros.

EXERCISES

Compute the determinant

1. $\begin{vmatrix} 2 & 1 \\ -1 & 3 \end{vmatrix}$ 2. $\begin{vmatrix} 4 & -3 \\ 2 & 1 \end{vmatrix}$ 3. $\begin{vmatrix} 8 & 1 \\ -3 & -2 \end{vmatrix}$ 4. $\begin{vmatrix} 2 & 3 \\ 3 & 5 \end{vmatrix}$ 5. $\begin{vmatrix} 1 & 2 \\ 2 & 4 \end{vmatrix}$

6. $\begin{vmatrix} 4 & -4 \\ 5 & -5 \end{vmatrix}$ 7. $\begin{vmatrix} a & -b \\ 0 & b \end{vmatrix}$ 8. $\begin{vmatrix} 0 & a \\ b & 0 \end{vmatrix}$ 9. $\begin{vmatrix} x & x-1 \\ x+1 & x \end{vmatrix}$ 10. $\begin{vmatrix} x & -3 \\ 3x & 1 \end{vmatrix}$

11–20. Use Cramer's rule to do Exercises 1–10 of Section 1.

Compute $\begin{vmatrix} -3 & 0 & 3 \\ 4 & -2 & 1 \\ 5 & 2 & 1 \end{vmatrix}$ by minors of

21. first row, second row

22. second column, third column

Compute the determinant

23. $\begin{vmatrix} 2 & 1 & 6 \\ 0 & 3 & 5 \\ 0 & 0 & 4 \end{vmatrix}$ **24.** $\begin{vmatrix} -1 & 0 & 0 \\ 7 & 2 & 0 \\ -6 & 3 & 5 \end{vmatrix}$ **25.** $\begin{vmatrix} 2 & 0 & 1 \\ 3 & 1 & 6 \\ 4 & 2 & 3 \end{vmatrix}$ **26.** $\begin{vmatrix} 2 & 3 & 0 \\ 4 & 6 & 9 \\ 7 & 1 & 2 \end{vmatrix}$

Use Cramer's rule to solve the system

27. $\begin{cases} x_1 + x_2 - 2x_3 = 0 \\ 3x_1 - 2x_2 = 0 \\ x_2 - 3x_3 = 4 \end{cases}$ **28.** $\begin{cases} x_1 - 2x_2 + x_3 = 2 \\ x_1 + x_2 = 1 \\ x_2 - 5x_3 = 0 \end{cases}$

29. $\begin{cases} 2x_1 + 5x_2 = 1 \\ 3x_2 - x_3 = -4 \\ x_1 + 2x_3 = 6 \end{cases}$ **30.** $\begin{cases} x_1 + x_2 + x_3 = 4 \\ 2x_1 - 3x_2 = 1 \\ x_1 - 4x_3 = -3 \end{cases}$

Verify that

31. $\begin{vmatrix} a & b \\ c & d \end{vmatrix} = - \begin{vmatrix} c & d \\ a & b \end{vmatrix}$ **32.** $\begin{vmatrix} ka & b \\ kc & d \end{vmatrix} = k \begin{vmatrix} a & b \\ c & d \end{vmatrix}$

33. $\begin{vmatrix} a & b \\ c + ka & d + kb \end{vmatrix} = \begin{vmatrix} a & b \\ c & d \end{vmatrix}$ **34.** $\begin{vmatrix} a & b + ka \\ c & d + kc \end{vmatrix} = \begin{vmatrix} a & b \\ c & d \end{vmatrix}$

35. $\begin{vmatrix} a & b & c \\ 0 & d & e \\ 0 & 0 & f \end{vmatrix} = adf$ **36.** $\begin{vmatrix} a & 0 & 0 \\ b & c & 0 \\ d & e & f \end{vmatrix} = acf$

37. $\begin{vmatrix} a & b & c \\ a & b & c \\ x & y & z \end{vmatrix} = 0$ **38.** $\begin{vmatrix} a & a & r \\ b & b & s \\ c & c & t \end{vmatrix} = 0$

Solve for x

39. $\begin{vmatrix} x & x+1 \\ 5 & 2 \end{vmatrix} = 7$ **40.** $\begin{vmatrix} 2 & x \\ -1 & 3 \end{vmatrix} = \begin{vmatrix} 3 & 2 \\ 4x & 1 \end{vmatrix}$

5 GENERAL DETERMINANTS AND THEIR PROPERTIES

Given a 4×4 matrix A, we define its minors M_{ij} as before. Delete the i-th row and j-th column; then M_{ij} is the determinant of the resulting matrix. We define the determinant of A by

$$|A| = a_{11}M_{11} - a_{12}M_{12} + a_{13}M_{13} - a_{14}M_{14}$$

Thus $|A|$ is expressed in terms of four 3×3 determinants.

Systems of Equations and Inequalities

The definition of the general $n \times n$ determinant follows the same pattern.

> **Determinant of an $n \times n$ Matrix** If A is an $n \times n$ matrix, then
> $$|A| = a_{11}M_{11} - a_{12}M_{12} + a_{13}M_{13} - + \cdots + (-1)^{1+n}a_{1n}M_{1n}$$

Notice that $|A|$ is *defined* as an expansion by minors of its first row. In courses on linear algebra, it is proved that $|A|$ can be evaluated by the minors of *any* row or column.

> The determinant of an $n \times n$ matrix ($n > 1$) can be evaluated by the minors of any row or column.

EXAMPLE 1 Evaluate $|A| = \begin{vmatrix} 3 & 0 & 1 & 0 \\ -2 & 0 & 5 & 4 \\ 8 & 2 & -9 & 7 \\ -1 & 0 & 2 & 6 \end{vmatrix}$

SOLUTION We expand by the second column, taking advantage of its three zeros:

$$|A| = -0 \cdot M_{12} + 0 \cdot M_{22} - 2M_{32} + 0 \cdot M_{42} = -2 \begin{vmatrix} 3 & 1 & 0 \\ -2 & 5 & 4 \\ -1 & 2 & 6 \end{vmatrix}$$

We expand the determinant M_{32} by its first row:

$$M_{32} = \begin{vmatrix} 3 & 1 & 0 \\ -2 & 5 & 4 \\ -1 & 2 & 6 \end{vmatrix} = 3 \begin{vmatrix} 5 & 4 \\ 2 & 6 \end{vmatrix} - \begin{vmatrix} -2 & 4 \\ -1 & 6 \end{vmatrix}$$

$$= 3(22) - (-8) = 74$$

Therefore, $|A| = -2M_{32} = \underline{-148}$

Properties of Determinants

Determinants have a number of useful general properties. We are going to state some of these, mostly without proof. They are all easy to check for 2×2 determinants. It will be understood that each statement refers to an $n \times n$ matrix A, where n is any integer greater than 1.

> (1) If all entries in a row (or column) are zero, then $|A| = 0$.

Just expand the determinant by that row. All terms are 0, so $|A| = 0$.

> (2) If two rows (or columns) are interchanged, then the sign of the determinant is reversed.

For example,

$$\begin{vmatrix} 1 & 2 & 3 \\ 5 & 7 & 9 \\ 0 & 3 & 4 \end{vmatrix} = -\begin{vmatrix} 5 & 7 & 9 \\ 1 & 2 & 3 \\ 0 & 3 & 4 \end{vmatrix} \qquad \begin{vmatrix} 2 & 0 & 6 \\ 2 & 1 & 7 \\ 2 & 3 & 8 \end{vmatrix} = -\begin{vmatrix} 6 & 0 & 2 \\ 7 & 1 & 2 \\ 8 & 3 & 2 \end{vmatrix}$$

(interchange Rows 1 and 2)   (interchange Columns 1 and 3)

> (3) If two rows (columns) are identical, then $|A| = 0$.

Interchanging the two identical rows produces a matrix A'. On the one hand, $A' = A$ so $|A'| = |A|$. On the other hand, Property (2) gives $|A'| = -|A|$. Hence $|A| = -|A|$, which forces $|A| = 0$.

> (4) If each entry of a row (column) is multiplied by k, then the determinant is multiplied by k.

For example,

$$\begin{vmatrix} 10 & 20 & -50 \\ 0 & 6 & 7 \\ 2 & 1 & 4 \end{vmatrix} = 10 \begin{vmatrix} 1 & 2 & -5 \\ 0 & 6 & 7 \\ 2 & 1 & 4 \end{vmatrix}$$

In practice, (4) allows us to factor out a common factor from a row or a column.

> (5) If two rows (columns) are proportional, then $|A| = 0$.

For example, suppose that A is a 3×3 matrix whose second and third columns are proportional. Then

$$|A| = \begin{vmatrix} a_{11} & a_{21} & ka_{21} \\ a_{12} & a_{22} & ka_{22} \\ a_{13} & a_{23} & ka_{23} \end{vmatrix} = k \begin{vmatrix} a_{11} & a_{21} & a_{21} \\ a_{12} & a_{22} & a_{22} \\ a_{13} & a_{23} & a_{23} \end{vmatrix}$$

The determinant on the right is zero because it has two identical columns.

For computation of determinants, probably the most useful property is:

> (6) The value of a determinant is unchanged when a multiple of one row (column) is added to another row (column).

This property allows us to create zeros in strategic positions and greatly reduce the work in evaluating a determinant.

You may have noticed that (2), (4), and (6) involve the elementary row operations discussed in Section 3. We can summarize these rules as follows:

Elementary Row Operation on A	Effect on $|A|$
$R_i \leftrightarrow R_j$	Reverses sign
$R_i \to kR_i$	Multiplies $|A|$ by k
$R_i \to R_i + kR_j$	None

240 Systems of Equations and Inequalities

The corresponding *column* operations produce the same effects.

EXAMPLE 2 Evaluate $\begin{vmatrix} 1 & -1 & 2 \\ 5 & 6 & 8 \\ -4 & 8 & 1 \end{vmatrix}$

SOLUTION Using Property (6), we create two zeros in the first row. We add the first column to the second, then subtract twice the first column from the third:

$$\begin{vmatrix} 1 & -1 & 2 \\ 5 & 6 & 8 \\ -4 & 8 & 1 \end{vmatrix} = \begin{vmatrix} 1 & 0 & 2 \\ 5 & 11 & 8 \\ -4 & 4 & 1 \end{vmatrix} = \begin{vmatrix} 1 & 0 & 0 \\ 5 & 11 & -2 \\ -4 & 4 & 9 \end{vmatrix}$$

$$= \begin{vmatrix} 11 & -2 \\ 4 & 9 \end{vmatrix} \qquad \text{Expanding by Row 1}$$

$$= 11 \cdot 9 - (-2)4 = \underline{107}$$

EXAMPLE 3 Evaluate $\begin{vmatrix} 2 & 3 & -2 & 0 \\ 1 & 2 & -5 & 1 \\ 3 & -1 & -7 & 2 \\ 2 & 0 & 13 & -1 \end{vmatrix}$

SOLUTION Column 4 already contains one zero. Let us use the 1 in that column to knock out the 2 and the -1 below it:

$$\begin{vmatrix} 2 & 3 & -2 & 0 \\ 1 & 2 & -5 & 1 \\ 3 & -1 & -7 & 2 \\ 2 & 0 & 13 & -1 \end{vmatrix} = \begin{vmatrix} 2 & 3 & -2 & 0 \\ 1 & 2 & -5 & 1 \\ 1 & -5 & 3 & 0 \\ 3 & 2 & 8 & 0 \end{vmatrix} \qquad \begin{array}{l} R_3 \to R_3 - 2R_1 \\ R_4 \to R_4 + R_1 \end{array}$$

$$= \begin{vmatrix} 2 & 3 & -2 \\ 1 & -5 & 3 \\ 3 & 2 & 8 \end{vmatrix} \qquad \text{Expanding by Col. 4}$$

$$= \begin{vmatrix} 0 & 13 & -8 \\ 1 & -5 & 3 \\ 0 & 17 & -1 \end{vmatrix} \qquad \begin{array}{l} R_1 \to R_1 - 2R_2 \\ R_3 \to R_3 - 3R_2 \end{array}$$

$$= - \begin{vmatrix} 13 & -8 \\ 17 & -1 \end{vmatrix} \qquad \text{Expanding by Col. 1}$$

$$= -[13(-1) - (-8)17] = \underline{-123}$$

EXAMPLE 4 With as little computation as possible, show that

$$\begin{vmatrix} 3 & 4 & 5 \\ 7 & 8 & 9 \\ 23 & 24 & 25 \end{vmatrix} = 0$$

SOLUTION Subtract the first row from the second and the third:

$$\begin{vmatrix} 3 & 4 & 5 \\ 4 & 4 & 4 \\ 20 & 20 & 20 \end{vmatrix}$$

This determinant is 0 because its second and third rows are proportional.

Remark on Cramer's rule We saw in Section 4 that Cramer's rule gives formulas for solution of 2×2 and 3×3 linear systems. There is also a natural extension of Cramer's rule to $n \times n$ systems. But it is a very poor method for computing solutions. As n increases, the amount of arithmetic required becomes enormous. Even a 4×4 system requires evaluating five 4×4 determinants. Gaussian elimination is a much more practical method.

EXERCISES

Evaluate

1. $\begin{vmatrix} 1 & 2 & 3 \\ 2 & 1 & 3 \\ 3 & 2 & 1 \end{vmatrix}$

2. $\begin{vmatrix} 1 & 2 & 3 \\ 1 & 4 & 9 \\ 1 & 8 & 27 \end{vmatrix}$

3. $\begin{vmatrix} 1 & 1 & 1 \\ 1 & -1 & 2 \\ 1 & 1 & 4 \end{vmatrix}$

4. $\begin{vmatrix} 2 & 3 & 6 \\ -1 & 2 & -9 \\ 1 & 3 & -2 \end{vmatrix}$

5. $\begin{vmatrix} 3 & 6 & -1 \\ 2 & 1 & -2 \\ 4 & -3 & 1 \end{vmatrix}$

6. $\begin{vmatrix} -2 & 4 & 3 \\ -3 & 2 & 5 \\ 5 & 1 & 2 \end{vmatrix}$

7. $\begin{vmatrix} 2 & 4 & 6 \\ -1 & 3 & 7 \\ 9 & 12 & 15 \end{vmatrix}$

8. $\begin{vmatrix} 7 & 14 & -28 \\ 12 & 60 & -12 \\ -5 & 15 & 20 \end{vmatrix}$

9. $\begin{vmatrix} 8 & 7 & 2 \\ 5 & 4 & 3 \\ 6 & 8 & 3 \end{vmatrix}$

10. $\begin{vmatrix} 1 & 2 & 3 \\ 4 & 5 & 6 \\ 7 & 8 & 9 \end{vmatrix}$

11. $\begin{vmatrix} x & 1 & 0 \\ -1 & x & 1 \\ 0 & -1 & x \end{vmatrix}$

12. $\begin{vmatrix} x & y & y \\ y & x & y \\ y & y & x \end{vmatrix}$

13. $\begin{vmatrix} 1 & 2 & 3 & 1 \\ 0 & 2 & 1 & -3 \\ -1 & -4 & 0 & 2 \\ -3 & 1 & 1 & 1 \end{vmatrix}$

14. $\begin{vmatrix} 5 & 0 & 1 & 2 \\ 2 & 4 & -4 & 3 \\ 1 & 3 & -3 & 2 \\ 0 & -2 & 2 & 1 \end{vmatrix}$

15. $\begin{vmatrix} 1 & -1 & 0 & 0 \\ -1 & 1 & -1 & 0 \\ 0 & -1 & 1 & -1 \\ 0 & 0 & -1 & 1 \end{vmatrix}$

16. $\begin{vmatrix} 0 & 0 & 0 & 4 \\ 0 & 0 & 3 & 0 \\ 0 & 2 & 0 & 0 \\ 1 & 0 & 0 & 0 \end{vmatrix}$

17. $\begin{vmatrix} 6 & 7 & 0 & 0 \\ 4 & 5 & 0 & 0 \\ 0 & 0 & 1 & 4 \\ 0 & 0 & 2 & 5 \end{vmatrix}$

18. $\begin{vmatrix} 1 & 1 & 1 & 1 \\ 1 & 1 & -1 & -1 \\ 1 & -1 & 1 & -1 \\ 1 & -1 & -1 & 1 \end{vmatrix}$

19. $\begin{vmatrix} 2 & 0 & 0 & 0 & 0 \\ 5 & 3 & 0 & 0 & 0 \\ 3 & 4 & 4 & 0 & 0 \\ 6 & 1 & 7 & 5 & 0 \\ 8 & 9 & 6 & 3 & 2 \end{vmatrix}$

20. $\begin{vmatrix} 1 & 2 & 0 & 0 & 0 \\ 2 & 6 & 0 & 0 & 0 \\ 0 & 0 & 2 & -1 & -2 \\ 0 & 0 & 1 & 3 & 5 \\ 0 & 0 & 4 & -7 & 1 \end{vmatrix}$

21–30. Use Cramer's rule to do Exercises 1–10 of Section 2.

True or false? Why?

31. Doubling each entry in a square matrix doubles its determinant.

32. If each entry in a square matrix is positive, its determinant is positive.

The solution set of the system is the intersection of the two half-planes (Fig. 3). The corner (6, 2) is the intersection of the lines. It is found by solving the system of equations

$$\begin{cases} x + y = 8 \\ x - 2y = 2 \end{cases}$$

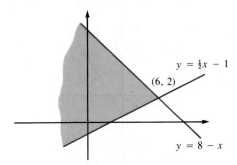

FIG. 3

EXAMPLE 2 Sketch the solution set

$$\begin{cases} x + y \le 8 & x - 2y \le 2 \\ x \ge 0 & y \ge 0 & y - x \le 5 \end{cases}$$

SOLUTION The top two inequalities determine the region in Figure 3. The inequalities $x \ge 0$ and $y \ge 0$ cut that region down to the first quadrant (Fig. 4a). Finally, the inequality $y - x \le 5$ determines the half-plane below and on the line $y = x - 5$. This cuts off the part of the region above that line, leaving the desired solution set shown in Figure 4b.

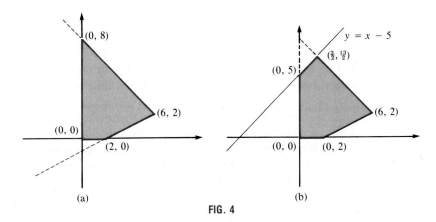

FIG. 4

Linear Programming

Linear programming (in two variables) is a technique for finding the maximum or minimum value of a linear function $f = ax + by$, where x and y must satisfy a system of linear inequalities. The function f is called the **objective function,** and the inequalities are called **constraints.**

6 Systems of Linear Inequalities and Linear Programming

Linear programming was developed in the 1940's and has become an important mathematical technique, especially in helping management make decisions. In a typical application, the objective function expresses profit to be maximized, and the constraints express various limitations such as availability of raw material, labor, time, machines, storage space, etc. The next example illustrates the type of mathematics involved.

EXAMPLE 3 Maximize $f = 3x + y$ subject to the constraints

$$\begin{cases} x + y \leq 8 & x - 2y \leq 2 \\ x \geq 0 & y \geq 0 \quad y - x \leq 5 \end{cases}$$

SOLUTION At first this seems like a very complicated problem in algebra. But looking at it geometrically leads to a solution.

The constraints determine the region discussed in Example 2 and shown in Figure 4b. This region is called the **constraint set** or set of **feasible solutions** of the problem. We denote it by S. Now the problem can be stated: Find the largest value of $f = 3x + y$ for all points (x, y) in S.

To solve this problem, we plot the lines $3x + y = c$ for various values of c, as in Figure 5.

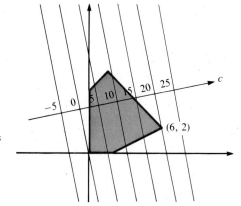

FIG. 5
The lines $3x + y = c$ for various values of c

We obtain a family of parallel lines moving in the direction indicated as c increases. Among those that intersect S, the line with the highest value of c is the one through the corner $(6, 2)$. That value of c is $3 \cdot 6 + 2 = 20$. Therefore, the largest value of $3x + y$ in S is 20.

Answer The maximum value is 20. It occurs only for $x = 6$ and $y = 2$.

Example 3 illustrates an important principle from the theory of linear programming: The maximum or minimum (if it exists) occurs at a vertex (corner) of the constraint set. When the constraint set is a polygon together with its interior (as in Example 3), the maximum and minimum always exist. Otherwise, they may not. For example, there is no maximum value of $x + y$ subject to $x \geq 0$, $y \geq 0$ (the first quadrant).

> If a linear programming problem has a solution, then the solution occurs at a vertex of the constraint set.
>
> A solution always exists when the constraint set is a polygon together with its interior.

EXAMPLE 4 A refinery produces gasoline and heating oil with a combined capacity of at most 30,000 barrels per day. It must produce at least 10,000 barrels of gasoline per day. Also it must produce at least half as much oil as gas, but no more than 15,000 barrels per day. The profit is \$1.50/barrel on gas and \$2.00/barrel on oil. How many barrels of each fuel should the refinery produce daily in order to maximize its profit?

SOLUTION Suppose the daily output is $1000x$ barrels of gas and $1000y$ barrels of oil. Then the problem is to maximize the profit function

$$P = 1.5(1000x) + 2(1000y) = 1500x + 2000y$$

subject to the constraints

$$\begin{cases} x + y \leq 30 & \text{(limitation on total capacity)} \\ y \geq \tfrac{1}{2}x & \text{(condition on oil production versus gas production)} \\ x \geq 10, \; y \leq 15 & \text{(restrictions on gas production and oil production)} \end{cases}$$

Plot the constraint set and find its vertices (Fig. 6).

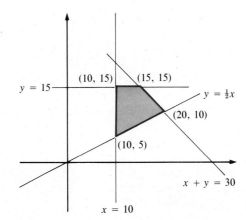

FIG. 6

The maximum must occur at one of the four vertices. So evaluate $P = 1500x + 2000y$ at each vertex and look for the largest result:

Vertex	(10, 5)	(10, 15)	(15, 15)	(20, 10)
Profit = $1500x + 2000y$	25,000	45,000	52,500	50,000

The maximum is 52,500, which occurs at (15, 15).

Answer
  15,000 barrels of gasoline and 15,000 barrels of heating oil

EXERCISES

Sketch the solution set

1. $y > x - 3$
2. $y < x + 1$
3. $y \leq 2x - 5$
4. $y \leq \frac{1}{2}x - 4$
5. $x - y > -1$
6. $x + y > -1$
7. $2x + 3y \leq 6$
8. $4x + 3y \geq 12$
9. $\frac{2}{5}x + \frac{3}{5}y \geq 2$
10. $\frac{1}{4}x + \frac{3}{4}y \leq 1$
11. $7x - 5y \leq 20$
12. $9x - 8y \geq 0$

Sketch the solution set and find all vertices

13. $\begin{cases} x \geq 1 \\ y \leq 2x \end{cases}$
14. $\begin{cases} x + y \leq 1 \\ y \leq 2 \end{cases}$
15. $\begin{cases} x + y \leq 3 \\ 1 \leq x \leq 2 \\ y \geq 0 \end{cases}$

16. $\begin{cases} 2x + y \leq 6 \\ x \geq 0 \\ y \geq 0 \end{cases}$
17. $\begin{cases} 3x - y \geq -2 \\ 2x + y \geq 0 \end{cases}$
18. $\begin{cases} 2x + y \leq 12 \\ 3x - 2y \leq 6 \end{cases}$

19. $\begin{cases} y \leq x + 1 \\ 2x - y \leq 3 \\ x \geq 0, \ y \geq 0 \end{cases}$
20. $\begin{cases} 3x + 2y \leq 12 \\ \frac{1}{3}x - y \geq -2 \\ x \geq 0, \ y \geq 0 \end{cases}$
21. $\begin{cases} x + y \geq 10 \\ 4x + y \geq 16 \\ 2x + 3y \geq 23 \\ x \geq 0, \ y \geq 0 \end{cases}$

22. $\begin{cases} x + y \geq 10 \\ x + 2y \geq 14 \\ x \geq 0, \ y \geq 1 \end{cases}$
23. $\begin{cases} |x - y| \leq 2 \\ 2x + 3y \leq 15 \\ x \geq 0, \ y \geq 0 \end{cases}$
24. $\begin{cases} 0 \leq y \leq x + 3 \\ 2x - 5y \leq 8 \\ 0 \leq x \leq 10, \\ 0 \leq y \leq 7 \end{cases}$

Find the maximum and minimum of the function f subject to the given constraints

25. $f = 4x + 6y$
$\begin{cases} 4x + 5y \leq 20 \\ x \geq 0, \ y \geq 0 \end{cases}$

26. $f = 5x + 4y$
$\begin{cases} x + 2y \leq 10 \\ x \geq 0, \ y \geq 0 \end{cases}$

27. $f = 10x - 3y$
$\begin{cases} x + 2y \leq 12 \\ 2 \leq x \leq 8 \\ y \geq 0 \end{cases}$

28. $f = x - 4y$
$\begin{cases} x \leq y \\ x + y \leq 20 \\ x \geq 0, \ y \geq 3 \end{cases}$

29. $f = 5x - 10y$
$\begin{cases} 0 \leq y \leq x \\ x + y \geq 8 \\ 3x + 2y \leq 30 \end{cases}$

30. $f = -2x + 3y$
$\begin{cases} x + 2y \leq 8 \\ y \leq \frac{1}{2}x + 2 \\ x \geq 1, \ y \geq 1 \end{cases}$

31. $f = x + y$
$\begin{cases} 4x + 3y \leq 12 \\ x - 2 \leq y \leq x - 1 \\ y \geq 0 \end{cases}$

32. $f = x + 5y$
$\begin{cases} 4x + y \leq 15 \\ x + y \geq 2 \\ 0 \leq y \leq x + 1 \end{cases}$

33. A manufacturer of stereo components makes two types of speakers, A and B. Each week he can produce up to 300 units of type A and 400 units of type B, but no more than 600 units combined. His profit is $50/unit on A and $40/unit on B. How many of each type should he produce weekly for the maximum profit?

34. A bakery has the capacity to make a combined maximum of 1000 pies and cakes per day. They need at least 150 cakes per day and at least three times as many pies as cakes. If their profit is 80¢/cake and 50¢/pie, how many of each should they bake daily for the maximum profit?

35. Do Example 4 of the text, assuming the profit is $2.00/barrel on gas and $1.50/barrel on oil.

36. Do Example 4 of the text, assuming a *loss* of 10¢/barrel on oil.

37. A firm manufactures two products, P and Q. The following table gives the costs per unit for raw materials and labor and the maximum funds available.

cost ($)/unit	P	Q	maximum funds available
materials	20	30	6,000
labor	60	40	10,000

The firm has already accepted orders for 100 units of Q. Their profit is $20/unit of P and $15/unit of Q. How many of each product should they produce for the maximum profit?

38. A company produces two products, X and Y. Both are manufactured using two machines. The following table gives the time (in hours) required on each machine to produce one unit, and the number of hours each machine is available.

	X	Y	hours available per week
Machine 1	4	2	120
Machine 2	1	3	40

The company has accepted orders for 25 units of X per week. If the profit per unit is the same for X and Y, how many units of each should they produce weekly for the maximum profit?

39. A nutritionist wishes to combine two foods, F and G, into a diet with a certain minimum number of vitamins and minerals.

per ounce	F	G	minimum requirement
vitamins	4 units	2 units	160 units
minerals	1 unit	2 units	100 units

Food F costs 20¢/ounce and food G costs 15¢/ounce. How many ounces of each will satisfy the minimum requirements at the least cost? (Assume a minimum cost exists.)

40. A farmer feeds his cattle with a blend of two types of feed, A and B.

per pound	A	B	minimum requirement
protein	3 units	1 unit	100 units
carbohydrate	1 unit	5 units	290 units

If feed A costs 20¢/pound and feed B costs 10¢/pound, what combination of A and B will satisfy the minimum requirements at the least cost? (Assume a minimum cost exists.)

REVIEW EXERCISES FOR CHAPTER 8

Solve the system

1. $\begin{cases} 3x + 4y = 10 \\ 7x + 5y = 17 \end{cases}$

2. $\begin{cases} x + 2y + z = 3 \\ x + y - z = 11 \\ 3x + 5y + 4z = -1 \end{cases}$

3. $\begin{cases} 2x - 3y + 5z = 6 \\ 4x - y - z = -1 \\ x + y + 3z = 0 \end{cases}$

4. $\begin{cases} 5x + 8y - z = -1 \\ 3x + 5y + 2z = 4 \end{cases}$

5. $\begin{cases} 2x + 5y - z = 4 \\ 4x + 9y - 4z = 0 \\ 2x + 6y + z = 13 \end{cases}$

6. $\begin{cases} 7x - 4y + z = 1 \\ -4x + 5y - z = 3 \\ 10x - 3y + z = 5 \end{cases}$

7. $\begin{cases} x - y - z = 0 \\ 2x + 3z - w = 2 \\ 3x - 2y - 4w = 4 \\ 5y + 6z + 2w = 7 \end{cases}$

8. $\begin{cases} x + y + z + w = 1 \\ 8x + 4y + 2z + w = 3 \\ 27x + 9y + 3z + w = 11 \\ 64x + 16y + 4z + w = 37 \end{cases}$

9. Evaluate the determinant $\begin{vmatrix} 3 & -1 & 4 \\ 6 & 2 & -3 \\ 5 & 8 & 1 \end{vmatrix}$

   using (a) minors of the second column (b) minors of the third row

10. Solve using Cramer's rule $\begin{cases} 4x - 3y + z = 5 \\ -x + y + z = 0 \\ 3x + 4y - 2z = -16 \end{cases}$

Evaluate the determinant in any way

11. $\begin{vmatrix} 3 & 5 & 7 \\ -1 & 0 & 4 \\ -2 & 6 & 1 \end{vmatrix}$

12. $\begin{vmatrix} 4 & -3 & -3 \\ 2 & 1 & 8 \\ 8 & -1 & 13 \end{vmatrix}$

13. $\begin{vmatrix} 1 & -5 & 0 & 1 \\ 3 & 2 & 2 & 1 \\ 1 & 4 & 0 & -3 \\ 2 & -2 & 4 & 5 \end{vmatrix}$

14. Verify that $\begin{vmatrix} a_1 & b_1 + c_1 & d_1 \\ a_2 & b_2 + c_2 & d_2 \\ a_3 & b_3 + c_3 & d_3 \end{vmatrix} = \begin{vmatrix} a_1 & b_1 & d_1 \\ a_2 & b_2 & d_2 \\ a_3 & b_3 & d_3 \end{vmatrix} + \begin{vmatrix} a_1 & c_1 & d_1 \\ a_2 & c_2 & d_2 \\ a_3 & c_3 & d_3 \end{vmatrix}$

15. A student averaged 77 on three exams. The average of the first two exams was 70, and the average of the second and third was 82. Find all three scores.

16. A distiller mixed three types of whisky, A, B, C, to make 100 gallons of a blend containing 43% alcohol. The cost was $1320. From the data in the following table, determine how many gallons of each type were used.

Type	A	B	C
Percentage of alcohol	50	45	40
Cost (dollars per gallon)	20	16	10

Sketch the solution set

17. $\begin{cases} 4x + 3y \le 24 \\ y \ge x + 2 \\ x \ge -3 \end{cases}$
18. $\begin{cases} 4x + 9y \le 36 \\ 0 \le y \le \frac{1}{3}x \\ 3x + y \ge 12 \end{cases}$

19. Without any computation show that the equation

$$\begin{vmatrix} x & 2 & 3 \\ 1 & x & 3 \\ 1 & 2 & 3 \end{vmatrix} = 0$$

has solutions $x = 1$ and $x = 2$, and no others.

20. A manufacturer makes two products, A and B. Each unit of A requires 10 minutes of work; each B requires 6 minutes. Total work time cannot exceed 720 minutes per day. He needs at least 40 units of A per day, and at least $\frac{1}{3}$ as many units of B as of A. If the profit is $4 per unit of A, and $2 per unit of B, how many of each should be made daily to maximize profit?

Zeros of Polynomials and Complex Numbers

1 DIVISION OF POLYNOMIALS; SYNTHETIC DIVISION

Recall some properties of division from arithmetic. Dividing 51 by 4 gives a quotient 12 and a remainder 3:

$$51 = 4 \cdot 12 + 3$$

dividend divisor quotient remainder

Note that the remainder is less than the divisor.

There is a similar division of polynomials. Dividing a polynomial $f(x)$ by a polynomial $d(x)$ gives a quotient polynomial $q(x)$ and a remainder polynomial $r(x)$:

$$f(x) = d(x)\,q(x) + r(x)$$

dividend divisor quotient remainder

The remainder is either 0 or a polynomial of degree less than the degree of the divisor.

Division Algorithm Let $f(x)$ and $d(x)$ be polynomials, $d(x) \neq 0$. Then there are unique polynomials $q(x)$ and $r(x)$ such that

$$f(x) = d(x)q(x) + r(x)$$

where $r(x) = 0$ or degree $r(x)$ < degree $q(x)$.

Generally, $d(x)$ and $r(x)$ are computed by a process of long division. An example will illustrate the process.

Zeros of Polynomials and Complex Numbers

EXAMPLE 1 Find the quotient and remainder when $f(x) = x^4 - 3x^3 + x - 2$ is divided by $d(x) = x^2 + x + 3$.

SOLUTION We set up the long division as in long division of integers, indicating the missing x^2 term in $f(x)$ by $0 \cdot x^2$:

$$x^2 + x + 3 \overline{\smash{\big)}\, x^4 - 3x^3 + 0 \cdot x^2 + x - 2}$$

The first term of the quotient is x^2 because $x^2(x^2 + x + 3)$ will cancel the x^4 in the dividend when subtracted. The first step of the long division is

$$\begin{array}{r} x^2 \\ x^2 + x + 3 \overline{\smash{\big)}\, x^4 - 3x^3 + 0x^2 + x - 2} \\ \underline{x^4 + x^3 + 3x^2 } \\ -4x^3 - 3x^2 + x \end{array}$$

The new dividend is $-4x^3 - 3x^2 + x$. The next term of the quotient is $-4x$ because $-4x(x^2 + x + 3)$ will cancel the $-4x^3$ in the new dividend when subtracted. The next step in the long division is

$$\begin{array}{r} x^2 - 4x \\ x^2 + x + 3 \overline{\smash{\big)}\, x^4 - 3x^3 + 0x^2 + x - 2} \\ \underline{x^4 + x^3 + 3x^2 } \\ -4x^3 - 3x^2 + x \\ \underline{-4x^3 - 4x^2 - 12x } \\ x^2 + 13x - 2 \end{array}$$

Now, 1 times $x^2 + x + 3$ will cancel the first term in $x^2 + 13x - 2$, so the next step is

$$\begin{array}{r} x^2 - 4x + 1 \leftarrow \text{quotient } q(x) \\ x^2 + x + 3 \overline{\smash{\big)}\, x^4 - 3x^3 + 0x^2 + x - 2} \\ \underline{x^4 + x^3 + 3x^2 } \\ -4x^3 - 3x^2 + x \\ \underline{-4x^3 - 4x^2 - 12x } \\ x^2 + 13x - 2 \\ \underline{x^2 + x + 3} \\ 12x - 5 \leftarrow \text{remainder } r(x) \end{array}$$

At this point we are done because the degree of $12x - 5$ is less than the degree of $x^2 + x + 3$. Let us check that $f(x) = d(x)q(x) + r(x)$:

$$\begin{aligned} d(x)q(x) + r(x) &= (x^2 + x + 3)(x^2 - 4x + 1) + (12x - 5) \\ &= (x^4 - 3x^3 - 11x + 3) + (12x - 5) \\ &= x^4 - 3x^3 + x - 2 = f(x) \end{aligned}$$

Answer   quotient $x^2 - 4x + 1$ remainder $12x - 5$

Synthetic Division

Especially important for this chapter is division by linear divisors $x - a$. Then the remainder is either 0 or a polynomial of degree 0, that is, a non-zero constant. Thus, in all cases, the remainder is just a real number.

1 Division of Polynomials; Synthetic Division

Division by x − a Let $f(x)$ be a polynomial of degree $n \geq 1$ and let a be a real number. Then there is a unique polynomial $q(x)$ of degree $n - 1$ and a unique real number r such that
$$f(x) = (x - a)q(x) + r$$

As an example, let $f(x) = 3x^3 - 5x^2 - 9x + 8$ and $a = 2$. To find $q(x)$ and r, use long division:

$$\begin{array}{r} 3x^2 + x - 7 \\ x - 2 \overline{\smash{\big)}\ 3x^3 - 5x^2 - 9x + 8} \\ \underline{3x^3 - 6x^2} \\ x^2 - 9x \\ \underline{x^2 - 2x} \\ -7x + 8 \\ \underline{-7x + 14} \\ -6 \end{array}$$

Hence $q(x) = 3x^2 + x - 7$ and $r = -6$ so
$$f(x) = 3x^3 - 5x^2 - 9x + 8 = (x - 2)(3x^2 + x - 7) + (-6)$$
$$= (x - 2)q(x) + r$$

There is a lot of repetition and unnecessary notation in this long division. Let us see how much we can shorten it. First of all, we can omit all the powers of x (below, left):

$$\begin{array}{r} 3 1 - 7 \\ 1 - 2 \overline{\smash{\big)}\ 3 - 5 - 9 8} \\ \rightarrow 3 - 6 \\ 1 - 9 \leftarrow \\ \rightarrow 1 - 2 \\ -7 8 \leftarrow \\ \rightarrow -7 14 \\ -6 \end{array} \qquad \begin{array}{r} 3 1 - 7 \\ -2 \overline{\smash{\big)}\ 3 - 5 - 9 8} \\ -6 \\ 1 \\ -2 \\ -7 \\ 14 \\ -6 \end{array}$$

The numbers indicated by arrows occur twice. We omit them and also the 1 in the divisor (above, right). This is fine, but we can do better. We move the last four lines up:

$$\begin{array}{r} 3 1 -7 \\ -2 \overline{\smash{\big)}\ 3 -5 -9 8} \\ -6 -2 14 \\ \hline 1 -7 -6 \end{array}$$

Much neater, but there is still room for improvement. The 1 and -7 occur both on the top and bottom rows. So we move the 3 from the top to the bottom, and eliminate the top row altogether:

$$\begin{array}{r} -2 | 3 -5 -9 8 \\ -6 -2 14 \\ \hline 3 1 -7 -6 \end{array}$$

Now all results appear on the bottom row. The numbers 3, 1, −7 represent the quotient, $3x^2 + x - 7$, and −6 is the remainder.

One last change. On the bottom row, 3, 1, −7 come from subtraction. To get the same results by *addition*, we reverse all signs on the second row and also the sign of −2⌋. The final scheme looks like this:

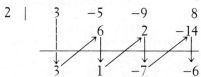

Start with the 3 on the top row and follow the arrows. The dotted arrow means that the 3 is brought down. The slanted arrows represent multiplication by 2, and the vertical arrows represent addition. The algorithm (process) described here is called **synthetic division**.

Synthetic Division To divide a polynomial $f(x)$ by $x - a$, follow these steps:

(1) Write the coefficients of $f(x)$ in descending order from left to right, using 0 for missing powers. Write $a⌋$ on the same line to the left.

(2) Bring down the first coefficient. Multiply it by a and add the product to the second coefficient. Multiply the result by a and add the product to the third coefficient, etc.

(3) The last number on the third row is the remainder. The other numbers on the third row are the coefficients of the quotient, in descending order.

EXAMPLE 2 Use synthetic division to find the quotient and remainder when $2x^4 - 8x^2 - 3x + 7$ is divided by (a) $x - 3$ (b) $x + 1$

SOLUTION The computations go like this:

```
3⌋2   0   −8   −3    7          −1⌋2    0   −8   −3    7
      6   18   30   81                 −2    2    6   −3
   ─────────────────────               ─────────────────────
   2   6   10   27   88                 2   −2   −6    3    4
```

Answer (a) $q(x) = 2x^3 + 6x^2 + 10x + 27$ $r = 88$
(b) $q(x) = 2x^3 - 2x^2 - 6x + 3$ $r = 4$

EXAMPLE 3 Express the polynomial $f(x) = 4x^5 - 3x^4 - x^3 - 5x^2 + 8x - 10$ in the form $(x - 1)q(x) + r$.

SOLUTION Divide $f(x)$ by $x - 1$ using synthetic division. The quotient is $q(x)$ and the remainder is r:

```
1⌋4   −3   −1   −5    8   −10
       4    1    0   −5    3
   ──────────────────────────
   4    1    0   −5    3   −7
```

From the last row, $q(x) = 4x^4 + x^3 - 5x + 3$, and $r = -7$.
Answer $f(x) = (x-1)(4x^4 + x^3 - 5x + 3) - 7$

Remainder and Factor Theorems

Suppose a polynomial $f(x)$ is divided by $x - a$, leaving a remainder r:
$$f(x) = (x-a)q(x) + r$$
Set $x = a$:
$$f(a) = 0 + r$$
Hence the remainder r is the value of $f(x)$ for $x = a$.

> **Remainder Theorem** When a polynomial $f(x)$ is divided by $x - a$, the remainder is $f(a)$.

The Remainder Theorem is useful for computing values of polynomials.

EXAMPLE 4 If $f(x) = 2x^4 - 7x^3 - 19x^2 + 8x - 3$, compute $f(5)$.

SOLUTION The value $f(5)$ is the remainder when $f(x)$ is divided by $x - 5$. Use synthetic division:

$$\begin{array}{r|rrrrr} 5 & 2 & -7 & -19 & 8 & -3 \\ & & 10 & 15 & -20 & -60 \\ \hline & 2 & 3 & -4 & -12 & -63 \end{array}$$

Answer -63

Remark Try doing Example 3 by substituting $x = 5$ in $f(x)$. You will see that synthetic division saves work.

We can write the Factor Theorem as
$$f(x) = (x-a)q(x) + f(a)$$
It follows that $f(x) = (x-a)q(x)$ if and only if $f(a) = 0$.

> **Factor Theorem** A polynomial $f(x)$ has the factor $x - a$ if and only if $f(a) = 0$.

EXAMPLE 5 Show that $f(x) = x^{15} + 2x^9 + x^4 - x - 3$ is divisible by $x - 1$.

SOLUTION Don't divide! Because of the Factor Theorem, it is enough to check that $f(1) = 0$:
$$f(1) = 1^{15} + 2 \cdot 1^9 + 1^4 - 1 - 3 = 1 + 2 + 1 - 1 - 3 = 0$$
Thus, $f(x) = 0$, so $f(x)$ has the factor $x - 1$.

EXAMPLE 6 Given that -2 is one root of the equation $x^3 + 7x^2 + 8x - 4 = 0$, find all other roots.

SOLUTION Call the left side of the equation $f(x)$. We are given that $f(-2) = 0$. Hence by the Factor Theorem, $f(x)$ has a factor $x + 2$. We divide $f(x)$ by $x + 2$ using synthetic division:

$$\begin{array}{r|rrrr} -2 & 1 & 7 & 8 & -4 \\ & & -2 & -10 & 4 \\ \hline & 1 & 5 & -2 & 0 \end{array}$$

The quotient is $x^2 + 5x - 2$, so the equation can be written as
$$(x + 2)(x^2 + 5x - 2) = 0$$
The remaining roots are roots of the quadratic equation $x^2 + 5x - 2 = 0$.

Answer $\dfrac{-5 \pm \sqrt{33}}{2}$

EXERCISES

Find the quotient and remainder when the first polynomial is divided by the second.

1. $x^2 - 2x + 5, \quad x - 3$
2. $x^2 + x + 4, \quad x + 2$
3. $2x^3 + 6x - 1, \quad x + 2$
4. $4x^3 - 7x, \quad x - 1$
5. $x^4 - 8x + 2, \quad x + 3$
6. $2x^4 - 9x^3 - 21, \quad x - 5$
7. $x^4 + 2x^3 - 4x^2 - x - 1, \quad x - 2$
8. $x^4 - 5x^3 + 6x^2 + 2x, \quad x + 3$
9. $2x^5 + x^4 + x^3 - 5x^2 - 8x + 2, \quad x + 1$
10. $x^5 - 3x^4 - 2x^3 + 10x^2 + 4x - 7, \quad x - 1$
11. $2x^3 - 5x + 1, \quad x^2 + 1$
12. $5x^3 - 4x^2 + 2, \quad x^2 - 4$
13. $x^4 + 2x^3 + 7x^2 + 3x + 4, \quad x^2 - x + 3$
14. $x^4 - x^3 + 5x^2 + 6x + 1, \quad x^2 + 2x - 1$
15. $x^5 - 4x^2 + 9, \quad x^2 + 2x$
16. $x^6, \quad x^2 + 1$

Express $f(x)$ in the form $(x - 3)q(x) + r$

17. $f(x) = x^3 - 7x^2 + 2x + 1$
18. $f(x) = x^3 + 6x^2 - 4x - 9$
19. $f(x) = x^4$
20. $f(x) = 5x^4 - 17x^3$

Compute $f(a)$ by synthetic division

21. $f(x) = 4x^3 - x^2 + 6x + 1, \quad a = -2$
22. $f(x) = 5x^3 + 2x^2 - 8x - 7, \quad a = 2$
23. $f(x) = -x^4 + 6x^3 + 4x^2 - 5x + 2, \quad a = -5$
24. $f(x) = 2x^4 - 9x^3 - x^2 + 4x + 1, \quad a = 5$
25. $f(x) = x^6 + x^5 + x^4 + x^3 + x^2 + x + 1, \quad a = 2$
26. $f(x) = 2x^3 - 5x^2 + x - 4, \quad a = \tfrac{1}{2}$

Without dividing, find the remainder when the first polynomial is divided by the second

27. $x^{10} + 2x^8$, $x - 1$
28. $x^{12} + x^7 - x^5$, $x + 1$
29. x^6, $x + 2$
30. x^5, $x - 3$
31. $x^7 - 50x$, $x - 2$
32. $x^6 - 3x^4$, $x - 2$
33. $x^5 + 2x^4 + 3x^3 + 4x^2 + 5x - 15$, $x - 1$
34. $x^6 - 2x^5 + 3x^4 - 4x^2 + 5x - 6$, $x - 1$

Show that the first polynomial is divisible by the second.

35. $x^{61} - x^{42} - x^6 + 1$, $x - 1$
36. $x^{43} - 3x^{31} + 2x^{15}$, $x + 1$
37. $x^n - a^n$, $x - a$
38. $x^n + a^n$, $x + a$ (n odd)

Factor the polynomial into linear factors

39. $x^3 - 6x^2 + 11x - 6$, given one factor $x - 3$
40. $x^3 + 6x^2 + 3x - 10$, given one factor $x - 1$
41. $x^4 + 4x^3 - 14x^2 - 36x + 45$, given factors $x + 3$ and $x + 5$
42. $x^4 - 2x^3 - 9x^2 + 2x + 8$, given factors $x - 1$ and $x + 2$

Find all real roots of the equation

43. $x^3 + 2x^2 - 4x + 1 = 0$, given one root $x = 1$
44. $x^3 - 5x^2 - 4x - 12 = 0$, given one root $x = 6$
45. $x^4 + 3x^3 - 2x^2 - 3x + 1 = 0$, given roots $x = 1$ and $x = -1$
46. $x^4 - 9x^3 + 28x^2 - 34x + 12 = 0$, given roots $x = 2$ and $x = 3$

2 POLYNOMIAL EVALUATION

We have evaluated polynomials using synthetic division. Let us examine that process. For example, suppose that $f(x) = 5x^3 - 4x^2 + 7x - 1$. To evaluate $f(r)$, the synthetic division is

$$
\begin{array}{c|cccc}
r & 5 & -4 & 7 & -1 \\
  &   & 5r & (5r-4)r & [(5r-4)r+7]r \\
\hline
  & 5 & 5r-4 & (5r-4)r+7 & \underline{[(5r-4)r+7]r - 1}
\end{array}
$$

The answer is correct because

$$[(5r - 4)r + 7]r - 1 = 5r^3 - 4r^2 + 7r - 1 = f(r)$$

Thus, evaluating $f(r)$ by synthetic division is equivalent to expressing $f(x)$ in this special factored form, then substituting $x = r$.

Every polynomial can be expressed this way. For example, take

$$f(x) = 2x^4 + 3x^3 - 5x^2 - 4x + 1$$

First factor x out of all terms but the last:
$$f(x) = \{2x^3 + 3x^2 - 5x - 4\}x + 1$$
Now repeat the process inside the braces:
$$f(x) = \{[2x^2 + 3x - 5]x - 4\}x + 1$$
$$= \{[(2x + 3)x - 5]x - 4\}x + 1$$
This factored form suggests a calculator routine for $f(r)$:

2 $\boxed{\times}$ r $\boxed{+}$ 3 $\boxed{=}$ $\boxed{\times}$ r $\boxed{-}$ 5 $\boxed{=}$ $\boxed{\times}$ r $\boxed{-}$ 4 $\boxed{=}$ $\boxed{\times}$ r $\boxed{+}$ 1 $\boxed{=}$

This routine is fine, but it requires keying in r *four* times. That wastes effort and increases the chances of a keying error. It is more efficient to key in r once, store it in the memory, and recall it as needed. Thus, an improved routine is

r $\boxed{STO}$ 2 $\boxed{\times}$ $\boxed{RCL}$ $\boxed{+}$ 3 $\boxed{=}$ $\boxed{\times}$ $\boxed{RCL}$ $\boxed{-}$ 5 $\boxed{=}$
$\boxed{\times}$ $\boxed{RCL}$ $\boxed{-}$ 4 $\boxed{=}$ $\boxed{\times}$ $\boxed{RCL}$ $\boxed{+}$ 1 $\boxed{=}$

This routine (or algorithm) can be adapted to any polynomial.

Polynomial Evaluation by Calculator (algebraic logic)   A calculator algorithm for evaluating
$$f(x) = a_n x^n + a_{n-1} x^{n-1} + \cdots + a_1 x + a_0$$
is the key sequence

x $\boxed{STO}$ a_n $\boxed{\times}$ $\boxed{RCL}$ $\boxed{+}$ a_{n-1} $\boxed{=}$ $\boxed{\times}$ $\boxed{RCL}$ $\boxed{+}$ a_{n-2} $\boxed{=}$ $\cdots$
$\cdots$ $\boxed{\times}$ $\boxed{RCL}$ $\boxed{+}$ a_1 $\boxed{=}$ $\boxed{\times}$ $\boxed{RCL}$ $\boxed{+}$ a_0 $\boxed{=}$

If $a_k < 0$, replace $\boxed{+}$ a_k by $\boxed{-}$ $|a_k|$. If $a_k = 0$, simply omit $\boxed{+}$ a_k.

EXAMPLE 1   Verify that $x = 29$ is a solution of the equation
$$2x^6 - 59x^5 + 26x^4 + 88x^3 - 24x^2 - 144x - 29 = 0$$
SOLUTION   By hand, the computation would be staggering. By calculator, it takes about one minute. Just evaluate the polynomial on the left-hand side:

2 9 $\boxed{STO}$ 2 $\boxed{\times}$ $\boxed{RCL}$ $\boxed{-}$ 5 9 $\boxed{=}$ $\boxed{\times}$ $\boxed{RCL}$ $\boxed{+}$ 2 6 $\boxed{=}$
$\boxed{\times}$ $\boxed{RCL}$ $\boxed{+}$ 8 8 $\boxed{=}$ $\boxed{\times}$ $\boxed{RCL}$ $\boxed{-}$ 2 4 $\boxed{=}$
$\boxed{\times}$ $\boxed{RCL}$ $\boxed{-}$ 1 4 4 $\boxed{=}$ $\boxed{\times}$ $\boxed{RCL}$ $\boxed{-}$ 2 9 $\boxed{=}$

The result is 0. Hence $x = 29$ is a solution.

If some coefficients of a polynomial are 0, evaluation can be streamlined by proper factoring. For instance, the polynomial
$$f(x) = x^6 - 7x^4 + 3$$
can be written in the form
$$f(x) = (x^2 - 7)x^4 + 3$$

which suggests the routine

$$x\ \boxed{STO}\ \boxed{x^2}\ \boxed{-}\ 7\ \boxed{=}\ \boxed{\times}\ \boxed{RCL}\ \boxed{x^2}\ \boxed{x^2}\ \boxed{+}\ 3\ \boxed{=}$$

There are other ways to evaluate a polynomial besides the algorithm given here. For example, suppose you want $f(3)$ where

$$f(x) = 2x^4 + 3x^3 - 5x^2 - 4x + 1$$

If your calculator has a $\boxed{y^x}$ key, then the following routine will work:

$$3\ \boxed{STO}\ 2\ \boxed{\times}\ \boxed{RCL}\ \boxed{y^x}\ 4\ \boxed{+}\ 3\ \boxed{\times}\ \boxed{RCL}\ \boxed{y^x}\ 3$$
$$\boxed{-}\ 5\ \boxed{\times}\ \boxed{RCL}\ \boxed{x^2}\ \boxed{-}\ 4\ \boxed{\times}\ \boxed{RCL}\ \boxed{+}\ 1\ \boxed{=}\qquad 187$$

Without a $\boxed{y^x}$ key, you can do it this way:

$$3\ \boxed{STO}\ 2\ \boxed{\times}\ \boxed{RCL}\ \boxed{x^2}\ \boxed{x^2}\ \boxed{+}\ 3\ \boxed{\times}\ \boxed{RCL}\ \boxed{x^2}\ \boxed{\times}\ \boxed{RCL}$$
$$\boxed{-}\ 5\ \boxed{\times}\ \boxed{RCL}\ \boxed{x^2}\ \boxed{-}\ 4\ \boxed{\times}\ \boxed{RCL}\ \boxed{+}\ 1\ \boxed{=}\qquad 187$$

These routines require a few more steps than the evaluation algorithm, but you may find them more natural. Be careful calculating powers of a stored number x if you don't have $\boxed{y^x}$. For instance, to find x^6 use

$$\boxed{RCL}\ \boxed{x^2}\ \boxed{x^2}\ \boxed{\times}\ \boxed{RCL}\ \boxed{x^2}\qquad \text{not}\qquad \boxed{RCL}\ \boxed{x^2}\ \boxed{x^2}\ \boxed{x^2}$$

Why?

EXERCISES

Evaluate $f(r)$ by synthetic division

1. $f(x) = 2x^3 + 4x^2 + 6x + 1$
2. $f(x) = 5x^3 - 2x^2 - x + 7$
3. $f(x) = x^4 + 2x^3 - 6x^2 - 8x + 9$
4. $f(x) = x^4 - x^3 + 5x^2 + 6x - 1$
5. $f(x) = 2x^5 + x^2 - 6$
6. $f(x) = x^6 - 4x^3 + 9$

Evaluate the polynomial for the given values of x

7. $x^3 + 5x^2 + 7x + 2$ $1.4, -32$
8. $x^3 + 15x^2 + 3x + 10$ $0.35, 13$
9. $0.3x^3 - 2.9x^2 + 3.7x + 4.1$ $167, -5.6$
10. $1.4x^3 + 3.8x^2 - 11.5x + 1.7$ $4.3, -21$
11. $854x^4 + 291x^3 - 4733x + 588$ $7, -3.8$
12. $7.1x^4 + 0.9x^3 + 1.8x^2 - 12.5x - 4.7$ $-12.4, 1.9$
13. $x^5 + 2x^3 - 17x^2 + 26x - 9$ $0.3, 26$
14. $3x^5 - 10x^4 + 2x^2 + 7$ $49, -0.7$
15. $x^8 + 9x^6 - 55x^4 - 107x^2 + 83$ $4, 5$
16. $x^{10} + 7x^4 + 5$ $3, 4$

Which of the numbers is a solution of the equation?

17. $18x^5 - 7x^4 + 3x^3 - 17x + 294 = 0 \quad -2, 2, 3$

18. $2x^6 + 3x^5 + 4x^4 + 5x^3 + 6x^2 + 7x - 951 = 0 \quad -4, -3, 3$

19. Find a routine for evaluating the factored polynomial
$$f(x) = (x - a_1)(x - a_2)(x - a_3) \cdots (x - a_n)$$
on a calculator with parentheses.

20. The TI-58 has a built-in program for polynomial evaluation, based on the following method. If $f(x) = a_n x^n + \cdots + a_1 x + a_0$, then $f(r)$ is evaluated by computing a sequence of numbers $b_n, b_{n-1}, \ldots, b_1, b_0$:

$$b_n = a_n$$
$$b_{n-1} = a_{n-1} + b_n r = a_{n-1} + a_n r$$
$$b_{n-2} = a_{n-2} + b_{n-1} r = a_{n-2} + (a_{n-1} + a_n r)r$$
$$\cdot$$
$$\cdot$$
$$\cdot$$
$$b_0 = a_0 + b_1 r$$

Then $f(r) = b_0$. Explain this method.

3 RATIONAL ZEROS OF POLYNOMIALS

A **zero** of a function $f(x)$ is a number r such that $f(r) = 0$. The number r is also called a **root** or **solution** of the equation $f(x) = 0$.

A polynomial may not have any real zeros. An example is $f(x) = x^2 + 1$, which is positive for every real x. On the other hand $f(x) = x(x - 1)(x - 2)$ has three real zeros, 0, 1, and 2. Now we ask what is the maximum number of real zeros a polynomial can have?

The answer comes from the Factor Theorem. Suppose that $f(x)$ is a polynomial of degree $n \geq 1$, and r_1 is a zero of $f(x)$. By the Factor Theorem,

$$f(x) = (x - r_1)q_1(x)$$

where $q_1(x)$ is a polynomial of degree $n - 1$. Now suppose r_2 is another zero of $f(x)$, that is, $r_2 \neq r_1$. Then

$$f(r_2) = (r_2 - r_1)q_1(r_2) = 0 \qquad r_2 - r_1 \neq 0$$

Hence $q_1(r_2) = 0$, so r_2 is a zero of $q_1(x)$. By the Factor Theorem again,

$$q_1(x) = (x - r_2)q_2(x)$$

where $q_2(x)$ has degree $n - 2$. Thus,

$$f(x) = (x - r_1)(x - r_2)q_2(x)$$

We continue this way. If $f(x)$ has k distinct real zeros $r_1, r_2, \ldots, r_k$, then

(1) $$f(x) = (x - r_1)(x - r_2) \cdots (x - r_k)q_k(x)$$

where $q_k(x)$ is a polynomial of degree $n - k$. Now $n - k \geq 0$ so $k \leq n$. Therefore the number of distinct real zeros is at most the degree of $f(x)$.

> A polynomial of degree $n \geq 1$ has at most n distinct real zeros.

Now suppose that $f(x)$ has n distinct real zeros (the maximum number possible). Then $k = n$ in equation (1). Hence $q_k(x)$ has degree 0, that is, $q_k(x)$ is a non-zero constant c.

> If $f(x)$ is a polynomial of degree $n \geq 1$ and has n distinct real zeros $r_1, r_2, \cdots, r_n$, then
> $$f(x) = c(x - r_1)(x - r_2) \cdots (x - r_n) \qquad c \neq 0$$

EXAMPLE 1 Find a cubic polynomial $f(x)$ with zeros 1, 2, and -3, and such that $f(0) = 5$.

SOLUTION $f(x)$ is of degree 3 and has three distinct real zeros, 1, 2, and -3. Therefore,
$$f(x) = c(x - 1)(x - 2)(x + 3)$$
Now, find the constant c that makes $f(0) = 5$. Set $x = 0$:
$$f(0) = c(0 - 1)(0 - 2)(0 + 3) = 6c$$
Therefore, $6c = 5$, so $c = \frac{5}{6}$.

Answer $f(x) = \frac{5}{6}(x - 1)(x - 2)(x + 3)$

The polynomial $f(x) = (x - 3)^2$ has only one real zero, 3, but it has *two factors* $(x - 3)$. We call 3 a zero of **multiplicity** 2 of $f(x)$. In general, if $f(x)$ is divisible by $(x - r)^m$, then r is a zero of **multiplicity** m. For example, let
$$f(x) = x(x - 1)(x - 5)^2(x - 6)^3(x + 10)^7$$
Then 0 and 1 are zeros of multiplicity 1 (simple zeros), 5 is a zero of multiplicity 2 (**double** zero), 6 is a zero of multiplicity 3 (**triple** zero), and -10 is a zero of multiplicity 7.

Rational Zeros

We come now to the practical matter of finding the zeros of a given polynomial. This is a complicated problem in general, so let us start with one of the easier cases: finding *rational* zeros of polynomials with *integer* coefficients. In this case there is a useful test for zeros.

> **Rational Zeros** Suppose
> $$f(x) = a_n x^n + \cdots + a_1 x + a_0 \qquad a_n \neq 0$$
> has *integer* coefficients. If r is a *rational* zero of $f(x)$, and $r = p/q$ in lowest terms, then p divides a_0 and q divides a_n.

Zeros of Polynomials and Complex Numbers

We leave the proof to more advanced courses.

EXAMPLE 2 Find all rational zeros of
$$f(x) = 5x^3 + 3x^2 - 7x + 2$$

SOLUTION Suppose $r = p/q$ is a rational zero expressed in lowest terms. Then p must divide 2 and q must divide 5. The possibilities for p and q are
$$p = \pm 1, \ \pm 2 \qquad q = \pm 1, \ \pm 5$$
Hence the possibilities for $r = p/q$ are
$$\pm 1, \ \pm 2, \ \pm \tfrac{1}{5}, \ \pm \tfrac{2}{5}$$

Let us check each of these for $f(r) = 0$. By synthetic division, we find
$$f(1) = 3 \qquad f(-1) = 7 \qquad f(2) = 40 \qquad f(-2) = -12$$
No zeros so far. Continuing, we find
$$f(\tfrac{1}{5}) = \tfrac{19}{25} \qquad f(-\tfrac{1}{5}) = \tfrac{87}{25} \qquad f(\tfrac{2}{5}) = 0 \qquad f(-\tfrac{2}{5}) = \tfrac{124}{25}$$
Thus, $\tfrac{2}{5}$ is a zero, and since there are no other possibilities, $\tfrac{2}{5}$ is the only rational zero.

Suppose we were lucky enough to try $\tfrac{2}{5}$ first. The synthetic division is

$$\tfrac{2}{5} \, \big| \begin{array}{rrrr} 5 & 3 & -7 & 2 \\ & 2 & 2 & -2 \\ \hline 5 & 5 & -5 & 0 \end{array}$$

This shows that $\tfrac{2}{5}$ is a zero, and also that
$$f(x) = (x - \tfrac{2}{5})(5x^2 + 5x - 5) = 5(x - \tfrac{2}{5})(x^2 + x - 1)$$
Hence the remaining zeros of $f(x)$ are zeros of $x^2 + x - 1$. But these are $\tfrac{1}{2}(-1 \pm \sqrt{5})$, irrational.

Answer $f(x)$ has one rational zero, $\tfrac{2}{5}$.

Suppose we are looking for rational zeros of $f(x) = x^n + \cdots + a_1 x + a_0$, a polynomial with integer coefficients and leading coefficient 1. If $r = p/q$ is a rational zero in lowest terms, then p divides a_0 and q divides a_n. But $a_n = 1$, so q divides 1. Therefore, $q = \pm 1$. It follows that $p/q = \pm p$, so r is an **integer**.

Integer Zeros Suppose
$$f(x) = x^n + \cdots + a_1 x + a_0$$
has integer coefficients and leading coefficient 1. Then each rational zero of $f(x)$ is an *integer* that divides a_0.

EXAMPLE 3 Find all rational zeros of $f(x) = x^4 - 2x^3 - 8x^2 + 19x - 6$

SOLUTION The coefficients are integers and the leading coefficient is 1. Therefore any rational zero is an integer that divides -6. The possibilities are
$$\pm 1, \ \pm 2, \ \pm 3, \ \pm 6$$

We start testing:
$$f(1) = 4 \qquad f(-1) = -30 \qquad f(2) = 0$$
Therefore, 2 is a zero, but 1 and -1 are not.

We could test the remaining possibilities. Instead, let us factor $f(x)$. Since 2 is a zero, $f(x)$ is divisible by $x - 2$. Synthetic division gives
$$f(x) = (x - 2)(x^3 - 8x + 3) = (x - 2)q(x)$$
Any rational zero of $q(x) = x^3 - 8x + 3$ is an integer that divides 3: hence ± 1 or ± 3. We have already ruled out ± 1. That leaves only 3 and -3. Testing them shows
$$q(3) = 6 \qquad q(-3) = 0$$

Answer 2 and -3

EXAMPLE 4 By the methods of this section, show that $\sqrt{2}$ is irrational.

SOLUTION The number $\sqrt{2}$ is a zero of $f(x) = x^2 - 2$. Now any rational zero of $f(x)$ must be an integer that divides -2. The only possibilities are ± 1 and ± 2, but none of these works. Hence all zeros of $f(x)$ are irrational. Since $\sqrt{2}$ is a zero, $\sqrt{2}$ is irrational.

EXERCISES

Find all real zeros and their multiplicities of $f(x) =$
1. $x(x - 1)(x - 5)$
2. $(x + 3)(x + 4)(x - 2)$
3. $(x^2 - 1)(x^2 + 4)$
4. $(x^2 - 9)(x^2 - 16)$
5. $3(x + 2)(x - 3)^2$
6. $-2(x - 1)^3(x + 4)$
7. $(x + 1)^3(x + 2)^4$
8. $x^2(x - 6)^3(x^2 + 1)^4$
9. $(x^2 + x - 2)(x^2 - 1)^3$
10. $(x^2 - 4)^2(x^3 - 8)$
11. $x^4 - 8x^2 + 16$
12. $x^7 - 3x^5$

Find the polynomial $f(x)$ described
13. quadratic, zeros -1, 3 and $f(0) = 6$
14. quadratic, zeros 2, -4 and $f(0) = 1$
15. cubic, $f(0) = f(5) = f(8) = 0$ and $f(10) = 17$
16. cubic, triple zero 4 and $f(1) = 9$

Find all rational zeros of the polynomial
17. $x^3 + x^2 + 2x + 2$
18. $x^3 + 3x^2 + 2x + 6$
19. $x^3 + 6x^2 + 5x - 12$
20. $x^3 + 2x^2 - 3x + 20$
21. $3x^4 - 8x^2 - 6x - 1$
22. $4x^3 + 3x^2 + 3x - 1$
23. $x^5 + 2x - 4$
24. $x^5 - 4x^3 + 10$

25. $2x^4 + 3x^3 + 8x^2 + 3x - 4$ **26.** $3x^4 + 5x^3 + x^2 + 5x - 2$

27. $x^{17} + x^{13} + 3x^4 - 1$ **28.** $2x^{10} + x^3 + x^2 + 1$

Show that the number is irrational

29. $\sqrt[3]{2}$ **30.** $\sqrt[5]{10}$ **31.** $\sqrt[4]{\tfrac{2}{3}}$ **32.** $\sqrt[3]{0.7}$

33. Explain why the graph of a polynomial of degree 7 intersects a given straight line at most seven times.

34. Explain why the graphs of a quadratic and a cubic have at most three points in common.

35. Explain why no three points of a parabola can lie on a straight line.

4 COMPLEX NUMBERS

The equation
$$x^2 + 1 = 0$$
has no real solution because there is no real number $\sqrt{-1}$. The equation
$$x^2 - 2x + 10 = 0$$
also has no real solution because the quadratic formula yields
$$x = \frac{2 \pm \sqrt{-36}}{2} = \frac{2 \pm 6\sqrt{-1}}{2} = 1 \pm 3\sqrt{-1}$$
Thus we run into the same trouble: there is no real number $\sqrt{-1}$.

Solving such equations requires an extended number system containing the real numbers and also a number $\sqrt{-1}$. There is such a system, called the **complex number system**, which we will now describe.

The complex number system consists of all formal expressions
$$a + bi$$
where a and b are real numbers, and i is a new symbol that plays the role of -1. Two such expressions, or complex numbers, are equal if and only if they are identical

$$\boxed{a + bi = c + di \quad \text{if and only if} \quad a = c \text{ and } b = d}$$

Addition and subtraction of complex numbers are defined in a natural way:
$$(a + bi) + (c + di) = (a + c) + (b + d)i$$
$$(a + bi) - (c + di) = (a - c) + (b - d)i$$
Multiplication is also defined in a natural way except that i^2 is replaced by -1:
$$(a + bi)(c + di) = ac + (ad + bc)i + (bd)i^2$$
$$= ac + (ad + bc)i - bd$$
$$= (ac - bd) + (ad + bc)i$$

> **Addition, Subtraction, and Multiplication of Complex Numbers** By definition
> $$(a + bi) + (c + di) = (a + c) + (b + d)i$$
> $$(a + bi) - (c + di) = (a - c) + (b - d)i$$
> $$(a + bi)(c + di) = (ac - bd) + (ad + bc)i$$

It is not necessary to memorize these definitions. Just work with $a + bi$ and $c + di$ as you would with $a + bx$ and $c + dx$, but replace i^2 by -1.

Examples
$$(4 + 3i) + (6 - 2i) = (4 + 6) + (3 - 2)i = 10 + i$$
$$(\tfrac{1}{2} + 4i) - (1 + \tfrac{3}{2}i) = (\tfrac{1}{2} - 1) + (4 - \tfrac{3}{2})i = -\tfrac{1}{2} + \tfrac{5}{2}i$$
$$(5 - 3i)(2 + 7i) = 5 \cdot 2 + [(-3) \cdot 2 + 5 \cdot 7]i + (-3) \cdot 7 \cdot i^2$$
$$= 10 + 29i + (-21) \cdot (-1) = 31 + 29i$$

Equipped with these operations (and division, to be defined shortly) the set of complex numbers form a system in which all the usual rules of arithmetic hold: the commutative, associative, and distributive laws, the existence of a zero element $0 + 0 \cdot i$, etc. We leave verification of some of these properties as exercises.

Each real number a can be considered as the complex number $a + 0 \cdot i$. Thus, the complex numbers contain the real numbers. Furthermore, their arithmetic is consistent with the arithmetic of real numbers. For example, if real numbers a and b are treated as complex numbers, then their product is
$$(a + 0 \cdot i)(b + 0 \cdot i) = ab + 0 \cdot i$$
which is consistent with their product as real numbers. In short, the complex number system is an extension of the real number system.

We denote $a + 0 \cdot i$ by a and $0 + bi$ by bi.

EXAMPLE 1 Express in the form $a + bi$
 (a) $4(2 + i) + (3 - 2i)$ (b) $(1 - 2i)(5 - 3i)$
 (c) $(1 + i)^4$ (d) $i + i^2 + i^3 + i^4$

SOLUTION

(a)
$$4(2 + i) + (3 - 2i) = (8 + 4i) + (3 - 2i)$$
$$= (8 + 3) + (4 - 2)i$$
$$= \underline{11 + 2i}$$

(b) $(1 - 2i)(5 - 3i) = 5 - 13i + 6i^2 = 5 - 13i - 6 = \underline{-1 - 13i}$

(c) First compute $(1 + i^2)$, then square the result:
$$(1 + i)^2 = 1 + 2i + i^2 = 1 + 2i - 1 = 2i$$
Hence, $(1 + i)^4 = (2i)^2 = 4i^2 = \underline{-4}$

(d) $i^2 = -1$, $i^3 = (i^2)i = -i$, $i^4 = (i^3)i = (-i)i = -i^2 = 1$
Hence, $i + i^2 + i^3 + i^4 = i - 1 - i + 1 = \underline{0}$

Conjugates and Division

The complex number $a - bi$ is called the **conjugate** of the complex number $a + bi$. Note that also $a + bi$ is the conjugate of $a - bi$. The conjugate of a *real* number is itself, since $a - 0 \cdot i = a + 0 \cdot i$.

It is a useful fact that the product of a pair of conjugate complex numbers is a non-negative real number:

$$\boxed{(a + bi)(a - bi) = a^2 - (bi)^2 = a^2 + b^2}$$

Clearly, $a^2 + b^2 > 0$ unless $a + bi = 0$.

We use this product to define division of complex numbers. For example, the quotient of $2 + 3i$ by $3 - 4i$ should be a complex number q such that

$$(3 - 4i)q = 2 + i$$

To find q, we multiply both sides by $3 + 4i$, the conjugate of the divisor:

$$(3 - 4i)(3 + 4i)q = (2 + i)(3 + 4i)$$

$$(3^2 + 4^2)q = 2 + 11i$$

$$q = \frac{1}{3^2 + 4^2} \cdot (2 + 11i) = \frac{2}{25} + \frac{11}{25}i$$

If we write $q = (2 + i)/(3 - 4i)$, then this computation amounts to multiplying numerator and denominator by the conjugate of the denominator:

$$\frac{2 + i}{3 - 4i} = \frac{2 + i}{3 - 4i} \cdot \frac{3 + 4i}{3 + 4i} = \frac{2 + 11i}{3^2 + 4^2} = \frac{2}{25} + \frac{11}{25}i$$

The same reasoning allows us to define the quotient of any two complex numbers, provided the divisor is not zero:

Division of Complex Numbers By definition the quotient

$$\frac{a + bi}{c + di} \qquad c + di \neq 0$$

is the complex number obtained by multiplying numerator and denominator by the conjugate of the denominator.

EXAMPLE 2 Compute (a) $\dfrac{1}{5 + 2i}$ (b) $\dfrac{2 + 3i}{1 - 4i}$ (c) $\dfrac{6 + 7i}{i}$

SOLUTION (a) Multiply numerator and denominator by $5 - 2i$:

$$\frac{1}{5 + 2i} = \frac{1}{5 + 2i} \cdot \frac{5 - 2i}{5 - 2i} = \frac{5 - 2i}{5^2 + 2^2} = \underline{\frac{5}{29} - \frac{2}{29}i}$$

(b) Multiply numerator and denominator by $1 + 4i$:

$$\frac{2 + 3i}{1 - 4i} = \frac{2 + 3i}{1 - 4i} \cdot \frac{1 + 4i}{1 + 4i} = \frac{-10 + 11i}{1^2 + 4^2} = \underline{\frac{10}{17} + \frac{11}{17}i}$$

(c) Multiply numerator and denominator by $-i$:
$$\frac{6+7i}{i} = \frac{6+7i}{i} \cdot \frac{-i}{-i} = \frac{7-6i}{1} = \underline{7-6i}$$

EXAMPLE 3 Compute $\dfrac{(2-i)^2}{(1+i)(1+6i)}$

SOLUTION First compute the numerator and denominator, then multiply by the conjugate of the denominator:

$$\frac{(2-i)^2}{(1+i)(1+6i)} = \frac{3-4i}{-5+7i}$$

$$= \frac{3-4i}{-5+7i} \cdot \frac{-5-7i}{-5-7i} = \frac{-43-i}{5^2+7^2} = \underline{-\frac{43}{74} - \frac{1}{74}i}$$

The operation of taking the conjugate of a complex number has some useful properties. Let us denote the conjugate of z by $\bar{z}$. For example, $\overline{8-5i} = 8+5i$.

Rules for Conjugates

(1) $\overline{z_1 + z_2} = \bar{z}_1 + \bar{z}_2$ (2) $\overline{z_1 z_2} = \bar{z}_1 \bar{z}_2$

(3) $\bar{z} = z$ if and only if z is real

We leave the verifications as exercises. These rules extend to any number of summands and factors, for instance,

$$\overline{z_1 + z_2 + z_3} = \bar{z}_1 + \bar{z}_2 + \bar{z}_3 \qquad \overline{z_1 z_2 z_3} = \bar{z}_1 \bar{z}_2 \bar{z}_3 \qquad \overline{z^n} = (\bar{z})^n$$

EXAMPLE 4 Show that $\overline{3z^2 + 7z + 4} = 3(\bar{z})^2 + 7\bar{z} + 4$

SOLUTION

$$\overline{3z^2 + 7z + 4} = \overline{3z^2} + \overline{7z} + \overline{4} \qquad [\text{Rule (1)}]$$
$$= \overline{3}\,\overline{z^2} + \overline{7}\bar{z} + \overline{4} \qquad [\text{Rule (2)}]$$
$$= 3\overline{z^2} + 7\bar{z} + 4 \qquad [\text{Rule (3)}]$$
$$= 3(\bar{z})^2 + 7\bar{z} + 4 \qquad [\text{Rule (2)}]$$

EXERCISES

Express in the form $a + bi$

1. $(1-i) + (7+6i)$
2. $(8+6i) - (3-2i)$
3. $(9+4i) - 2(3-i)$
4. $3(2-2i) + (-5+8i)$
5. $(5+4i)(3-i)$
6. $(1-6i)(5+i)$
7. $(2+5i)^2$
8. $(-1+3i)^2$
9. $(6+8i)(6-8i)$
10. $(9-2i)(9+2i)$
11. $(1-i)^4$
12. $(1+i)(1-i)^3$
13. $(4i)(5i)$
14. $(2i)^3$
15. i^{101}

16. $1 + i + i^2 + \cdots + i^{11}$
17. $\dfrac{1}{2 + 5i}$
18. $\dfrac{2}{3 + 1}$
19. $\dfrac{i}{3 + 4i}$
20. $\dfrac{2 - 3i}{i}$
21. $\dfrac{-7 + i}{1 - 2i}$
22. $\dfrac{6 - 5i}{5 - 4i}$
23. $\dfrac{1}{(2 + 3i)(5 - 4i)}$
24. $\dfrac{1}{(1 - i)(8 + 3i)}$
25. $\dfrac{1 - 3i}{(2 + i)(2 + 5i)}$
26. $\dfrac{3 - 5i}{(1 + 3i)^2}$
27. $\dfrac{(1 + i)(1 + 2i)}{(3 - 2i)(4 - 3i)}$
28. $\dfrac{(2 - i)(3 - i)}{(1 + 2i)(1 + 3i)}$
29. $\dfrac{i}{2 + i} + \dfrac{3 + i}{4 + i}$
30. $\dfrac{1}{4 - 3i} + \dfrac{5 + 3i}{2 - i}$

31. Let $f(z) = z^3 + 3z^2 + z + 3$. Compute $f(i)$, $f(1 + i)$.
32. Let $f(z) = z^2 - 4z + 13$. Compute $f(-i)$, $f(2 + 3i)$.

Check the Associative Law $(z_1 z_2) z_3 = z_1 (z_2 z_3)$
33. for $z_1 = i$, $z_2 = 3 + 2i$, $z_3 = 1 - i$
34. for $z_1 = 1 + i$, $z_2 = 1 - i$, $z_3 = 1 + 4i$

Verify the relation for complex numbers
35. $z_1 + z_2 = z_2 + z_1$
36. $z_1 z_2 = z_2 z_1$
37. $z_1(z_2 + z_3) = z_1 z_2 + z_1 z_3$
38. $(z_1 + z_2) + z_3 = z_1 + (z_2 + z_3)$
39. $\overline{z_1 + z_2} = \overline{z_1} + \overline{z_2}$
40. $\overline{z_1 z_2} = \overline{z_1}\,\overline{z_2}$
41. $\overline{3z + i} = 3\overline{z} - i$
42. $\overline{5z^3 + 2z - 1} = 5(\overline{z})^3 + 2\overline{z} - 1$

43. Verify that $(\sqrt{3} + i)^3 = 8i$, then find $(\sqrt{3} - i)^3$ without computing.
44. Compute $(1 + i)(1 + 2i)(1 - 3i)$, then find $(1 - i)(1 - 2i)(1 + 3i)$ without computing.

Solve for z

45. $2iz + 5i = 6 - 7z$
46. $\dfrac{1}{z - 3i} = 1 + 4i$
47. $z^2 = 8 - 6i$
48. $z^2 = 15 - 8i$

(Hint In Exercises 47–48, set $z = a + bi$, then solve for a and b.)

5 COMPLEX ZEROS OF POLYNOMIALS

In the complex number system, a negative real number has two complex square roots. For example,

$$\sqrt{-4} = \sqrt{4i^2} = \pm 2i$$

It follows that a polynomial such as $x^2 + x + 4$, which has no real zeros, does have complex zeros. By the quadratic formula, they are
$$\frac{-1 \pm \sqrt{-15}}{2} = \frac{-1 \pm \sqrt{15}\,i}{2}$$
Similarly, every quadratic polynomial $ax^2 + bx + c$ has complex zeros. If $b^2 - 4ac \geq 0$ those zeros are real; if $b^2 - 4ac < 0$, they are non-real. (Remember that real numbers are complex numbers.)

What about polynomials of higher degree? If a new quantity i is needed for quadratics, is another new quantity needed for cubics, another for quartics, etc.?

Fortunately, the answer is no. Once the real number system is enlarged to contain zeros for quadratics, that is enough. It then contains zeros for every polynomial, even if the coefficients are complex. This remarkable result was proved by Gauss in 1799 and is called the **Fundamental Theorem of Algebra.**

> **Fundamental Theorem of Algebra** If $f(z)$ is a polynomial of degree ≥ 1 with complex coefficients, then there exists a complex number c such that $f(c) = 0$.

Stated briefly, a complex polynomial has a complex zero. The proof of this theorem requires advanced theory.

The same reasoning as that used in Section 1 gives the following version of the Factor Theorem:

> **Factor Theorem** Let $f(z)$ be a polynomial of degree $n \geq 1$ with complex coefficients. A complex number c is a zero of $f(z)$ if and only if
> $$f(z) = (z - c)q(z)$$
> where $q(z)$ is a polynomial of degree $n - 1$.

For example, the number i is a zero of $f(z) = z^2 + 1$ since $i^2 + 1 = 0$. Therefore $z - i$ must be a factor of $f(z)$. It *is*:
$$z^2 + 1 = (z - i)(z + i)$$

Complete Factorization

The polynomial $z^2 + 1$ cannot be factored if only real numbers are allowed, but it can if complex numbers are allowed. In fact, *every* polynomial of degree n factors into n complex linear factors. For suppose $f(z) = a_n z^n + \cdots + a_1 z + a_0$, where the coefficients are complex and $a_n \neq 0$. By the Fundamental Theorem of Algebra, $f(z)$ has a complex zero c_1. Hence, by the Factor Theorem,
$$f(z) = (z - c_1)q_1(z)$$
where $q_1(z)$ has degree $n - 1$. If $n - 1 > 0$, then the same reasoning gives
$$q_1(z) = (z - c_2)q_2(z)$$

so that
$$f(z) = (z - c_1)(z - c_2)q_2(z)$$
where $q_2(z)$ has degree $n - 2$.

Repeating this procedure n times, we reach
$$f(z) = (z - c_1)(z - c_2) \cdots (z - c_n)q_n(z)$$
Here $q_n(z)$ is a polynomial of degree 0, that is, a non-zero constant q_n. In fact, $q_n = a_n$, the leading coefficient of $f(z)$.

> **Complete Factorization** Let
> $$f(z) = a_n z^n + \cdots + a_1 z + a_0$$
> be a polynomial of degree $n \geq 1$ with complex coefficients. Then there exist complex numbers $c_1, c_2, \ldots, c_n$ such that
> $$f(z) = a_n(z - c_1)(z - c_2) \cdots (z - c_n)$$

Example
$$\begin{aligned} f(z) &= 3z^4 - 48 \\ &= 3(z^4 - 16) = 3(z^2 - 4)(z^2 + 4) \\ &= 3(z - 2)(z + 2)(z - 2i)(z + 2i) \end{aligned}$$

The numbers $c_1, c_2, \cdots, c_n$ the zeros of $f(z)$. They may all be distinct, or there may be only k distinct zeros ($k < n$) due to repetitions. For instance, c_1 may occur m_1 times, c_2 may occur m_2 times, etc. Then the complete factorization of $f(z)$ takes the form
$$f(z) = a_n(z - c_1)^{m_1}(z - c_2)^{m_2} \cdots (z - c_k)^{m_k}$$
We call c_j a **zero of multiplicity** m_j (a **simple zero** if $m_j = 1$). Notice that $m_1 + m_2 + \cdots + m_k = n$, the degree of $f(z)$. Hence $f(z)$ has exactly n zeros if we count c_j as m_j zeros.

> A polynomial of degree $n \geq 1$ has at most n distinct complex zeros. It has exactly n complex zeros if a zero of multiplicity m is counted m times.

For example,
$$f(z) = (z - 1)(z - 2)(z - 1 - i)^2(z - 2i)^3$$
is a polynomial of degree 7 with 4 distinct zeros. The numbers 1 and 2 are simple zeros, $1 + i$ is a zero of multiplicity 2, and $2i$ is a zero of multiplicity 3. There are 7 zeros if they are counted this way:
$$1, 2, 1 + i, 1 + i, 2i, 2i, 2i$$

Polynomials with Real Coefficients

The polynomial $x^2 + x + 4$ has two complex zeros,
$$-\tfrac{1}{2} + \tfrac{1}{2}\sqrt{15}\, i \quad \text{and} \quad -\tfrac{1}{2} - \tfrac{1}{2}\sqrt{15}\, i$$

The fact that they are conjugates is not an accident but a general property of polynomials with *real coefficients*. For brevity, we call such a polynomial a **real polynomial**.

> **Conjugate Zeros** Let $f(z)$ be a polynomial with real coefficients. If $f(c) = 0$, then $f(\bar{c}) = 0$.

Stated briefly, non-real complex zeros of real polynomials come in conjugate pairs. Thus, you get two zeros for the price of one.

Let us justify this statement. (We use the same kind of reasoning as in Example 4, page 267.) Let
$$f(z) = a_n z^n + a_{n-1} z^{n-1} + \cdots + a_1 z + a_0$$
where $a_n, \ldots, a_0$ are real. If $f(c) = 0$, then
$$a_n c^n + a_{n-1} c^{n-1} + \cdots + a_1 c + a_0 = 0$$
Take the conjugates of both sides:
$$\overline{a_n c^n + a_{n-1} c^{n-1} + \cdots + a_1 c + a_0} = \bar{0} = 0$$
Now, by the rules for conjugates (page 267),
$$\overline{a_n c^n} + \overline{a_{n-1} c^{n-1}} + \cdots + \overline{a_1 c} + \overline{a_0} = 0$$
$$\overline{a_n}\,\overline{c^n} + \overline{a_{n-1}}\,\overline{c^{n-1}} + \cdots + \overline{a_1}\,\overline{c} + \overline{a_0} = 0$$
But a_j is real, so $\overline{a_j} = a_j$. Therefore, $\overline{a_j c^j} = a_j \overline{c^j} = a_j (\bar{c})^j$. Hence
$$a_n (\bar{c})^n + a_{n-1} (\bar{c})^{n-1} + \cdots + a_1 \bar{c} + a_0 = 0$$
But this says that $f(\bar{c}) = 0$, as was to be shown.

EXAMPLE 1 Find a real polynomial of degree 4 having zeros i and $3 + 2i$.

SOLUTION Since the coefficients are real, $-i$ and $3 - 2i$ are also zeros. A logical choice is the polynomial
$$f(z) = (z - i)(z + i)(z - 3 - 2i)(z - 3 + 2i)$$
Let us check that it has real coefficients. We note that
$$(z - i)(z + i) = z^2 + 1$$
and
$$(z - 3 - 2i)(z - 3 + 2i) = [(z - 3) - 2i][(z - 3) + 2i]$$
$$= (z - 3)^2 - (2i)^2 = z^2 - 6z + 13$$
All coefficients are real and
$$f(z) = (z^2 + 1)(z^2 - 6z + 13) = \underline{z^4 - 6z^3 + 14z^2 - 6z + 13}$$

Remark Actually the most general polynomial decribed in Example 1 is $f(z) = a(z^4 - 6z^3 + 14z^2 - 6z + 13)$, where a is any real number.

EXAMPLE 2 Find all zeros of $f(z) = z^3 - 5z^2 + 8z - 6$ given that one zero is $1 + i$.

SOLUTION Since the coefficients are real, $1 - i$ is also a zero. Now a cubic has 3 zeros, so there is one zero left to find. (It must be real because non-real zeros come in *pairs*.)

Since $1 + i$ and $1 - i$ are zeros, $f(z)$ is divisible by
$$(z - 1 - i)(z - 1 + i) = (z - 1)^2 - i^2 = z^2 - 2z + 2$$
By long division,
$$f(z) = (z^2 - 2z + 2)(z - 3)$$
The remaining zero is 3.

Answer $1 + i,\ 1 - i,\ 3$

EXAMPLE 3 Find all zeros of
$$f(z) = z^6 + z^5 + 11z^4 + 8z^3 + 40z^2 + 16z + 48$$
given that $2i$ is a double zero.

SOLUTION The coefficients are real, so $-2i$ is also a double zero. Hence $f(z)$ is divisible by
$$(z - 2i)^2(z + 2i)^2 = [(z - 2i)(z + 2i)]^2$$
$$= (z^2 + 4)^2 = z^4 + 8z^2 + 16$$
By long division,
$$f(z) = (z^4 + 8z^2 + 16)(z^2 + z + 3)$$
The remaining zeros are zeros of $z^2 + z + 3$. We find them by the quadratic formula.

Answer $2i,\ 2i,\ -2i,\ -2i,\ -\tfrac{1}{2} + \tfrac{1}{2}\sqrt{11}\,i,\ -\tfrac{1}{2} - \tfrac{1}{2}\sqrt{11}\,i$

Let us examine the complete factorization of a real polynomial:
$$f(z) = a_n(z - c_1)(z - c_2) \cdots (z - c_n)$$
Each real zero c_j contributes a *real* linear factor $z - c_j$. Each conjugate pair of zeros, $a + bi$ and $a - bi$, contributes a *real* quadratic factor
$$(z - a - bi)(z - a + bi) = (z - a)^2 - (bi)^2 = z^2 - 2az + a^2 + b^2$$
This quadratic is called **irreducible** because it cannot be factored into real linear factors (otherwise it would have real zeros).

Complete Factorization of Real Polynomials A real polynomial of degree ≥ 1 is the product of real linear and real irreducible quadratic factors.

EXAMPLE 4 Factor $z^3 + 8$ into real linear and real irreducible quadratic factors.

SOLUTION Factor as the sum of two cubes:
$$z^3 + 8 = z^3 + 2^3$$
$$= (z + 2)(z^2 - 2z + 4)$$
The factors are real, and the quadratic $z^2 - 2z + 4$ is irreducible because it has no real zeros.

EXERCISES

Find a real polynomial as described
1. degree 2, zero $1 - 3i$
2. degree 2, zero $-4i$
3. degree 3, zeros 2, $3 + i$
4. degree 3, zeros -3, $1 - 2i$
5. degree 4, zeros $1 + i$, $2 + i$
6. degree 4, zeros 1, 2, $5 + i$
7. degree 4, double zero, $1 - 3i$
8. degree 5, zeros 1, $2 + i$, $3 + i$

Factor completely into linear factors using complex numbers if necessary
9. $z^2 + 7$
10. $z^2 - 2z + 5$
11. $z^2 + z + 3$
12. $z^3 + z$
13. $z^4 - z^2 - 6$
14. $z^4 + 10z^2 + 9$
15. $z^6 - 16z^2$
16. $z^4 + 10z^2 + 25$

Factor into real linear and quadratic factors
17. $z^3 - 27$
18. $z^4 - 81$
19. $z^6 - 9z^3 + 8$
20. $z^8 - 17z^4 + 16$

Find all zeros of the polynomial, using the information given
21. $z^3 - 3z^2 + z + 5$, one zero is $2 - i$
22. $z^3 - 4z^2 - 2z + 20$, one zero is $3 - i$
23. $z^2 - (3 + i)z + 3i$, one zero is i
24. $z^2 - (3 + i)z + 2 + 2i$, one zero is 2
25. $z^4 - 5z^3 + 7z^2 - 5z + 6$, one zero is i
26. $z^4 - z^3 - 2z^2 - 2z + 4$, one zero is $-1 + i$
27. $z^5 - z^4 + 4z^3 + 4z^2 + 3z + 5$, two zeros are i, $1 + 2i$
28. $z^6 - 2z^5 + 10z^4 - 10z^3 + 29z^2 - 8z + 20$, two zeros are i, $2i$
29. If a polynomial has zeros 1, -1, i, $-i$, show that it is divisible by $z^4 - 1$.
30. Show by the methods of this section that a real polynomial of odd degree has a real zero.
31. Show that the zeros of $z^2 - 4iz$ are not conjugates. Does this contradict the text?
32. Divide $z^4 - 1$ by $z - i$, using synthetic division.

True or false? Why?
33. A real polynomial of odd degree has a non-real complex zero.
34. A real polynomial of even degree cannot have a zero of odd multiplicity.
35. A real polynomial of degree 4 cannot have exactly 3 simple real zeros.

36. A real cubic polynomial cannot have a double non-real zero.

37. If all zeros of a polynomial $f(z)$ have even multiplicity, show that $f(z) = [g(z)]^2$ for some polynomial $g(z)$.

38. If $az^2 + bz + c$ is irreducible, what condition do the real coefficients a, b, c satisfy?

39. If $f(z)$ is a real polynomial, show that $f(\bar{z}) = \overline{f(z)}$.

40. How does it follow from Exercise 39 that complex zeros of real polynomials come in conjugate pairs?

6 REAL ZEROS OF POLYNOMIALS AND APPROXIMATION OF ZEROS

Given a polynomial, we may not know whether it has any real zeros. So let us begin with an important test for the existence of real zeros.

Existence of Real Zeros

Geometrically, a real zero of $f(x)$ is a point $x = r$ where the graph of $f(x)$ intersects the x-axis. For a polynomial, the graph is a continuous curve. If the graph is below the x-axis at $x = a$ and above at $x = b$, or vice versa, then it must cross the x-axis in between (Fig. 1). Hence $f(x)$ has at least one zero between a and b.

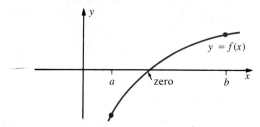

FIG. 1

Existence of Real Zeros If $f(x)$ is a polynomial and $f(a)$ and $f(b)$ have opposite signs, then $f(x)$ has a zero between a and b.

EXAMPLE 1 Show that $f(x) = x^4 - 5x^3 - 2x^2 + x + 1$ has a zero in the interval $[0, 1]$ and another in the interval $[5, 6]$.

SOLUTION By inspection, $f(0) = 1$ and $f(1) = -4$. The signs are opposite, so $f(x)$ has a zero between 0 and 1.

Next, check the signs of $f(5)$ and $f(6)$. Use synthetic division:

$$
\begin{array}{r|rrrrr}
5 & 1 & -5 & -2 & 1 & 1 \\
  &   & 5 & 0 & -10 & -45 \\
\hline
  & 1 & 0 & -2 & -9 & -44
\end{array}
$$

$$
\begin{array}{r|rrrrr}
6 & 1 & -5 & -2 & 1 & 1 \\
  &   & 6 & 6 & 24 & 150 \\
\hline
  & 1 & 1 & 4 & 25 & 151
\end{array}
$$

Hence $f(5) = -44$ and $f(6) = 151$. The signs are opposite, so $f(x)$ has a zero between 5 and 6.

A polynomial of *odd* degree always has a real zero. For suppose that $f(x) = a_n x^n + \cdots + a_0$, where n is odd. If $a_n > 0$, we know that $f(x) \to \infty$ as $x \to \infty$ and $f(x) \to -\infty$ as $x \to -\infty$. Therefore, $f(x)$ has opposite signs at x and $-x$ if $|x|$ is large enough. The same conclusion holds if $a_n < 0$.

> A polynomial of odd degree has at least one real zero.

EXAMPLE 2 Does there exist a real number whose seventh power exceeds its square by 1000?

SOLUTION The question is whether the equation
$$x^7 = x^2 + 1000$$
has a real solution. In other words, does the polynomial $x^7 - x^2 - 1000$ have a real zero? It does, because its degree is odd.

Answer Yes

Approximating Real Zeros [Optional]

Let us return to the practical problem of finding real zeros of polynomials. Only in special cases can we find real zeros *exactly*: if the polynomial is quadratic, or if it factors, or if the zeros are rational. Generally, however, factoring is impossible and the zeros are irrational. In such cases we use numerical methods for locating real zeros *approximately*. Let us describe an "educated" trial-and-error method.

Given a polynomial $f(x)$, we locate a zero between two consecutive integers n and $n + 1$. Then we divide the interval $[n, n + 1]$ into 10 equal subintervals and locate the zero in one of these. Next we divide that subinterval into 10 parts and locate the zero in one of these, etc.

For instance, suppose $f(3) < 0$ and $f(4) > 0$. Then there is a zero in $[3, 4]$. We test the values $f(3.0), f(3.1), f(3.2), \ldots, f(3.9), f(4.0)$. Somewhere the sign changes from $-$ to $+$, say $f(3.2) < 0$ but $f(3.3) > 0$. That means the zero is trapped in $[3.2, 3.3]$. Next we test $f(3.20), f(3.21), \ldots, f(3.29), f(3.30)$. Say $f(3.27) < 0$ but $f(3.28) > 0$. Then the zero is trapped in $[3.27, 3.28]$. Continuing this way, we gain one more decimal place accuracy at each stage. Since the method requires evaluating $f(x)$ for many values of x, a calculator helps enormously.

EXAMPLE 3 Estimate to three places the positive zero of $f(x) = x^4 + x - 5$.

SOLUTION A little trial and error shows that $f(1) = -3$ and $f(2) = 13$. Hence the zero is in $[1, 2]$.

Next we tabulate $f(1.0), f(1.1), f(1.2), \ldots$ until the sign changes. A calculator routine for $f(x)$ is

x STO x^2 x^2 $+$ RCL $-$ 5 $=$ or (faster) x $+$ x^2 x^2 $-$ 5 $=$

We find the values

x	1.0	1.1	1.2	1.3	1.4
$f(x)$	−3.0	−2.4359	−1.7264	−0.8439	0.2416

The signs of $f(1.3)$ and $f(1.4)$ are opposite. Hence the zero is between 1.3 and 1.4. Probably it is closer to 1.4, because $f(1.4) \approx 0.24$ which is closer to 0 than $f(1.3) \approx -0.84$.

Next we examine $f(1.30)$, $f(1.31)$, . . . , $f(1.39)$, $f(1.40)$. Since the zero may be nearer 1.40, let's start at 1.40 and work *backward* until the sign changes:

x	1.37	1.38	1.39	1.40	
$f(x)$	−0.1072	0.0067	0.1230	0.2416	(approx.)

The zero is between 1.37 and 1.38, probably close to 1.38. (Starting at 1.40 required computing 3 new values of $f(x)$. Starting at 1.30 would have required 8 new values.)

Next we start at 1.380 and work backward in steps of 0.001:

x	1.379	1.380
$f(x)$	−0.0048	0.0067

The zero is between 1.379 and 1.380, probably nearer 1.379. To be sure, we check the midpoint: $f(1.3795) \approx 0.0010$. Hence the zero is between 1.379 and 1.3795. To 3 places, $x = \underline{1.379}$.

Linear Interpolation

We can make our trial-and-error more "educated." In Example 3, the first table showed $f(1.3) \approx -0.84$ and $f(1.4) \approx 0.24$. We concluded that the zero was probably closer to 1.4 than to 1.3. But we can do better. Look at the evidence:

$$\underbrace{-0.84 \qquad\qquad 0 \qquad 0.24}_{1.08}^{0.84}$$

The data suggest going $84/108 \approx 0.77$ of the way from 1.3 to 1.4. So we start the next stage at 1.38, not at 1.40.

x	1.30	1.38
$f(x)$	−0.8439	0.0067

The signs of $f(1.30)$ and $f(1.38)$ are opposite, so we move down in steps of 0.01. The first step gives $f(1.37) < 0$. Hence the zero is between 1.37 and 1.38.

The technique here is **linear interpolation.** Between $x = 1.3$ and 1.4, we pretend that the graph of $f(x)$ is a straight line. Then we approximate the zero of $f(x)$ by the x-intercept of that line (Fig. 2, next page).

Under a microscope, a continuous graph looks nearly straight. Hence linear interpolation can be quite accurate on small intervals. To illustrate, suppose Example 2 had asked for 5-place accuracy. At one point we had

$$f(1.379) \approx 0.0048 \qquad f(1.380) \approx 0.0067$$

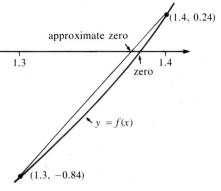

FIG. 2

Since [1.379, 1.380] is a small interval, we try guessing the next 2 decimal places by linear interpolation:

$$\underbrace{\overbrace{-0.0048 \qquad\qquad 0 \qquad\qquad 0.0067}^{48}}_{115}$$

Now, $48/115 \approx 0.42$, so we jump to 1.37942 and start testing there:

x	1.37941	1.37942
$f(x)$	-0.000049	0.000066

To 5 places, $x = 1.37941$ or 1.37942. We could test the midpoint or go on and find a few more digits.

Feeling bold, we interpolate again and try jumping 3 decimal places. Since $49/115 \approx 0.426$, we start at $x = 1.37941\ 426$.

x	1.37941 425	1.37941 426
$f(x)$	-4×10^{-8}	7×10^{-8}

To 7 places, $x = 1.37941\ 43$. To 8 places, $x = 1.37941\ 425$ or $1.37941\ 426$.

Solution of Equations

The approximation method described here does not require $f(x)$ to be a polynomial, only a function with a continuous graph. Therefore, it applies in many situations.

EXAMPLE 4 Find a solution of the equation $e^x = 2 - x$ to 3 decimal places.

SOLUTION First of all, does the equation have a solution? A rough sketch (Fig. 3, next page) shows that it does. The graphs of $y = e^x$ and $y = 2 - x$ intersect in exactly one point (x, y), with $x \approx 0.4$. Hence there is one solution, and it is near $x = 0.4$.

Let us write the equation as $f(x) = 0$, where $f(x) = e^x + x - 2$. Now $f(x)$ has a zero near $x = 0.4$. Since the function has a continuous graph, we can use our approximation method to squeeze down on that zero, just as we did for polynomials.

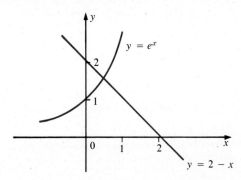

FIG. 3

We start at $x = 0.4$ and move in steps of 0.1. A calculator routine for $f(x)$ is $x\boxed{+}\boxed{e^x}\boxed{-}2\boxed{=}$.

x	0.0		0.4	0.5
$f(x)$	−1.0		−0.11	0.15

The zero is between 0.4 and 0.5. Linear interpolation suggests $x \approx 0.44$. We find

x	0.44	0.45
$f(x)$	−0.007	0.018

Another linear interpolation (jumping 2 places) suggests $x \approx 0.4428$:

x	0.4428	0.4429
$f(x)$	−0.00014	0.00012

Thus, $0.4428 < x < 0.4429$. To 3 places, $x = \underline{0.443}$.

EXERCISES

1. Is there a real number that exceeds its cube by 2?
2. Is there a real number that is 1 more than half its fifth power?

Show that the polynomial has a zero in the given interval

3. $x^3 + 4x - 1$, $[0, 1]$
4. $x^3 - 12x + 3$, $[3, 4]$
5. $2x^4 - x^2 - 6x - 3$, $[-1, 0]$
6. $x^7 - x^2 - 1000$, $[2, 3]$
7. $x^3 + x - 1$, $[0.6, 0.7]$
8. $x^4 + 100x - 1$, $[0.00, 0.01]$

The polynomial has exactly one real zero. Locate it between two consecutive integers

9. $x^3 + 2x - 4$
10. $x^3 + x + 116$
11. $\tfrac{1}{4}x^5 + x + 12$
12. $\tfrac{1}{2}x^5 + 2x^3 - 70$
13. $12x^3 - 8x^2 + 7x - 10$
14. $x^3 + 4x^2 + 10x + 15$

Estimate the positive solution to 3 decimal places

15. $2x^3 + x - 10 = 0$
16. $x^3 + 5x - 1 = 0$
17. $x^4 + x^3 - 3x^2 - 1 = 0$
18. $2x^4 - 4x^3 - 7 = 0$
19. $x^5 + x^2 = 100$
20. $x^7 = x^2 + 1000$

Estimate the negative solution to 5 decimal places

21. $x^3 + 5x^2 + 1 = 0$ **22.** $x^4 + 6x - 1 = 0$

23. $x^5 + 3x^3 - x^2 + 6 = 0$ **24.** $x^5 + 4x^4 + 10 = 0$

25. $x^7 = \frac{1}{2}x - 8$ **26.** $x^6 + x^2 - 2x = 5$

Estimate the solution to 3 decimal places

27. $e^x + 3x = 41$ **28.** $10^x = 6 - x$

29. $x^4 = 5\sqrt{x} + 2$ **30.** $xe^x = 100$

31. $\ln x + x = 0$ **32.** $\log x = \dfrac{1}{x-3}$

33. If $f(x)$ is a polynomial whose coefficients are all positive, show that $f(x)$ has no *positive* zeros.

34. If $f(x)$ is a polynomial whose coefficients alternate in sign, show that $f(x)$ has no *negative* zeros.

35. Suppose that $f(x)$ is a polynomial with even degree and positive leading coefficient. If $f(0) < 0$ why must $f(x)$ have at least two real zeros?

REVIEW EXERCISES FOR CHAPTER 9

Compute $f(a)$ by synthetic division

1. $f(x) = x^4 + 2x^3 - 7x^2 - x + 5$, $a = 3$

2. $f(x) = x^5 - 3x^4 + x^2 - 7x - 1$, $a = -2$

Find all rational zeros

3. $f(x) = 2x^3 + 3x^2 + 4x + 6$

4. $f(x) = 3x^4 + 5x^3 + x^2 + 5x - 2$

Express in the form $a + bi$

5. $\dfrac{1}{1+2i} + \dfrac{3i}{4-i}$ **6.** $\left(\dfrac{4+i}{2-3i}\right)^2$

7. Express $f(x) = x^5 + 2x^2 - 6x + 3$ in the form $(x-2)q(x) + r$.

8. Compute $i^{27} + i^{18} + i^{13} + 1$.

9. Find all complex numbers z for which $z\bar{z} = 0$.

10. Find the cubic polynomial $f(x)$ for which -1 is a zero, 1 is a double zero, and $f(0) = 5$.

Find all real zeros and their multiplicities

11. $f(x) = (x^2 - 1)^2(x^2 + 3)^4(x^2 - 4x + 3)$

12. $f(x) = (x^4 - 4x^2)^3$

Find all zeros of the polynomial using the information given

13. $x^3 + 2x^2 + 6x + 5$, one zero is -1

14. $x^4 - 5x^3 + 5x^2 - 20x + 4$, one zero is $2i$

The polynomial has one positive zero. Locate it between two consecutive integers

15. $x^3 - 5x^2 - 2x - 9$

16. $2x^5 + x^4 - 23x - 8$

17. Explain why the equation $x^5 + 6x^2 = 1 + \dfrac{1}{x^2}$ has a real solution.

18. Under what conditions on a, b, c is $ax^2 + bx + c$ irreducible?

19. Show that $f(x) = x^6 + 7x^4 + 6x^3 + x^2 + 9x - 1$ has at least two real zeros.

20. Express $f(x) = x^6 - 1$ as a product of real linear and irreducible quadratic factors. (Hint Factor first as the difference of squares.)

10

Discrete Algebra

1 SEQUENCES AND SERIES

We often deal with sets of numbers that occur in a definite order, for instance the hourly temperature readings over a day,

$$t_1, t_2, t_3, \ldots, t_{24}$$

or the integer powers of 2,

$$2^1, 2^2, 2^3, 2^4, \ldots$$

These are examples of **sequences.** The first is a **finite sequence** because it has a last **term.** Technically, a finite sequence is any set of objects in one-to-one correspondence with a block of integers $\{1, 2, 3, \ldots, n\}$. In this case, the block is $\{1, 2, 3, \ldots, 24\}$. We will always take the "objects" to be real numbers.

The second example above is an **infinite sequence** because it has no last term. Technically, an infinite sequence is a set of objects in one-to-one correspondence with the set of positive integers $\{1, 2, 3, \ldots\}$.

It is common to specify an infinite sequence $a_1, a_2, a_3, \ldots$ by giving a formula for its n-th term. For example, the formula $a_n = 1/n$ describes the sequence

$$1, \frac{1}{2}, \frac{1}{3}, \frac{1}{4}, \ldots, \frac{1}{n}, \ldots$$

For a finite sequence, the range of n must also be given. For example, $a_n = \sqrt{n}$, $1 \leq n \leq 6$ describes the sequence

$$\sqrt{1}, \sqrt{2}, \sqrt{3}, \sqrt{4}, \sqrt{5}, \sqrt{6}$$

When no restriction on n is given, we assume the sequence is infinite.

Examples

$a_n = 2n$: $\quad 2, 4, 6, 8, \ldots \qquad\qquad a_n = (-1)^n n^2$: $\quad -1^2, 2^2, -3^2, 4^2, \ldots$

$a_n = \dfrac{n}{n+1}$: $\quad \dfrac{1}{2}, \dfrac{2}{3}, \dfrac{3}{4}, \dfrac{4}{5}, \ldots$

EXAMPLE 1 Write out the first 5 terms of the sequence whose n-th term is a_n

(a) $a_n = 3n + 1$ (b) $a_n = 3 \cdot 2^n$ (c) $a_n = \dfrac{1}{(n+1)(n+2)}$

SOLUTION Just substitute $n = 1, 2, 3, 4, 5$.
Answer
(a) 4, 7, 10, 13, 16   (b) 6, 12, 24, 48, 96   (c) $\frac{1}{6}, \frac{1}{12}, \frac{1}{20}, \frac{1}{30}, \frac{1}{42}$

EXAMPLE 2 Find a formula for the n-th term

(a) 1, 3, 5, 7, ...   (b) $\frac{1}{4}, \frac{1}{9}, \frac{1}{16}, \frac{1}{25}, \ldots$

(c) $17\sqrt{3}, 27\sqrt{6}, 37\sqrt{9}, 47\sqrt{12}, \ldots$

SOLUTION (a) The terms are 1 less than 2, 4, 6, ..., $2n$, ... Hence, $a_n = 2n - 1$.

(b) The denominators are $2^2, 3^2, 4^2, \ldots$, where the number squared is 1 more than the number of the term. Hence, $a_n = 1/(n+1)^2$.

(c) The numbers inside the radicals are 3, 6, 9, ..., $3n$, ... So a_n has the factor $\sqrt{3n}$. The factor multiplying it is 7 more than $10n$. Hence, $a_n = (10n + 7)\sqrt{3n}$.

Not every sequence can be described by a formula. An example is

3, 1, 4, 1, 5, 9, 2, 6, 5, ...

the decimal digits of π. This is a well-defined sequence, but there is no formula for computing its n-th term.

Summation Notation

A sum of terms of a sequence is sometimes called a **series**. We use the symbol

$$\sum_{j=1}^{n} a_j$$

to abbreviate the sum

$$a_1 + a_2 + \cdots + a_n$$

of the terms of the sequence $a_1, a_2, \ldots, a_n$.

The Greek sigma Σ stands for *sum*. The letter j is called the **summation index**; it labels the terms to be added. For example, if $a_j = j^2$, then

$$\sum_{j=1}^{10} a_j = 1^2 + 2^2 + \cdots + 10^2$$

In practice, we usually write this sum as

$$\sum_{j=1}^{10} j^2$$

The notation means the sum of all terms of the form j^2, where j takes the values 1, 2, ..., 10.

The letter j is a **dummy variable** because any other letter will do just as well. Thus

$$\sum_{i=1}^{10} i^2 = \sum_{j=1}^{10} j^2 = \sum_{k=1}^{10} k^2 = 1^2 + 2^2 + \cdots + 10^2$$

The index need not start at 1. For example,

$$\sum_{j=5}^{9} a_j = a_5 + a_6 + a_7 + a_8 + a_9$$

Examples

$$\sum_{i=6}^{10} i(i+1) = 6 \cdot 7 + 7 \cdot 8 + 8 \cdot 9 + 9 \cdot 10 + 10 \cdot 11 = 370$$

$$\sum_{j=2}^{6} \frac{(-1)^j}{j} = \frac{1}{2} - \frac{1}{3} + \frac{1}{4} - \frac{1}{5} + \frac{1}{6} = \frac{23}{60}$$

$$\sum_{k=0}^{n} a_k x^k = a_0 + a_1 x + a_2 x^2 + \cdots + a_n x^n$$

EXAMPLE 3 Express in summation notation

(a) $5^3 + 6^3 + 7^3 + \cdots + n^3$ (b) $\dfrac{1}{8} + \dfrac{1}{16} + \dfrac{1}{32} + \dfrac{1}{64} + \dfrac{1}{128}$

(c) $x^9 y + x^8 y^2 + x^7 y^3 + \cdots + x y^9$

SOLUTION (a) The typical term is j^3, where j runs from 5 to n. Hence

$$5^3 + 6^3 + 7^3 + \cdots + n^3 = \sum_{j=5}^{n} j^3$$

(b) The typical term is $1/2^j$, where j runs from 3 to 7. Hence

$$\frac{1}{8} + \frac{1}{16} + \frac{1}{32} + \frac{1}{64} + \frac{1}{128} = \sum_{j=3}^{7} \frac{1}{2^j}$$

(c) Each term has the form $x^p y^q$, where $p + q = 10$. The j-th term has the factor y^j, so $q = j$ and $p = 10 - j$. The index j runs from 1 to 9. Hence

$$x^9 y + x^8 y^2 + x^7 y^3 + \cdots + x y^9 = \sum_{j=1}^{9} x^{10-j} y^j$$

EXERCISES

Find the first 5 terms of the sequence whose n-th term is a_n

1. $a_n = 4n$
2. $s_n = \dfrac{1}{3n}$
3. $a_n = 5 - \tfrac{1}{2}n$
4. $a_n = 7 - 2n$
5. $a_n = \dfrac{n}{n+3}$
6. $a_n = \dfrac{1}{n^2 + 1}$

Discrete Algebra

7. $a_n = 1 + (-1)^n$
8. $a_n = (-1)^{n+1} \log n$
9. $a_n = 3 \cdot 2^{n-1}$
10. $a_n = 2^{8-3n}$
11. $a_n = n\sqrt{n+2}$
12. $a_n = n^2 - n + 4$

Compute the sum

13. $\sum_{i=1}^{4} i$
14. $\sum_{i=1}^{3} i^2$
15. $\sum_{j=1}^{6} (2j + 5)$
16. $\sum_{j=4}^{8} (3j - 7)$
17. $\sum_{k=6}^{10} (10 - k)$
18. $\sum_{k=5}^{9} (15 - 2k)$
19. $\sum_{n=1}^{3} \frac{n+1}{2n}$
20. $\sum_{n=1}^{4} \frac{2^n}{n}$
21. $\sum_{j=2}^{5} (-1)^j j(j+2)$
22. $\sum_{j=3}^{7} \frac{j(j+1)}{2}$
23. $\sum_{n=1}^{6} n \cdot 10^{n-1}$
24. $\sum_{n=0}^{3} 2^{-(n+1)}$

Find a formula for the n-th term of the sequence

25. $8, 16, 24, 32, \ldots$
26. $3, 9, 27, 81, \ldots$
27. $\sqrt{1}, 2\sqrt{3}, 3\sqrt{5}, 4\sqrt{7}, \ldots$
28. $9^2, 19^2, 29^2, 39^2, \ldots$
29. $\frac{1}{8}, \frac{1}{16}, \frac{1}{32}, \frac{1}{64}, \ldots$
30. $\frac{1}{2}, \frac{1}{8}, \frac{1}{32}, \frac{1}{128}, \ldots$
31. $-\frac{5}{6}, \frac{6}{11}, -\frac{7}{16}, \frac{8}{21}, \ldots$
32. $\frac{3}{2}, -\frac{9}{4}, \frac{27}{8}, -\frac{81}{16}, \ldots$
33. $6, 12, 20, 30, \ldots$
34. $10, 18, 28, 40, \ldots$

(Hint for Exercises 33–34 Factor the terms)

Express using summation notation

35. $2^4 + 2^5 + 2^6 + 2^7 + 2^8$
36. $1 + 4 + 7 + 10 + 13 + 16$
37. $\frac{1}{3} + \frac{2}{4} + \frac{3}{5} + \cdots + \frac{15}{17}$
38. $\frac{1}{5^2} + \frac{1}{7^2} + \frac{1}{9^2} + \cdots + \frac{1}{21^2}$
39. $1 - \frac{1}{2} + \frac{1}{4} - \frac{1}{8} + \frac{1}{16}$
40. $\frac{4}{5 \cdot 6} + \frac{5}{6 \cdot 7} + \frac{6}{7 \cdot 8} + \cdots + \frac{15}{16 \cdot 17}$
41. $a_0 + a_1 x^2 + a_2 x^4 + \cdots + a_6 x^{12}$
42. $x^2 y^{10} + x^4 y^9 + x^6 y^8 + \cdots + x^{20} y$

43. the sum of all odd integers between 200 and 400
44. the sum of all even integers between 21 and 77

Is the statement true or false?

45. $\sum_{j=1}^{10} j(j+1) = \sum_{j=2}^{11} (j-1)j$
46. $\sum_{j=1}^{16} (2j - 1) = \sum_{j=0}^{15} (2j + 1)$
47. $\sum_{k=1}^{8} (1 + k^3) = 1 + \sum_{k=1}^{8} k^3$
48. $\sum_{k=1}^{3} k^2 = \left(\sum_{k=1}^{3} k\right)^2$

49. $\sum_{i=1}^{9} x^{10-i} y^i = \sum_{i=1}^{9} x^i y^{10-i}$

50. $\sum_{i=-4}^{4} 2^i = \sum_{i=-4}^{4} 2^{-i}$

Justify the rule for summations

51. $\sum_{i=1}^{n} c a_i = c \sum_{i=1}^{n} a_i$

52. $\sum_{i=1}^{n} (a_i + b_i) = \sum_{i=1}^{n} a_i + \sum_{i=1}^{n} b_i$

2 ARITHMETIC SEQUENCES

An **arithmetic sequence** or **progression** is a sequence of equally spaced numbers.

Examples

(1) The sequence of odd integers: 1, 3, 5, 7, . . .
(2) The yearly values of a $25,000 property that depreciates $1000 per year:
$$\$25{,}000,\ 24{,}000,\ 23{,}000,\ \ldots,\ 1000,\ 0$$
(3) Successive readings on a taxi meter if the fare is $1 for the first $\frac{1}{6}$ mile plus 10¢ for each additional $\frac{1}{6}$ mile:
$$\$1.00,\ 1.10,\ 1.20,\ 1.30,\ \ldots$$

In an arithmetic progression, consecutive terms differ by a fixed number d, called the **common difference**. In the preceding examples, $d = 2$, -1000, and 0.10, respectively. In general, an arithmetic sequence has the form
$$a_1,\ a_2 = a_1 + d,\ a_3 = a_1 + 2d,\ a_4 = a_1 + 3d,\ \ldots$$
In each term, d is multiplied by 1 less than the number of the term. Hence

$$\boxed{a_n = a_1 + (n-1)d}$$

EXAMPLE 1 Find the 20-th term of the arithmetic progression
$$47,\ 44,\ 41,\ 38,\ \ldots$$

SOLUTION Here $a_1 = 47$ and $d = -3$. Therefore,
$$a_{20} = a_1 + (20-1)d = 47 + 19(-3) = \underline{-10}$$

EXAMPLE 2 If $a_1, a_2, a_3, \ldots$ is the arithmetic progression
$$2.6,\ 2.8,\ 3.0,\ 3.2,\ \ldots$$
and $a_n = 15.8$, find n.

SOLUTION We have
$$a_n = a_1 + (n-1)d = 15.8$$
For this sequence, $a_1 = 2.6$ and $d = 0.2$. Therefore,
$$2.6 + (n-1)(0.2) = 15.8$$

Solve for n:
$$n - 1 = \frac{15.8 - 2.6}{0.2} = 66$$
$$n = \underline{67}$$

Let us find a formula for the sum of an arithmetic progression. In Chapter 1 (page 4) we computed $1 + 2 + 3 + \cdots + 100$ by observing that
$$1 + 100 = 2 + 99 = 3 + 98 = \cdots = 50 + 51 = 101$$
Now we apply the same idea to the sum of any arithmetic progression with n terms. We write the sum twice, first forward, then backward:
$$S = a_1 + (a_1 + d) + (a_1 + 2d) + \cdots + [a_1 + (n - 1)d]$$
$$S = a_n + (a_n - d) + (a_n - 2d) + \cdots + [a_n - (n - 1)d]$$
Adding corresponding terms gives
$$2S = (a_1 + a_n) + (a_1 + a_n) + \cdots + (a_1 + a_n)$$
There are n terms on the right, hence
$$2S = n(a_1 + a_n) \qquad S = n\left(\frac{a_1 + a_n}{2}\right)$$
Since $a_n = a_1 + (n - 1)d$, we can also write
$$S = \frac{n}{2}[a_1 + a_1 + (n - 1)d] = \frac{n}{2}[2a_1 + (n - 1)d]$$

Sum of an Arithmetic Progression

$$a_1 + (a_1 + d) + (a_1 + 2d) + \cdots + [a_1 + (n - 1)d]$$
$$= \frac{n}{2}[2a_1 + (n - 1)d] = n\left(\frac{a_1 + a_n}{2}\right)$$

The form
$$n\left(\frac{a_1 + a_n}{2}\right)$$
shows that the sum is the number of terms times the average of the first and last terms.

EXAMPLE 3 Compute (a) the sum of the first 16 terms of the arithmetic sequence $7.4, 7.1, 6.8, 6.5, \ldots$

(b) $3 + 13 + 23 + 33 + \cdots + 983 + 993$.

SOLUTION (a) We have $a_1 = 7.4$, $d = -0.3$, and $n = 16$. Hence the sum is
$$\frac{n}{2}[2a_1 + (n - 1)d] = \frac{16}{2}[2(7.4) + 15(-0.3)]$$
$$= 8(14.8 - 4.5) = \underline{82.4}$$

(b) This is the sum of an arithmetic progression with first term $a_1 = 3$ and last term $a_n = 993$. To apply the formula, we need n, the number of terms. Writing the progression as

$$3,\ 3 + 10,\ 3 + 2 \cdot 10,\ \ldots,\ 3 + 99 \cdot 10$$

we see that $n = 100$. (Or, we note the that the progression contains exactly $\frac{1}{10}$ of all positive integers up to 1000.) Hence the sum is

$$n\left(\frac{a_1 + a_n}{2}\right) = 100\left(\frac{3 + 993}{2}\right) = 100 \cdot 498 = \underline{49{,}800}$$

It is useful to know the sum of the first n positive integers. Just put $a_1 = 1$ and $a_n = n$ in the formula for the sum of an arithmetic progression.

$$\sum_{j=1}^{n} j = 1 + 2 + 3 + \cdots + n = \frac{n(n+1)}{2}$$

EXAMPLE 4 On the first day of Christmas my true love gave to me, a partridge in a pear tree. On the second day of Christmas, my true love gave to me, two turtle doves and a partridge in a pear tree. On the third day: three French hens, two turtle doves, and a partridge in a pear tree, etc. How many gifts did she give me on the 12-th day?

SOLUTION

$$1 + 2 + 3 + \cdots + 12 = \frac{12 \cdot 13}{2} = \underline{78}$$

EXAMPLE 5 To study the motion of falling objects, Galileo first observed a ball rolling down a gently inclined plane (Fig. 1). He found that if the ball rolls a distance k in the first second, it rolls $3k$ in the next second, then $5k$, then $7k$, etc. How far does the ball roll in n seconds?

FIG. 1

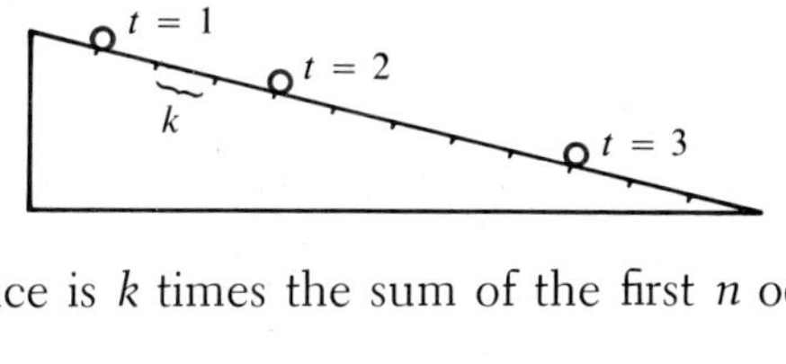

SOLUTION The distance is k times the sum of the first n odd integers, an arithmetic sequence:

$$k[1 + 3 + 5 + \cdots + (2n - 1)] = kn\left(\frac{1 + 2n - 1}{2}\right)$$

$$= kn\left(\frac{2n}{2}\right) = \underline{kn^2}$$

Remark Thus, the distance traveled in n seconds is proportional to n^2. (The proportionality constant k depends on the slope of the plane.) As the plane approaches vertical position, the motion approaches free fall. Hence, Galileo concluded that a free-falling object falls a distance kt^2 in t seconds, for some suitable constant k.

EXERCISES

Find the 12-th term and the n-th term of the arithmetic progression
1. 4, 7, 10, 13, . . .
2. 23, 19, 15, 11, . . .
3. 5.1, 4.6, 4.1, 3.6, . . .
4. 2.2, 2.9, 3.6, 4.3, . . .
5. $3, 5 + \pi, 7 + 2\pi, 9 + 3\pi, \ldots$
6. $\frac{1}{2}\sqrt{2} - 1, \sqrt{2} - 1, \frac{3}{2}\sqrt{2} - 1, 2\sqrt{2} - 1, \ldots$
7. $a + b, 2a + 5b, 3a + 9b, 4a + 13b, \ldots$
8. $x, 3x - y, 5x - 2y, 7x - 3y, \ldots$

Given the term a_n in the arithmetic progression, find n
9. $-30, -41, -52, -63, \ldots \quad a_n = -1229$
10. $6, 49, 92, 135, \ldots \quad a_n = 1511$
11. $7.09, 7.16, 7.23, 7.30, \ldots \quad a_n = 12.76$
12. $-4.31, -4.29, -4.27, -4.25, \ldots \quad a_n = 27.03$

Find the missing terms in the arithmetic progression
13. $6, a_2, 20$
14. $7, a_2, -9$
15. $10.5, a_2, a_3, 3.0$
16. $9, a_2, a_3, a_4, 33$

Compute the sum of the arithmetic progression
17. $1.01 + 1.02 + 1.03 + \cdots + 1.50$
18. $1 + 5 + 9 + 13 + \cdots + 41$
19. $8 + 11 + 14 + \cdots + 53$
20. $11 + 12 + 13 + \cdots + 35$
21. $x + (x + 3) + (x + 6) + \cdots + (x + 24)$
22. $(3x - 1) + (9x - 1) + (15x - 1) + \cdots + (57x - 1)$
23. $(n + 1) + (n + 2) + (n + 3) + \cdots + 2n$
24. $(2n + 1) + (2n + 2) + (2n + 3) + \cdots + 3n$
25. $\sum_{j=1}^{10} (5j - 4)$
26. $\sum_{j=1}^{20} (\frac{2}{3}j + 1)$

Find the sum of all integers
27. between 200 and 400, and odd
28. between 21 and 77, and even
29. between 0 and 500, and divisible by 3
30. between 500 and 1000, and leaving a remainder of 4 when divided by 7

31. A class picture taken on the steps of the school shows 17 students standing on the ground, 16 on the first step, 15 on the second step, etc., up to and including the eighth step. How many students are in the picture?

32. An employee has received a $750 increase each year. In 8 years on the job, she has earned a total of $109,000. What was her starting yearly salary?

33. In 1970, a company had sales of $7,000,000. Each year the sales increased by the same amount, reaching $15,000,000 in 1979. Compute the company's total sales for the decade.

34. A rich uncle gave his niece $500 on her first birthday and increased the gift by $100 on each successive birthday. How much had he given her just after her 21st birthday?

35. The sum of the first n positive integers is 595. Find n.

36. Show that the sum of any 20 consecutive positive integers is divisible by 10.

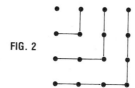
FIG. 2

37. Prove the formula $1 + 3 + 5 + \cdots + (2n - 1) = n^2$ by counting the dots as indicated in Figure 2.

38. Numbers of the form $1 + 2 + 3 + \cdots + n = \frac{1}{2}n(n + 1)$ are called **triangular numbers** (Fig. 3). Show that the sum of two consecutive triangular numbers is a perfect square.

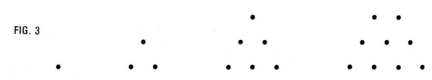
FIG. 3

39. Use Figure 4 to obtain the result of Exercise 38.

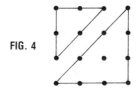
FIG. 4

40. Show that $(k + 1) + (k + 2) \cdots + n = \frac{1}{2}n(n + 1) - \frac{1}{2}k(k + 1)$, is the difference of two triangular numbers.

41. Suppose that $a_1, a_2, \ldots, a_{99}$ and $b_1, b_2, \ldots, b_{99}$ are arithmetic progressions. Show that their sums are equal if and only if $a_{50} = b_{50}$.

42. Suppose that $a_1, a_2, a_3, \ldots$ and $b_1, b_2, b_3, \ldots$ are arithmetic progressions. Which of the following sequences are also arithmetic progressions?
    (a) $a_1 + b_1, a_2 + b_2, a_3 + b_3, \ldots$
    (b) $5a_1 - 2b_1, 5a_2 - 2b_2, 5a_3 - 2b_3, \ldots$
    (c) $a_1 b_1, a_2 b_2, a_3 b_3, \ldots$

43. Suppose that $a_1, a_2, a_3, \ldots$ is an arithmetic progression. Show that there is a linear function $f(x)$ for which $f(1) = a_1$, $f(2) = a_2$, etc.

44. Find all strings of consecutive positive integers with sum 303. (Hint See Exercise 40.)

3 GEOMETRIC SEQUENCES

In an arithmetic sequence, you *add* a fixed number to each term to get the next term. In a **geometric sequence** or **progression**, you *multiply* by a fixed number.

Examples

(1) Suppose $1000 is deposited in an account paying 6% interest, compounded yearly. The yearly values of the account are

   $1000, \quad 1000(1.06), \quad 1000(1.06)^2, \quad 1000(1.06)^3, \ldots$

(2) A radioactive material has a half-life of one day. A 5-gm sample disintegrates daily according to the sequence

   $5, \quad 5(\tfrac{1}{2}), \quad 5(\tfrac{1}{2})^2, \quad 5(\tfrac{1}{2})^3, \ldots$

(3) A chemical solution contains 2 gm of impurities. Filtering removes 80% of the impurities. Repeated filtering reduces the amount of impurities according to the sequence

   $2, \quad 2(0.8), \quad 2(0.8)^2, \quad 2(0.8)^3, \ldots$

The general geometric progression $a_1, a_2, a_3, \ldots$ has the form

$$a_1, \quad a_2 = a_1 r, \quad a_3 = a_1 r^2, \quad a_4 = a_1 r^3, \ldots$$

The number r is called the **common ratio** of the progression. In each term, the power of r is 1 less than the number of the term, hence:

$$\boxed{a_n = a_1 r^{n-1}}$$

EXAMPLE 1 The numbers 16, a_2, a_3, 54 form a geometric progression. Find a_2 and a_3.

SOLUTION We need the value of r. Since $a_4 = a_1 r^3$,

$$54 = 16 r^3$$
$$r^3 = \tfrac{54}{16} = \tfrac{27}{8}$$
$$r = \tfrac{3}{2}$$

It follows that

$$a_2 = 16(\tfrac{3}{2}) = \underline{24} \qquad a_3 = 24(\tfrac{3}{2}) = \underline{36}$$

Let us find a formula for the sum of a finite geometric progression with n terms,
$$\sum_{j=0}^{n-1} a_1 r^j = a_1 + a_1 r + a_1 r^2 + \cdots + a_1 r^{n-1} = a_1(1 + r + r^2 + \cdots + r^{n-1})$$

Now the sum in parentheses is hidden in the factoring formula (page 40),
$$x^n - y^n = (x - y)(x^{n-1} + x^{n-2}y + x^{n-3}y^2 + \cdots + xy^{n-2} + y^{n-1})$$
For $x = 1$ and $y = r$, this says
$$1 - r^n = (1 - r)(1 + r + r^2 + \cdots + r^{n-1})$$
Dividing by $1 - r$ gives
$$1 + r + r^2 + \cdots + r^{n-1} = \frac{1 - r^n}{1 - r} = \frac{r^n - 1}{r - 1} \qquad r \neq 1$$

Sum of a Geometric Progression If $r \neq 1$, then
$$a_1 + a_1 r + a_1 r^2 + \cdots + a_1 r^{n-1} = a_1 \frac{1 - r^n}{1 - r} = a_1 \frac{r^n - 1}{r - 1}$$
If $r = 1$, the sum is just
$$a_1 + a_1 + a_1 + \cdots + a_1 = na_1$$

Note The form $a_1(1 - r^n)/(1 - r)$ is more convenient when $|r| < 1$ because the numerator and denominator are positive.

EXAMPLE 2 Compute the sum of the geometric progression
(a) $2 + 6 + 18 + 54 + 162$ (b) $3 - \frac{3}{2} + \frac{3}{4} - \frac{3}{8} + \frac{3}{16} - \frac{3}{32} + \frac{3}{64}$

SOLUTION (a) Here $a_1 = 2$, $r = 3$, and $n = 5$. Hence the sum is
$$2 \frac{3^5 - 1}{3 - 1} = 2 \frac{242}{2} = \underline{242}$$
(b) In this case, $a_1 = 3$, $r = -\frac{1}{2}$, and $n = 7$. Hence the sum is
$$3 \frac{1 - \left(-\frac{1}{2}\right)^7}{1 - \left(-\frac{1}{2}\right)} = 3 \frac{1 + \frac{1}{128}}{\frac{3}{2}} = 2\left(1 + \frac{1}{128}\right) = \frac{129}{64}$$

EXAMPLE 3 Players A and B bet on heads-or-tails. Player **A** has a pet strategy: he bets \$1 on heads and quits if a head turns up. Otherwise he keeps doubling the bet and quits when the first head appears. Show that *theoretically* **A** will always come out a winner.

SOLUTION If a head turns up on the first toss, **A** wins \$1. If not, suppose the first head appears on the n-th toss. Then **A** loses \$1, 2, 4, . . . , 2^{n-2} on the first $n - 1$

tosses, but wins 2^{n-1} on the n-th toss. At this point, he quits with net gain

$$2^{n-1} - (1 + 2 + 4 + \cdots + 2^{n-2}) = 2^{n-1} - \frac{2^{n-1} - 1}{2 - 1}$$
$$= 2^{n-1} - (2^{n-1} - 1) = 1$$

Conclusion No matter when the game ends, **A** wins one dollar.

Question **A**'s strategy is fine *in theory*; where does it break down *in practice*?

In Chapter 7, Section 5, we gave a formula for annuities. Suppose n periodic deposits P are made into an account paying interest i per period. Then the value of the account just after the n-th deposit is

$$A_n = P + P(1 + i) + P(1 + i)^2 + \cdots + P(1 + i)^{n-1}$$

EXAMPLE 4 Derive the formula (stated on page 213)

$$A_n = P \frac{(1 + i)^n - 1}{i}$$

SOLUTION A_n is the sum of a geometric progression with $a_1 = P$ and $r = 1 + i$. Hence the sum is

$$a_1 \frac{r^n - 1}{r - 1} = P \frac{(1 + i)^n - 1}{(1 + i) - 1} = P \frac{(1 + i)^n - 1}{i}$$

Infinite Geometric Series

The expression

$$\sum_{j=0}^{\infty} a_1 r^j = a_1 + a_1 r + a_1 r^2 + a_1 r^3 + \cdots$$

is called an **infinite geometric series**. Can such an *infinite* sum make any sense? It can. For example, take the series

$$\tfrac{1}{2} + \tfrac{1}{4} + \tfrac{1}{8} + \tfrac{1}{16} + \cdots$$

We add up blocks of terms to see what happens:

$$\tfrac{1}{2} + \tfrac{1}{4} = \tfrac{3}{4} \qquad \tfrac{1}{2} + \tfrac{1}{4} + \tfrac{1}{8} = \tfrac{7}{8} \qquad \tfrac{1}{2} + \tfrac{1}{4} + \tfrac{1}{8} + \tfrac{1}{16} = \tfrac{15}{16}$$

In general, the sum of n terms is

$$S_n = \frac{1}{2} + \frac{1}{4} + \frac{1}{8} + \cdots + \frac{1}{2^n} = \frac{1}{2} \cdot \frac{1 - \frac{1}{2^n}}{1 - \frac{1}{2}} = 1 - \frac{1}{2^n}$$

As n increases, we see that S_n comes closer and closer to 1. Therefore, we *define* the sum of the infinite series to be 1.

For a general infinite series, the n-th **partial sum** is

$$S_n = a_1 + a_1 r + a_1 r_2 + \cdots + a_1 r^{n-1} = a_1 \frac{1 - r^n}{1 - r}$$

If $|r| < 1$, then r^n approaches 0 as n increases. Hence S_n approaches

$$a_1 \frac{1-0}{1-r} = \frac{a_1}{1-r}$$

We *define* the sum of the series to be this number.

Sum of an Infinite Geometric Series If $|r| < 1$, then

$$\sum_{j=0}^{\infty} a_1 r^j = a_1 + a_1 r + a_1 r^2 + \cdots = \frac{a_1}{1-r}$$

EXAMPLE 5 Compute the sum of the infinite geometric series

(a) $1 - \frac{1}{3} + \frac{1}{9} - \frac{1}{27} + \cdots$ (b) $\frac{2}{5} + (\frac{2}{5})^3 + (\frac{2}{5})^5 + (\frac{2}{5})^7 + \cdots$

SOLUTION (a) Use the formula with $a_1 = 1$ and $r = -\frac{1}{3}$. The sum is

$$\frac{a_1}{1-r} = \frac{1}{1-(-\frac{1}{3})} = \frac{3}{4}$$

(b) Here $a_1 = \frac{2}{5}$ and $r = (\frac{2}{5})^2 = \frac{4}{25}$. The sum is

$$\frac{a_1}{1-r} = \frac{\frac{2}{5}}{1-\frac{4}{25}} = \underline{\frac{10}{21}}$$

EXAMPLE 6 The bob at the end of a pendulum swings through an arc of length 2 ft. The next swing is 5% shorter, the next 5% shorter again. Assuming this pattern continues indefinitely, find the total distance traveled by the bob.

SOLUTION The sum of all distances is an infinite geometric series:

$$2 + 2(0.95) + 2(0.95)^2 + \cdots = \frac{2}{1-0.95} = \frac{2}{0.05} = \underline{40 \text{ ft}}$$

Every real number that has a repeating decimal is a rational number, that is, expressible as a fraction. We can find that fraction using infinite geometric series.

EXAMPLE 7 Express the repeating decimal $x = 0.61616161 \cdots$ as a fraction.

SOLUTION

$$x = 0.61 + 0.0061 + 0.000061 + \cdots = \frac{61}{10^2} + \frac{61}{10^4} + \frac{61}{10^6} + \cdots$$

$$= \frac{61}{10^2}\left[1 + \frac{1}{10^2} + \frac{1}{10^4} + \cdots\right]$$

The expression in brackets is an infinite geometric series with $r = \frac{1}{100}$. Its sum is

$$\frac{1}{1-\frac{1}{100}} = \frac{100}{99}$$

Hence

$$x = \frac{61}{100} \cdot \frac{100}{99} = \underline{\frac{61}{99}}$$

EXERCISES

Find the 7th term and the n-th term of the geometric progression

1. $1, 3, 9, 27, \ldots$

2. $5, 20, 80, 320, \ldots$

3. $32, 16, 8, 4, \ldots$

4. $9, 3, 1, \frac{1}{3}, \ldots$

5. $3, 3\sqrt{2}, 6, 6\sqrt{2}, \ldots$

6. $12\pi, 3\pi, \frac{3}{4}\pi, \frac{3}{16}\pi, \ldots$

7. $x, -x^3, x^5, -x^7, \ldots$

8. $\frac{1}{x}, -\frac{1}{x^2}, \frac{1}{x^3}, -\frac{1}{x^4}, \ldots$

9. $a, a^4b, a^7b^2, a^{10}b^3, \ldots$

10. $ab, ab^2c^2, ab^3c^4, ab^4c^6, \ldots$

Find the missing terms in the geometric sequence

11. $2, a_2, a_3, 250$

12. $27, a_2, a_3, 64$

13. $2, a_2, a_3, 4\sqrt{2}$

14. $9, a_2, a_3, 8/3$

15. $0.35, a_2, a_3, -350$

16. $3, a_2, a_3, a_4, a_5, -96$

Compute the sum

17. $2^4 + 2^5 + 2^6 + \cdots + 2^9$

18. $\frac{1}{3} + \frac{1}{3^2} + \frac{1}{3^3} + \frac{1}{3^4} + \frac{1}{3^5}$

19. $\sum_{i=1}^{8} \left(-\frac{1}{2}\right)^i$

20. $\sum_{i=1}^{5} \left(-\frac{1}{10}\right)^i$

21. $\sum_{i=0}^{7} (\sqrt{2})^i$

22. $\sum_{i=0}^{7} \left(\frac{1}{\sqrt{3}}\right)^i$

Find the sum of the infinite geometric series

23. $\frac{9}{10} + \left(\frac{9}{10}\right)^2 + \left(\frac{9}{10}\right)^3 + \left(\frac{9}{10}\right)^4 + \cdots$

24. $\frac{2}{3} + \left(\frac{2}{3}\right)^3 + \left(\frac{2}{3}\right)^5 + \left(\frac{2}{3}\right)^7 + \cdots$

25. $\frac{1}{5} - \frac{1}{25} + \frac{1}{125} - \frac{1}{625} + - \cdots$

26. $0.2 - 0.04 + 0.008 - 0.0016 + - \cdots$

27. $\sum_{j=0}^{\infty} \left(\frac{x}{2}\right)^j \quad |x| < 2$

28. $\sum_{j=0}^{\infty} \frac{1}{x^j} \quad |x| > 1$

29. A ball bounces to $\frac{1}{2}$ the height from which it falls. If dropped from 10 feet, how far does the ball travel until it hits the ground for the 8-th time?

30. How far does the pendulum in Example 6 move during its first 10 swings?

31. A rajah once demanded of his subjects 1 grain of rice on the first square of his chess board, 2 grains on the second square, 4 on the third, and so on until the 64-th square. Assuming about 500 grains of rice to the cubic inch, show that this amounts to about 145 *cubic miles* of rice.

32. Everyone has 2 parents, 4 grandparents, 8 great grandparents, etc. How many ancestors do you have in the 10 generations preceding you?

33. Two gossips spread a rumor. On the first day, each one tells it to 4 people. On the next day, each of those 4 tell it to 4 more people, etc. After a week, how many people have heard the rumor?

34. A person writes a chain letter. On the first round, he sends it to 3 people. On the second round, each of these send it to 3 more people, etc. Assuming the chain is not broken, after how many rounds will 100,000 letters have been sent?

Express the repeating decimal as a fraction

35. $0.484848 \cdots$
36. $0.757575 \cdots$
37. $0.9121212 \cdots$
38. $1.8636363 \cdots$
39. $0.415415415 \cdots$
40. $0.027027027 \cdots$

Given that $a_1, a_2, a_3, \ldots$ is a geometric sequence with positive terms, show that the sequence is also geometric

41. $\sqrt{a_1}, \sqrt{a_2}, \sqrt{a_3}, \ldots$
42. $\dfrac{1}{a_1}, \dfrac{1}{a_2}, \dfrac{1}{a_3}, \ldots$

43. If $a_1, a_2, a_3, \ldots$ is an arithmetic sequence, show that $2^{a_1}, 2^{a_2}, 2^{a_3}, \ldots$ is geometric.

44. If $a_1, a_2, a_3, \ldots$ is a geometric sequence, show that $\log|a_1|, \log|a_2|, \log|a_3|, \ldots$ is arithmetic.

45. A polygonal line starts at the origin and moves alternately to the right and upward (Fig. 1). The first segment has length 1; then each segment has length $\tfrac{3}{4}$ of the preceding length. What point is at the "end" of this line?

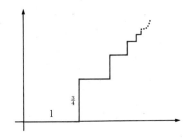

FIG. 1

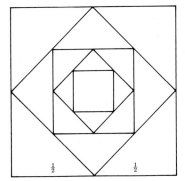

FIG. 2

46. Figure 2 (preceding page) suggests an infinite sequence of nested squares. The midpoints of the sides of each square are the vertices of the next square. Compute the total of the perimeters of all the squares.

47. (Brain teaser) Two trains are on the same straight track 100 miles apart. At noon, they start toward each other, train **A** at 60 mph and train **B** at 40 mph. Also starting at noon, a bird flies back and forth between the trains at 100 mph until they meet. How far does the bird fly?

4 MATHEMATICAL INDUCTION

The word "induction" means discovery of a general principle from special cases. For instance, the special facts

$$1 = 1^2 \qquad 1 + 3 = 2^2 \qquad 1 + 3 + 5 = 3^2 \qquad 1 + 3 + 5 + 7 = 4^2$$

suggest that the general formula

$$1 + 3 + 5 + \cdots + (2n - 1) = n^2$$

may hold for *all* positive integers n. But can we be sure? Is it possible that the formula might fail for, say, $n = 9837$? For instance, the formula $s_n = n^2 - n + 41$ yields

$$s_1 = 41 \qquad s_2 = 43 \qquad s_3 = 47 \qquad s_4 = 53 \qquad s_5 = 61$$

all prime numbers. More arithmetic shows that $s_6, s_7, \ldots, s_{40}$ are also primes. The "obvious" conjecture is that each s_n is a prime number. But that is wrong, because

$$s_{41} = 41^2 - 41 + 41 = 41^2$$

which is not a prime. After 40 correct cases, the conjecture fails.

As a more extreme example, take the statement "all positive integers n are less than 1,000,001." The first million cases are correct, yet the statement is false. Thus, no number of special cases can prove a conjecture containing infinitely many cases. We need a way to verify all cases at once without checking each one separately (life is too short). This is what mathematical induction does.

To understand the idea, imagine a row of dominoes standing on end (Fig. 1). If they are spaced close enough to each other, and the first one is knocked down, they will all fall. We know this for two reasons:

(1) The first domino *will* fall.

(2) *If* a domino falls, then it will knock down the next one.

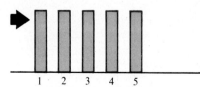
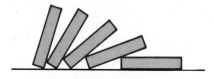

Domino 1 is knocked down.   Each one knocks down the next.

FIG. 1

Be careful of (2). We do not assert a given domino *will* fall, only that *if* it does, it will knock down the next.

The idea leads to a general principle called *mathematical* induction:

Principle of Mathematical Induction Let T_1, T_2, T_3, ... be an infinite sequence of statements. Suppose that

(1) T_1 is true.

(2) Whenever T_n is true, T_{n+1} is also true.

Then all the statements are true.

EXAMPLE 1 Prove the formula

$$1 + 3 + 5 + \cdots + (2n - 1) = n^2 \quad \text{for all } n \geq 1$$

SOLUTION This formula consists of a sequence of statements T_1, T_2, T_3, ..., where

T_1: $1 = 1^2$ $\qquad T_2$: $1 + 3 = 2^2 \qquad T_3$: $1 + 3 + 5 = 3^2 \quad$ etc.

We can prove them *all* in two steps using the principle of mathematical induction.

(1) Show that T_1 is true. But T_1 is the statement $1 = 1^2$, which is true.

(2) Show that whenever T_n is true, T_{n+1} is also true:

$\qquad T_n$ states: $\qquad 1 + 3 + 5 + \cdots + (2n - 1) = n^2$

$\qquad T_{n+1}$ states: $\quad 1 + 3 + 5 + \cdots + (2n - 1) + (2n + 1) = (n + 1)^2$

Now, the left side of the equation in T_{n+1} is just like the left side in T_n, except that it contains one more term, $2n + 1$. Assuming T_n is true, we can add $2n + 1$ to both sides and obtain another true statement:

$$\begin{array}{ll} T_n: \quad 1 + 3 + 5 + \cdots + (2n - 1) & = n^2 \\ \quad \text{(add)} \qquad\qquad\qquad\qquad\quad 2n + 1 & = 2n + 1 \\ \hline \quad 1 + 3 + 5 + \cdots + (2n - 1) + (2n + 1) & = n^2 + 2n + 1 = (n + 1)^2 \end{array}$$

But this statement is T_{n+1}. Therefore, whenever T_n is true, T_{n+1} is also true.

By the principle of mathematical induction, it follows that all the statements T_1, T_2, T_3, ... are true. In other words, the given formula holds for all $n \geq 1$.

Remark Study the solution of Example 1 carefully. The crucial step is (2). Note that we did not *prove* T_n; we showed that *if* T_n is true, then T_{n+1} is also true.

This boot-strap argument is the basis of induction. You assume that T_n is true, and then use it to prove T_{n+1}. Usually, most of what you need for T_{n+1} is already contained in T_n.

It may help to think of climbing an infinite ladder. If you can (1) get to the first step, (2) get from any step you reach to the next, then you will reach all steps.

EXAMPLE 2 Prove the formula

$$1 \cdot 2 + 2 \cdot 3 + 3 \cdot 4 + \cdots + n(n + 1) = \tfrac{1}{3}n(n + 1)(n + 2) \qquad n \geq 1$$

SOLUTION The formula consists of a sequence of statements $T_1, T_2, T_3, \ldots$, where

$$T_1: \quad 1 \cdot 2 = \tfrac{1}{3}(1 \cdot 2 \cdot 3) \qquad T_2: \quad 1 \cdot 2 + 2 \cdot 3 = \tfrac{1}{3}(2 \cdot 3 \cdot 4) \text{ etc.}$$

The proof by mathematical induction consists of two steps:

(1) Show that T_1 is true. Here T_1 is the statement $1 \cdot 2 = \tfrac{1}{3}(1 \cdot 2 \cdot 3)$, which is true.

(2) Show that whenever T_n is true, T_{n+1} is also true:

$T_n: \quad 1 \cdot 2 + 2 \cdot 3 + \cdots + n(n+1) = \tfrac{1}{3}n(n+1)(n+2)$

$T_{n+1}: \quad 1 \cdot 2 + 2 \cdot 3 + \cdots + n(n+1) + (n+1)(n+2)$
$$= \tfrac{1}{3}(n+1)(n+2)(n+3)$$

Assume that T_n is true. Now add $(n+1)(n+2)$ to both sides in T_n:

$1 \cdot 2 + 2 \cdot 3 + \cdots + n(n+1) + (n+1)(n+2)$
$$= \tfrac{1}{3}n(n+1)(n+2) + (n+1)(n+2)$$
$$= (n+1)(n+2)(\tfrac{1}{3}n + 1)$$
$$= (n+1)(n+2)\left(\frac{n+3}{3}\right)$$
$$= \tfrac{1}{3}(n+1)(n+2)(n+3)$$

But this statement is T_{n+1}, which completes step (2). By mathematical induction, the formula is proved for all $n \geq 1$.

EXAMPLE 3 Prove that $2^n > n + 1$ for $n \geq 2$.

SOLUTION Note that the inequality is not true for $n = 1$. We give a two-step proof by mathematical induction, starting at $n = 2$ instead of $n = 1$:

(1) For $n = 2$, the assertion is that $2^2 > 2 + 1$, which is correct.

(2) We must show that whenever

$$2^n > n + 1 \quad \text{then also} \quad 2^{n+1} > (n+1) + 1 = n + 2$$

Assuming $2^n > n + 1$, we multiply both sides of the inequality by 2:

$$2 \cdot 2^n > 2(n+1)$$
$$2^{n+1} > 2n + 2 > n + 2$$

This completes step (2). By mathematical induction, the inequality holds for all $n \geq 2$.

Inductive Definitions

Suppose a sequence of numbers $a_1, a_2, a_3, \ldots$ is defined this way: $a_1 = 1$ and $a_{n+1} = 2a_n + 1$. The sequence is not given explicitly, only that it starts with 1 and that each number is one more than twice the preceding one. We say that the sequence is defined **inductively** or **recursively**. Another example is the sequence

$$a_1 = 0 \qquad a_2 = 1 \qquad a_{n+1} = \tfrac{1}{2}(a_n + a_{n-1})$$

where the first two numbers are 0 and 1; after that, each number is the average of the two preceding ones.

The method of mathematical induction is the natural way of proving properties of inductively defined sequences.

EXAMPLE 4 Let $a_1 = 1$ and $a_{n+1} = 2a_n + 1$ for $n > 1$. Find a formula for a_n and prove it by mathematical induction.

SOLUTION Trying a few cases, we see

$$a_1 = 1 \qquad a_2 = 3 \qquad a_3 = 7 \qquad a_4 = 15 \qquad a_5 = 31$$

These numbers are 1 less than 2^1, 2^2, 2^3, 2^4, 2^5. We conjecture that $a_n = 2^n - 1$ for $n \geq 1$. Let us prove this by mathematical induction:

(1) Since $a_1 = 1 = 2^1 - 1$, the conjecture is true for $n = 1$.

(2) Now we assume $a_n = 2^n - 1$. If so, then

$$a_{n+1} = 2a_n + 1 = 2(2^n - 1) + 1 = 2^{n+1} - 1$$

Hence, whenever the conjecture is true for n, it is also true for $n + 1$. By mathematical induction, the conjecture is proved.

EXAMPLE 5 A sequence of snowflake figures is shown in Fig. 2. The first is an equilateral triangle of side 1. At each step, the middle third of each side is replaced by an equilateral triangle. Find the number of sides of the n-th figure.

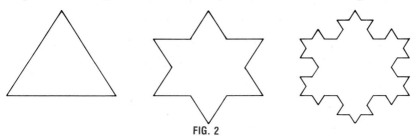

FIG. 2

SOLUTION Let s_n denote the number of sides of the n-th figure. Clearly, $s_1 = 3$. Now consider what happens when the n-th figure is transformed into the $(n + 1)$-th figure (Fig. 3).

FIG. 3

Side of n-th figure $(n + 1)$-st figure

Each of the s_n sides becomes 4 sides. Hence

$$s_{n+1} = 4s_n$$

It follows that $\underline{s_n = 3 \cdot 4^{n-1}}$, as is easily confirmed by induction.

EXERCISES

Prove by mathematical induction for $n \geq 1$

1. $2 + 5 + 8 + \cdots + (3n - 1) = \frac{1}{2}n(3n + 1)$
2. $2 + 4 + 6 + \cdots + 2n = n(n + 1)$
3. $9 + 17 + 25 + \cdots + (8n + 1) = n(4n + 5)$
4. $13 + 23 + 33 + \cdots + (10n + 3) = n(5n + 8)$
5. $1^2 + 2^2 + 3^2 + \cdots + n^2 = \frac{1}{6}n(n + 1)(2n + 1)$
6. $1^3 + 2^3 + 3^3 + \cdots + n^3 = [\frac{1}{2}n(n + 1)]^2$
7. $\dfrac{1}{1 \cdot 2} + \dfrac{1}{2 \cdot 3} + \dfrac{1}{3 \cdot 4} + \cdots + \dfrac{1}{n(n + 1)} = \dfrac{n}{n + 1}$
8. $\dfrac{1}{1 \cdot 3} + \dfrac{1}{3 \cdot 5} + \dfrac{1}{5 \cdot 7} + \cdots + \dfrac{1}{(2n - 1)(2n + 1)} = \dfrac{n}{2n + 1}$
9. $1 \cdot 3 + 2 \cdot 4 + \cdots + n(n + 2) = \frac{1}{6}n(n + 1)(2n + 7)$
10. $1 \cdot 2 \cdot 3 + 2 \cdot 3 \cdot 4 + 3 \cdot 4 \cdot 5 + \cdots + n(n + 1)(n + 2) = \frac{1}{4}n(n + 1)(n + 2)(n + 3)$
11. $a + (a + d) + (a + 2d) + \cdots + [a + (n - 1)d] = \frac{1}{2}n[2a + (n - 1)d]$
12. $1 + r + r^2 + \cdots + r^{n-1} = \dfrac{r^n - 1}{r - 1} \qquad r \neq 1$

Prove by mathematical induction

13. The sum of n even integers is even.
14. The product of n odd integers is odd.
15. $\log(a_1 a_2 a_3 \cdots a_n) = \log a_1 + \log a_2 + \log a_3 + \cdots + \log a_n \qquad n \geq 2$
16. $\sqrt{a_1 a_2 a_3 \cdots a_n} = \sqrt{a_1} \cdot \sqrt{a_2} \cdot \sqrt{a_3} \cdots \sqrt{a_n} \qquad n \geq 2$
17. To tie n pieces of string, $n \geq 2$, into one long string requires $n - 1$ knots.
18. If n people stand in line, $n \geq 2$, and if the first person in line is a woman and the last person is a man, then somewhere in the line there is a man directly behind a woman.
19. The number of committees that can be formed from a group of n people (allowing committees of 1, 2, 3, ..., n members) is $2^n - 1$.
20. An n-piece jigsaw puzzle requires $n - 1$ joinings (fitting of a piece to another piece, or a block of pieces to another block).

Prove the inequality by mathematical induction

21. $(\frac{3}{2})^n > n \qquad n \geq 1$
22. $3^n \geq 2n + 1 \qquad n \geq 1$
23. $2^n > 7n \qquad n \geq 6$
24. $5^n > 4^n + 3^n \qquad n \geq 3$
25. $1 \cdot 2 \cdot 3 \cdots n > 10n \qquad n \geq 5$
26. $(1 + a)^n > 1 + na \qquad a > 0, \quad n \geq 2$

Find a formula and prove it by induction for

27. $\left(1 - \dfrac{1}{4}\right)\left(1 - \dfrac{1}{9}\right)\left(1 - \dfrac{1}{16}\right) \cdots \left(1 - \dfrac{1}{n^2}\right) \qquad n \geq 2$

28. $(1 - x)(1 + x)(1 + x^2)(1 + x^4) \cdots (1 + x^{2^n}) \qquad n \geq 0$

29. the perimeter of the n-th figure in Figure 2, page 299

30. the length of each side of the n-th figure in Figure 2, page 299

31. Consider the sequence
$$a_1 = \sqrt{2}, \quad a_2 = \sqrt{2 + \sqrt{2}}, \quad a_3 = \sqrt{2 + \sqrt{2 + \sqrt{2}}}, \quad \text{etc.}$$
Prove that $a_n < 2$ for $n \geq 1$.

32. Consider the sequence defined by
$$a_0 = 1, \quad a_{n+1} = \dfrac{1}{2}\left(a_n + \dfrac{2}{a_n}\right)$$
Prove that $1 \leq a_n \leq 2$ for $n \geq 0$.

Find a formula for a_n and prove it by induction

33. $a_0 = 0, \; a_1 = 1,$ and $a_n = 2a_{n-1} - a_{n-2}$ for $n \geq 2$

34. $a_0 = 0, \; a_1 = 2,$ and $a_n = $ average of $a_0, a_1, \ldots, a_{n-1}$ for $n \geq 2$

35. $a_0 = 1, \; a_1 = 1,$ and $a_n = a_0 + a_1 + \cdots + a_{n-1}$ for $n \geq 2$

36. $a_0 = 3$ and $a_n = a_{n-1}^2$ for $n \geq 1$

37. You have 3^n coins and a balance. The coins are identical except for one which is slightly heavy. Prove that you can find the heavy coin using the balance n times.

38. Suppose the odd coin in Exercise 37 may be either heavy or light. (You are not told which.) Prove that you can find the odd coin using the balance $n + 1$ times.

39. A pyramid of n decreasing rings is placed on one of three pegs as shown in Fig. 4. The game (tower of Hanoi puzzle) is to transfer the pyramid to peg 2 in as few moves as possible. There are two rules: (1) only one ring can be moved at a time (2) a ring may be moved from one peg to another but may not be placed on top of a ring smaller than itself. Find the least number of moves required and confirm your answer by induction.

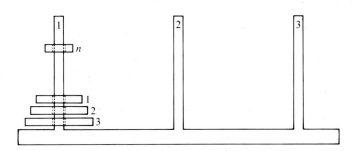

FIG. 4

40. (See Example 4, page 287) What is the total number of gifts my true love gave to me during all 12 days of Christmas? (Hint Example 2)

5 PERMUTATIONS

Factorials

In the rest of this chapter, we will often need the product $1 \cdot 2 \cdot 3 \cdots n$. We abbreviate this product by $n!$ and call it n-**factorial**.

> **n-Factorial** $\quad n! = 1 \cdot 2 \cdot 3 \cdots n$

For example,
$$5! = 1 \cdot 2 \cdot 3 \cdot 4 \cdot 5 = 120 \qquad 10! = 1 \cdot 2 \cdot 3 \cdots 10 = 3{,}628{,}800$$
We also define $0! = 1$.

Products of consecutive integers can be expressed in terms of factorials. For example,
$$4 \cdot 5 \cdot 6 \cdot 7 = \frac{1 \cdot 2 \cdot 3 \cdot 4 \cdot 5 \cdot 6 \cdot 7}{1 \cdot 2 \cdot 3} = \frac{7!}{3!}$$

A Counting Principle

Lil's Lunch offers a soup-and-sandwich special: your choice of either bean or vegetable soup, and either a cheese, ham, or tuna sandwich.

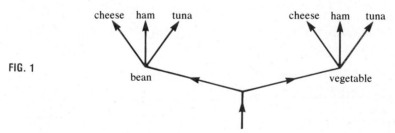

FIG. 1

There are $2 \times 3 = 6$ possible choices (Fig. 1). Each of the 2 choices of soups leads to 3 choices of sandwiches, so you multiply 2×3. This illustrates an important principle:

> **Counting Principle** Suppose that in a sequence of k events, the first event has n_1 possible outcomes, the second has n_2 possible outcomes, . . . , and the k-th has n_k possible outcomes. Then the number of ways in which all k events can occur is the product $n_1 n_2 \cdots n_k$.

Generally an "event" will mean a decision, and an "outcome" a possible choice. For instance, choosing a soup at Lil's is an event with 2 possible outcomes.

EXAMPLE 1 How many 3-digit numbers are there less than 500 with all 3 digits different?

SOLUTION For the hundreds digit d_2 there are 4 choices: 1, 2, 3, 4. Once d_2 is chosen, the tens digit d_1 can be any of the digits 0, 1, . . . , 9 except d_2. Hence there are 9 choices for d_1. Once d_2 and d_1 are chosen, the units digit d_0 can be any of 0, 1, . . . , 9 except d_1 and d_2. Hence there are 8 choices for d_0. By our counting principle, the number of ways to choose all 3 digits is $4 \cdot 9 \cdot 8 = \underline{288}$.

Permutations

In how many different orders can 8 horses finish a race? Let us apply our counting principle. There are 8 possibilities for the winner. Then there are 7 possibilities for second place, 6 possibilities for third place, etc. Hence the number of different orders is

$$8 \cdot 7 \cdot 6 \cdot 5 \cdot 4 \cdot 3 \cdot 2 \cdot 1 = 8! = 40{,}320$$

By the same reasoning, the number of orderings (or arrangements) of n objects is

$$n(n-1)(n-2) \cdots 3 \cdot 2 \cdot 1 = n!$$

Each ordering of a set of objects is called a **permutation**.

> There are $n!$ permutations of n objects.

Examples

(1) There are $4! = 24$ ways to hang 4 pictures on a wall from left to right.
(2) There are $9! = 362{,}880$ batting orders for the 9 players on a baseball team.
(3) There are $52! \approx 8.066 \times 10^{67}$ arrangements of a deck of 52 cards.

Next, suppose we have n objects and ask how many ways we can select k of them in order. (Last time we took all n of them.) There are n choices for the first, $n - 1$ for the second, . . . , and $n - (k - 1)$ choices for the k-th. By the counting principle, the total number of selections possible is

$$n(n-1)(n-2) \cdots (n-k+1) \qquad (k \text{ factors})$$

We denote this number by $P(n, k)$. It can be expressed in terms of factorials:

$$n(n-1) \cdots (n-k+1)$$
$$= n(n-1) \cdots (n-k+1) \frac{(n-k)(n-k-1) \cdots 2 \cdot 1}{(n-k)(n-k-1) \cdots 2 \cdot 1} = \frac{n!}{(n-k)!}$$

> **Permutation of n Objects k at a Time** The number of permutations of k objects selected from a set of n objects is
> $$P(n, k) = n(n-1)(n-2) \cdots (n-k+1) = \frac{n!}{(n-k)!}$$

Examples

(1) The number of choices for the first, second, and third batters from 9 baseball players is

$$P(9, 3) = 9 \cdot 8 \cdot 7 = \frac{9!}{6!} = 504$$

(2) The number of ways the first 5 cards can be dealt in order from a deck of 52 cards is

$$P(52, 5) = 52 \cdot 51 \cdot 50 \cdot 49 \cdot 48 = \frac{52!}{47!} = 311{,}875{,}200$$

EXAMPLE 2 I have 10 posters, but there is room for only 5 on my wall. How many arrangements from left to right can I make?

SOLUTION Since the order of the posters counts, this is the number of permutations of 10 objects, 5 at a time:

$$P(10, 5) = 10 \cdot 9 \cdot 8 \cdot 7 \cdot 6 = \underline{30{,}240}$$

EXAMPLE 3 In how many ways can 4 people from a group of 12 be seated side by side in a row of 20 seats?

SOLUTION Suppose the seats are numbered 1 to 20. The 4 people can occupy the block of seats $\{1, 2, 3, 4\}$ or $\{2, 3, 4, 5\}$, or $\cdots$ $\{17, 18, 19, 20\}$. Hence there are 17 choices for the block of seats.

Once the seats are chosen, there are $P(12, 4)$ arrangements of 4 people from the 12. By our counting principle, the answer to the problem is

$$17 \cdot P(12, 4) = 17 \cdot (12 \cdot 11 \cdot 10 \cdot 9) = \underline{201{,}960}$$

Distinguishable Permutations

A multiple-choice test has 10 questions, each with choices A, B, C. How many ways can you construct a test so that four of the answers are A, three are B, and three are C? For example, one possible sequence of answers is

$$B \quad A \quad A \quad C \quad A \quad C \quad B \quad B \quad C \quad A$$

To answer this question, let us pretend for a moment that we can distinguish among the A's, B's, and C's. Then we could write the sequence above as

$$B_1 \quad A_1 \quad A_2 \quad C_1 \quad A_3 \quad C_2 \quad B_2 \quad B_3 \quad C_3 \quad A_4$$

Without changing this answer sequence, we can permute its four A's in 4! ways, its three B's in 3! ways, and its three C's in 3! ways. Hence, there are (4!)(3!)(3!) permutations that give the same answer sequence.

Altogether, there are 10! permutations of the 10 letters. But among these, each **distinguishable permutation** is counted (4!)(3!)(3!) times. Hence the number of actual answer sequences is

$$\frac{10!}{4!\,3!\,3!} = 4200$$

This example illustrates a general principle:

> **Distinguishable Permutations** Suppose a set of n objects contains k types of objects: n_1 of the first type (all alike), ..., n_k of the k-th type (all alike). Then the number of distinguishable permutations of the n objects is
> $$\frac{n!}{n_1!\, n_2! \cdots n_k!} \qquad n_1 + n_2 + \cdots + n_k = n$$

EXAMPLE 4 How many 8-digit numbers contain the digits 3, 4, 4, 7, 7, 7, 7, 7?

SOLUTION The 8-digit numbers we want are distinguishable permutations of the given digits. Since the two 4's are alike and the five 7's are alike, the number of distinguishable permutations is

$$\frac{8!}{1!\,2!\,5!} = \frac{8 \cdot 7 \cdot 6 \cdot 5 \cdot 4 \cdot 3 \cdot 2 \cdot 1}{1 \cdot 2 \cdot 5 \cdot 4 \cdot 3 \cdot 2 \cdot 1} = \frac{8 \cdot 7 \cdot 6}{2} = \underline{168}$$

EXERCISES

Express in factorial notation

1. $13 \cdot 12 \cdot 11 \cdot 10 \cdot 9$
2. $8 \cdot 9 \cdot 10 \cdots 14$
3. $\dfrac{52 \cdot 51 \cdot 50 \cdot 49 \cdot 48}{120}$
4. $7(6!)$
5. $4 \cdot 5 \cdot 6 \cdots n$
6. $(n + 1)(n + 2)(n + 3)$

Compute

7. $\dfrac{9!}{7!}$
8. $\dfrac{15!}{12!}$
9. $\dfrac{11!}{3!\,8!}$
10. $\dfrac{8!}{4!\,4!}$
11. $\dfrac{14!}{9!\,3!\,2!}$
12. $\dfrac{13!}{7!\,4!\,2!}$
13. $\dfrac{n!}{(n-2)!}$
14. $\dfrac{(2n+1)!}{(2n)!}$

How many are there? (Leave very large answers in terms of factorials)

15. lunch menus; choice of 3 soups, 4 sandwiches, 2 desserts
16. car models; choice of 4 body types, 2 transmissions, 6 colors, 3 upholsteries
17. license plates with 2 letters followed by 3 digits, such as RG074 (repetitions allowed)
18. 4-letter names for radio stations, starting with K or W; no repeated letters
19. 4-digit numbers; no repeated digits
20. 3-digit area codes where the first digit is not 0 or 1 and the second digit can be only 0 or 1
21. 7-digit telephone numbers where the first digit is not 0
22. ways of trying 7 breakfast cereals on 7 consecutive mornings

Discrete Algebra

EXAMPLE 2 From a committee of 7 Democrats and 8 Republicans, how many subcommittees of 3 Democrats and 3 Republicans are possible?

SOLUTION There are $\binom{7}{3}$ ways to choose 3 Democrats out of 7 and $\binom{8}{3}$ ways to choose 3 Republicans out of 8. By our counting principle, the number of ways to make both choices is

$$\binom{7}{3} \cdot \binom{8}{3} = \frac{7 \cdot 6 \cdot 5}{1 \cdot 2 \cdot 3} \cdot \frac{8 \cdot 7 \cdot 6}{1 \cdot 2 \cdot 3} = 35 \cdot 56 = \underline{1960}$$

EXAMPLE 3 Seven points on a circle are connected in all possible ways by chords (Fig. 1). How many chords are there?

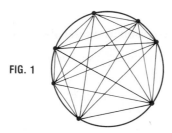

FIG. 1

SOLUTION Each chord connects 2 of the points. Hence there are as many chords as pairs of points chosen from 7 points:

$$\binom{7}{2} = \frac{7 \cdot 6}{1 \cdot 2} = \underline{21}$$

EXAMPLE 4 A man's office is 8 blocks east and 8 blocks north of his apartment. He walks to work along city streets, walking only eastward or northward (Fig. 2). How many possible routes can he take?

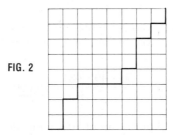

FIG. 2

SOLUTION Each route is a sequence of 16 one-block segments, 8 to the east and 8 to the north. Conversely, each such sequence defines a route. A route is determined once the positions of the 8 eastward segments are chosen from the 16 positions in the sequence. Hence the number of routes is

$$\binom{16}{8} = \frac{16!}{8!\,8!} = \frac{16 \cdot 15 \cdot 14 \cdot 13 \cdot 12 \cdot 11 \cdot 10 \cdot 9}{1 \cdot 2 \cdot 3 \cdot 4 \cdot 5 \cdot 6 \cdot 7 \cdot 8} = \underline{12{,}870}$$

EXAMPLE 5 Of the 12,870 possible routes in Example 4, how many start out with 4 northward segments?

SOLUTION Suppose a route starts with 4 northward segments. Then the rest of the route consists of a sequence of 12 segments, 4 north and 8 east. It is determined by choosing the positions of the 4 northward (or 8 eastward) segments in the 12 slots. The number of ways this can be done is

$$\binom{12}{4} = \binom{12}{8} = \frac{12 \cdot 11 \cdot 10 \cdot 9}{1 \cdot 2 \cdot 3 \cdot 4} = \underline{495}$$

EXERCISES

How many are there? (Leave very large answers in terms of factorials)
1. possible handshakes at a meeting of 15 people
2. games between basketball teams in the Big Ten (each team plays every other team)
3. choices of 3 flavors of ice cream from 31 flavors
4. choices of 3 sweaters from a rack of 25 sweaters on sale
5. choices by an officer of 5 out of his 18-man platoon for a patrol
6. subcommittees of 4 from a committee of 13
7. 5-card poker hands (from a 52-card deck)
8. 13-card bridge hands (from a 52-card deck)
9. triangles in Figure 1 with vertices on the circle
10. quadrilaterals in Figure 1 with vertices on the circle
11. committees of 4 Conservatives and 4 Liberals from a cabinet of 9 Conservatives and 11 Liberals
12. committees of 3 Radicals, 2 Loyalists, and 2 Progressives from a cabinet of 7 Radicals, 6 Loyalists, and 5 Progressives
13. from a group of 12 people, choices of two committees, one with 2 members and another with 3 members, no common members
14. ways to visit 3 exhibits tonight, and 4 different ones tomorrow, of 20 exhibits at a county fair
15. ways of dividing 10 players into two teams of 5 players
16. ways of dividing 20 people into two baseball teams of 9 players plus 2 umpires
17. sequences of 10 heads and tails containing 6 heads and 4 tails
18. 10-digit binary numbers such as 0110100011 having 5 zeros and 5 ones
19. paths in Figure 2 whose first and last steps are vertical
20. paths in Figure 2 whose first 5 steps are horizontal

21. 13-card bridge hands containing 1 club, 2 diamonds, 4 hearts, and 6 spades

22. full-house poker hands (3 of a kind plus a pair of another kind such as King, King, King, 4, 4)

23. doubles matches between two tennis teams, one with 6 members and one with 8

24. choices of players to start a hockey game between a team with 10 members and another with 11 members (a hockey team has 6 players)

25. dinner menus in a Chinese restaurant with choice of 3 dishes from group A (9 dishes) and 2 from group B (6 dishes)

26. choices of 4 rock, 3 jazz, and 2 classical records from a selection containing 10 records of each type

27. Verify the relation $\binom{n+1}{k} = \binom{n}{k} + \binom{n}{k-1}$

28. Explain the relation in Exercise 27 by counting combinations instead of computing.

29. Show that $1 + 2 + 3 + \cdots + n = \binom{n+1}{2}$. Verify this relation by computing, in two different ways, the number of handshakes possible in a group of n people.

30. Ten points on a circle are connected in all possible ways by chords. How many points of intersection of all these chords are there inside the circle? Assume that no three chords intersect in a point. (Hint Exercise 10)

7 THE BINOMIAL THEOREM

The binomial theorem is a formula for the expanded product $(x + y)^n$. Let us compute a few cases:

$$(x + y)^1 = x + y$$
$$(x + y)^2 = x^2 + 2xy + y^2$$
$$(x + y)^3 = x^3 + 3x^2y + 3xy^2 + y^3$$
$$(x + y)^4 = x^4 + 4x^3y + 6x^2y^2 + 4xy^3 + y^4$$

The general formula for $(x + y)^n$ will have terms in

$$x^n, \; x^{n-1}y, \; x^{n-2}y^2, \; \ldots, \; x^{n-k}y^k, \; \ldots, \; xy^{n-1}, \; y^n$$

The question is, what are the coefficients of these terms? The answer comes from examining how we multiply out

$$(x + y)^n = (x + y)(x + y)(x + y) \cdots (x + y) \qquad (n \text{ factors})$$

The product is the sum of all terms obtained as follows. From each of the n factors $(x + y)$, choose either the x or the y and multiply these. If k of the n choices are y's, the result is a term $x^{n-k}y^k$. The number of terms of this type is the number of combinations of n objects taken k at a time. Thus $x^{n-k}y^k$ occurs in $(x + y)^n$ with the coefficient $\binom{n}{k}$. For this reason, $\binom{n}{k}$ is called a **binomial coefficient**.

Binomial Theorem For each positive integer n,

$$(x + y)^n = x^n + \binom{n}{1} x^{n-1}y + \binom{n}{2} x^{n-2}y^2 + \binom{n}{3} x^{n-3}y^3$$
$$+ \cdots + \binom{n}{k} x^{n-k}y^k + \cdots + \binom{n}{n-1} xy^{n-1} + y^n$$

If we define

$$\binom{n}{0} = 1 \quad \text{and} \quad \binom{n}{n} = 1$$

then we can write the binomial theorem in compact summation notation:

$$(x + y)^n = \sum_{k=0}^{n} \binom{n}{k} x^{n-k}y^k$$

EXAMPLE 1 Expand $(x + y)^5$

SOLUTION By the binomial theorem,

$$(x + y)^5 = \binom{5}{0} x^5 + \binom{5}{1} x^4y + \binom{5}{2} x^3y^2 + \binom{5}{3} x^2y^3 + \binom{5}{4} xy^4 + \binom{5}{5} y^5$$
$$= x^5 + \frac{5}{1} x^4y + \frac{5 \cdot 4}{1 \cdot 2} x^3y^2 + \frac{5 \cdot 4 \cdot 3}{1 \cdot 2 \cdot 3} x^2y^3 + \frac{5 \cdot 4 \cdot 3 \cdot 2}{1 \cdot 2 \cdot 3 \cdot 4} xy^4 + y^5$$

Answer

$x^5 + 5x^4y + 10x^3y^2 + 10x^2y^3 + 5xy^4 + y^5$

Remark We could have saved work by using the symmetry of the binomial coefficients

$$\binom{5}{0} = \binom{5}{5} \qquad \binom{5}{1} = \binom{5}{4} \qquad \binom{5}{2} = \binom{5}{3}$$

EXAMPLE 2 Expand $(2a - b)^6$

SOLUTION Use the binomial theorem with $x = 2a$, $y = -b$, and $n = 6$:

$$(2a - b)^6 = \binom{6}{0} (2a)^6 + \binom{6}{1} (2a)^5(-b) + \binom{6}{2} (2a)^4(-b)^2$$
$$+ \binom{6}{3} (2a)^3(-b)^3 + \binom{6}{4} (2a)^2(-b)^4 + \binom{6}{5} (2a)(-b)^5$$
$$+ \binom{6}{6} (-b)^6$$

The coefficients are

$$\binom{6}{0} = \binom{6}{6} = 1 \qquad \binom{6}{1} = \binom{6}{5} = 6 \qquad \binom{6}{2} = \binom{6}{4} = 15 \qquad \binom{6}{3} = 20$$

Therefore,
$$(2a - b)^6 = 64a^6 - 6(32a^5)b + 15(16a^4)b^2 - 20(8a^3)b^3 + 15(4a^2)b^4 \\ - 6(2a)b^5 + b^6$$

Answer
$$64a^6 - 192a^5b + 240a^4b^2 - 160a^3b^3 + 60a^2b^4 - 12ab^5 + b^6$$

EXAMPLE 3 Find the coefficient of x in
$$\left(x^2 + \frac{2}{x}\right)^8$$

SOLUTION The typical term is
$$\binom{8}{k}(x^2)^{8-k}\left(\frac{2}{x}\right)^k = \binom{8}{k}x^{16-2k}x^{-k}2^k$$
$$= \binom{8}{k}2^k x^{16-3k}$$

The term involving x is the one for which $16 - 3k = 1$, that is, $k = 5$. Its coefficient is
$$\binom{8}{5}2^5 = \binom{8}{3}2^5 = \frac{8 \cdot 7 \cdot 6}{1 \cdot 2 \cdot 3} \cdot 32 = \underline{1792}$$

EXAMPLE 4 Without a calculator or logarithms, compute $(1.01)^{10}$ to 6-place accuracy.

SOLUTION Write $1.01 = 1 + 0.01$ and use the binomial theorem:
$$(1 + 0.01)^{10} = \binom{10}{0} + \binom{10}{1}(.01) + \binom{10}{2}(.01)^2 + \cdots + \binom{10}{10}(.01)^{10}$$
$$= 1 + 10(.01) + 45(.0001) + 120(.000001) + 210(.00000001)$$
$$+ \text{remaining terms}$$
$$= 1.1046221 + \text{remaining terms}$$

The 6 "remaining terms" are very small. The largest of them is
$$\binom{10}{5}(.01)^5 = 252 \times 10^{-10} = 2.52 \times 10^{-8}$$

All 6 of them together amount to at most 6 times as much:
$$6 \times 2.52 \times 10^{-8} < 2 \times 10^{-7}$$

This is too small to affect the 6-th decimal place. Hence $(1.01)^{10} = \underline{1.104622}$ to 6 places.

Pascal's Triangle

An interesting way of displaying binomial coefficients is an array called **Pascal's Triangle**. On the n-th row (starting with $n = 0$) we write the binomial coefficients

$\binom{n}{k}$, for $k = 0, 1, \ldots, n$:

$$
\begin{array}{ccccccccccc}
 & & & & & 1 & & & & & \\
 & & & & 1 & & 1 & & & & \\
 & & & 1 & & 2 & & 1 & & & \\
 & & 1 & & 3 & & 3 & & 1 & & \\
 & 1 & & 4 & & 6 & & 4 & & 1 & \\
1 & & 5 & & 10 & & 10 & & 5 & & 1 \\
& 6 & & 15 & & 20 & & 15 & & 6 & & 1 \\
\cdot & \cdot & \cdot & \cdot & \cdot & \cdot & \cdot & \cdot & \cdot
\end{array}
$$

Each interior entry is the sum of the two numbers above it because

$$\binom{n+1}{k} = \binom{n}{k} + \binom{n}{k-1}$$

(See Exercise 27, last section.)

The sum of the numbers in the n-th row is 2^n. This is because

$$2^n = (1+1)^n = \binom{n}{0} + \binom{n}{1} + \cdots + \binom{n}{n}$$

For instance ($n = 5$),

$$2^5 = 1 + 5 + 10 + 10 + 5 + 1$$

The alternating sum of each row is 0, that is,

$$\binom{n}{0} - \binom{n}{1} + \binom{n}{2} - + \cdots + \binom{n}{n} = 0$$

This is because $(1 - 1)^n = 0$. For instance ($n = 6$),

$$1 - 6 + 15 - 20 + 15 - 6 + 1 = 0$$

EXERCISES

Expand
1. $(x + y)^7$
2. $(x + y)^8$
3. $(x + 2)^5$
4. $(y - 1)^9$
5. $(2x - y)^4$
6. $(x + 2y)^4$
7. $(a^2 - b)^6$
8. $(2a - 3b)^3$
9. $\left(x - \dfrac{1}{x}\right)^{10}$
10. $\left(x + \dfrac{1}{x}\right)^7$
11. $(2x - \tfrac{1}{2})^5$
12. $(\tfrac{1}{3}x + 3)^4$

Find the coefficient of
13. x^4 in $(x - \tfrac{1}{3})^7$
14. y^3 in $(2y - 1)^6$
15. u^6 in $(u^2 + 2)^5$
16. $t^{5/2}$ in $(\sqrt{t} + 1)^9$
17. $x^5 y^3$ in $(x + 2y)^8$
18. $x^2 y^3$ in $(2x - 3y)^5$
19. $x^{10} y^4$ in $(x^2 - 2y)^9$
20. $x^4 y^{18}$ in $(3x - y^3)^{10}$
21. $\dfrac{1}{x^3}$ in $\left(2x + \dfrac{1}{x^2}\right)^6$
22. x^3 in $\left(\sqrt{x} + \dfrac{1}{x}\right)^{12}$

Find the term

23. without x in $\left(x^3 + \dfrac{1}{x}\right)^{12}$ **24.** without x in $\left(x - \dfrac{1}{x}\right)^{10}$

25. without y in $\left(xy - \dfrac{1}{\sqrt{y}}\right)^9$ **26.** without z in $\left(z^2 - \dfrac{1}{z^3}\right)^{15}$

Compute by the binomial theorem
27. $(102)^3$ **28.** $(101)^5$ **29.** $(99)^4$ **30.** $(999)^3$

Compute by the binomial theorem to 6-place accuracy
31. $(1.01)^8$ **32.** $(1.002)^{10}$ **33.** $(0.99)^{10}$ **34.** $(0.999)^{12}$

Prove by the binomial theorem
35. $(1.001)^{50} > 1.05$ **36.** $(1.03)^{10} > 1.3405$

37. Look at lines 1–4 of Pascal's triangle. Notice that $11 = 11^1$, $121 = 11^2$, $1331 = 11^3$, and $14{,}641 = 11^4$. Explain.

38. How many terms are there when the product of n factors $(x + y)(x + y) \cdots (x + y)$ is expanded? What does that say about $\binom{n}{0} + \binom{n}{1} + \cdots + \binom{n}{n}$?

REVIEW EXERCISES FOR CHAPTER 10

Find a formula for the n-th term
1. 8, 11, 14, 17, 20, $\cdots$ **2.** 4.04, 4.00, 3.96, 3.92, 3.88, $\cdots$

Compute the sum
3. $61 + 62 + 63 + \cdots + 99 + 100$

4. $9 + 20 + 31 + 42 + \cdots + 119 + 130$

5. $2 + \dfrac{4}{3} + \dfrac{8}{9} + \dfrac{16}{27} + \cdots + \dfrac{2^{n+1}}{3^n}$

6. $x + x^4 + x^7 + x^{10} + \cdots + x^{25} + x^{28}$

7. Find the sum of the infinite geometric series
$$\tfrac{3}{5} - (\tfrac{3}{5})^2 + (\tfrac{3}{5})^3 - (\tfrac{3}{5})^4 + - \cdots$$

8. Express $1 + 2x + 3x^2 + 4x^3 + \cdots + 10x^9$ in summation notation.

Prove by mathematical induction
9. $1^2 + 3^2 + 5^2 + \cdots + (2n - 1)^2 = \tfrac{1}{3}n(2n - 1)(2n + 1)$

10. $a(b_1 + b_2 + \cdots + b_n) = ab_1 + ab_2 + \cdots + ab_n$ $n \geq 2$

11. Define a sequence by $a_1 = 6$, $a_2 = 2$, and $a_{n+1} = a_n + a_{n-1}$. Prove by induction that all elements of this sequence are even integers.

12. Filtering removes 60% of the impurities in a chemical solution. How many filterings are needed to reduce 10 gm of impurities to less than 0.001 gm?

13. Expand $(a - 2b)^7$.

14. Find the term without x and y in $\left(\dfrac{x}{y} + \dfrac{y^2}{2x^2}\right)^9$.

15. A professor of piano has 4 undergraduate and 5 graduate students. In how many orders can he schedule one lesson for each student if he teaches all the undergraduates first?

16. Besides teaching, the pianist in Exercise 15 will appear as soloist with an orchestra in 6 concerts, playing 3 concertos by Beethoven and 3 by Mozart. How many orders are possible alternating the two composers?

17. In how many ways can a hotel assign a single room to each of 6 people if 10 single rooms are available?

18. Bruno's pizza comes in 3 sizes with 8 possible toppings. In how many ways can you order a pizza with 5 toppings?

19. In how many ways can a group of 10 players be divided into two teams of 5 players?

20. If 5 identical red pennants, 4 identical blue pennants, and 3 identical white pennants are flown from a vertical mast, how many distinguishable patterns are possible?

Appendix on Computation

1 USE OF CALCULATORS

Our object is to develop the know-how necessary for basic computations with a hand-held calculator. So many models are available, each with its own special features, that we cannot discuss them all. Instead, we shall discuss standard techniques common to most scientific calculators. We suggest that you read the instructions to your calculator carefully and do enough numerical experiments to familiarize yourself with it.

Logic

The logic of a calculator is its system for doing the operations of arithmetic. The three common logics are algebraic, left-to-right, and reverse Polish. We'll stick to algebraic and left-to-right logics (abbreviated Alg and LR).

In both Alg and LR, you do addition, subtraction, multiplication, and division in the same way:

$a + b$: $a\boxed{+}b\boxed{=}$ $\quad$ $a - b$: $a\boxed{-}b\boxed{=}$

$a \times b$: $a\boxed{\times}b\boxed{=}$ $\quad$ $a \div b$: $a\boxed{\div}b\boxed{=}$

(The keys $\boxed{+}$, $\boxed{-}$, $\boxed{\times}$, and $\boxed{\div}$ are called **binary** keys because they operate on pairs of numbers.)

It is easy to do strings of additions and subtractions. For example,

$477 + 92 + 1651$: $\quad 4\ 7\ 7\boxed{+}9\ 2\boxed{+}1\ 6\ 5\ 1\boxed{=}\quad\quad 2220$

$138.2 - 10.7 + 8.03 - 0.55$: $\quad 1\ 3\ 8\ .\ 2\boxed{-}1\ 0\ .\ 7\boxed{+}8\ .\ 0\ 3\boxed{-}.\ 5\ 5\boxed{=}\quad 134.98$

Similarly, you can do strings of multiplications and divisions. For example,

$19 \times 8.6 \times 0.712$: $\quad 1\ 9\boxed{\times}8\ .\ 6\boxed{\times}.\ 7\ 1\ 2\boxed{=}\quad\quad 116.3408$

$2157 \times 61.5 \div 4$: $\quad 2\ 1\ 5\ 7\boxed{\times}6\ 1\ .\ 5\boxed{\div}4\boxed{=}\quad\quad 33163.875$

If a string of operations mixes additions and subtractions with multiplications and divisions, then algebraic and left-to-right logics can disagree. In algebraic logic, multiplication and division take priority over addition and subtraction. Thus,

$$a\boxed{+}b\boxed{\times}c\boxed{=} \quad \text{produces} \quad a + bc \quad \text{(Alg)}$$

because the calculator multiplies first, then adds. In left-to-right logic, operations are done as they are keyed in. Thus,

$$a\boxed{+}b\boxed{\times}c\boxed{=} \quad \text{produces} \quad (a + b) \times c \quad \text{(LR)}$$

Two further examples:

Key Sequence	Algebraic	Left-to-Right
$a\boxed{\times}b\boxed{-}c\boxed{\times}d\boxed{=}$	$ab - cd$	$(ab - c)d$
$a\boxed{-}b\boxed{\times}c\boxed{+}d\boxed{\div}e\boxed{=}$	$a - bc + \dfrac{d}{e}$	$\dfrac{(a - b)c + d}{e}$

Check the logic of your calculator. If $1\boxed{+}2\boxed{\times}5\boxed{=}$ gives 11, you have algebraic logic; if it gives 15, you have left-to-right logic. For a further discussion of Alg vs. LR, see Chapter 1, Section 4.

Unary Keys

The keys $\boxed{x^2}, \boxed{1/x}, \boxed{\sqrt{}}, \boxed{+/-}$ are called **unary** keys because they operate on *single* numbers. Each unary key applies only to the number displayed. For example,

$$3\boxed{x^2}: \quad 3^2 = 9 \qquad 10\boxed{1/x}: \quad \frac{1}{10} = 0.1 \qquad 4\boxed{\sqrt{}}: \quad \sqrt{4} = 2$$

The key $\boxed{+/-}$ changes the sign of the number displayed:

$$6\boxed{+/-}: \quad -6 \qquad -8.45\boxed{+/-}: \quad 8.45$$

You can use several unary keys in succession. For example,

$$25\boxed{\sqrt{}}\boxed{1/x}: \quad \frac{1}{\sqrt{25}} = \frac{1}{5} = 0.2$$

$$3\boxed{x^2}\boxed{x^2}\boxed{+/-}: \quad -(3^2)^2 = -3^4 = -81$$

Unary keys take priority over binary keys. Thus, $\boxed{x^2}$ is done before $\boxed{+}$, so

$$1\boxed{+}3\boxed{x^2}\boxed{=} \quad \text{produces} \quad 1 + 3^2 = 10 \quad \text{not} \quad (1 + 3)^2 = 16$$

Similarly,

$$5\boxed{+}20\boxed{1/x}\boxed{=} \quad \text{produces} \quad 5 + \frac{1}{20} = 5.05 \quad \text{not} \quad \frac{1}{5 + 20} = 0.04$$

EXAMPLE 1 Calculate $\sqrt{18^2 + 37^2}$

SOLUTION First calculate $18^2 + 37^2$, then use $\boxed{\sqrt{}}$:

$$18\boxed{x^2}\boxed{+}37\boxed{x^2}\boxed{=}\boxed{\sqrt{}} \qquad \qquad \underline{41.146081}$$

Question Suppose you omit the $\boxed{=}$ in the solution of Example 1. What do you get?

EXAMPLE 2 Calculate $\dfrac{129}{7.7 + 25.6 - 1.8}$

SOLUTION Calculate the denominator first, take its reciprocal, and then multiply by the numerator:

7 . 7 $\boxed{+}$ 2 5 . 6 $\boxed{-}$ 1 . 8 $\boxed{=}$ $\boxed{1/x}$ $\boxed{\times}$ 1 2 9 $\boxed{=}$ <u>4.095238</u>

EXAMPLE 3 Calculate $\dfrac{1}{\dfrac{1}{7} + \dfrac{1}{\sqrt{5}}}$

SOLUTION Calculate the denominator, and then take its reciprocal:

7 $\boxed{1/x}$ $\boxed{+}$ 5 $\boxed{\sqrt{}}$ $\boxed{1/x}$ $\boxed{=}$ $\boxed{1/x}$ <u>1.694712</u>

Scientific calculators have other unary keys, such as $\boxed{10^x}$ or $\boxed{\log^x}$. These are discussed in Chapter 7.

Memory

How do you calculate $(a + b) \times (c + d)$? If your calculator has parentheses, it's easy:

$\boxed{(}$ a $\boxed{+}$ b $\boxed{)}$ $\boxed{\times}$ $\boxed{(}$ c $\boxed{+}$ d $\boxed{)}$ $\boxed{=}$

But without parentheses you must store $a + b$ somewhere while you compute $c + d$. That requires using the memory.

The key $\boxed{STO}$ (also called $\boxed{x \rightarrow M}$) stores the displayed number in the memory (without changing the display). The key $\boxed{RCL}$ (also called $\boxed{MR}$ or $\boxed{RM}$) recalls the stored number to the display (without changing the contents of the memory). For example,

5 $\boxed{STO}$ $\boxed{x^2}$ produces   25 in the display and 5 in the memory

$\boxed{+}$ $\boxed{RCL}$ $\boxed{=}$ adds the stored number to the display

Therefore,

5 $\boxed{STO}$ $\boxed{x^2}$ $\boxed{+}$ $\boxed{RCL}$ $\boxed{=}$ produces $5^2 + 5 = 30$

4 $\boxed{STO}$ $\boxed{+}$ $\boxed{RCL}$ $\boxed{1/x}$ $\boxed{+}$ $\boxed{RCL}$ $\boxed{\sqrt{}}$ $\boxed{=}$ produces $4 + \tfrac{1}{4} + \sqrt{4} = 6.25$

Storing a number in the memory replaces whatever number is already there. A "clear" memory contains 0. There are three ways to clear the memory: (1) by a key such as $\boxed{CM}$ which does just that; (2) by storing 0 in the memory; and (3) by shutting off the calculator.

EXAMPLE 4 Without parentheses calculate $(83 + 79)(436 - 87)$.

SOLUTION First calculate 83 + 79 and store it. Then calculate 436 − 87, recall the stored number, and multiply:

$$8\ 3\ \boxed{+}\ 7\ 9\ \boxed{=}\ \boxed{\text{STO}}\quad 4\ 3\ 6\ \boxed{-}\ 8\ 7\ \boxed{=}\quad \boxed{\times}\ \boxed{\text{RCL}}\ \boxed{=}\qquad \underline{56538}$$

Remark Suppose you forget the first $\boxed{=}$ in the solution of Example 4. Then only 79 goes into the memory and you get (436 − 87) × 79 = 27571.

This illustrates a danger in using calculators. A small mistake may throw the answer way off. Therefore, in practice, it's good to have a rough estimate of what the answer should be and mistrust an answer that seems out of line.

EXAMPLE 5 Without using parentheses calculate

$$\frac{185}{7.06} + 3\left(\frac{551}{816 - 479}\right)^2$$

SOLUTION Compute the first term and store it. Then compute the second term and add the stored term to it:

$$1\ 8\ 5\ \boxed{\div}\ 7\ .\ 0\ 6\ \boxed{=}\ \boxed{\text{STO}}$$

$$8\ 1\ 6\ \boxed{-}\ 4\ 7\ 9\ \boxed{=}\ \boxed{1/x}\ \boxed{\times}\ 5\ 5\ 1\ \boxed{=}\ \boxed{x^2}\ \boxed{\times}\ 3$$

$$\boxed{+}\ \boxed{\text{RCL}}\ \boxed{=}\qquad \underline{34.223786}$$

The Key $\boxed{y^x}$

This key computes powers. For example,

$$3\ \boxed{y^x}\ 1\ .\ 8\ \boxed{=}: \qquad 3^{1.8} = 7.224674$$

$$1\ .\ 0\ 9\ \boxed{y^x}\ 5\ \boxed{+/-}\ \boxed{=}: \qquad (1.09)^{-5} = 0.649931$$

(For further examples and definitions of negative and rational exponents, see Sections 2.1 and 2.7.)

EXAMPLE 6 Calculate $5.22 \times (4.013 - 1.968)^6$

SOLUTION With parentheses, the natural key sequence is

$$5\ .\ 2\ 2\ \boxed{\times}\ \boxed{(}\ 4\ .\ 0\ 1\ 3\ \boxed{-}\ 1\ .\ 9\ 6\ 8\ \boxed{)}\ \boxed{y^x}\ 6\ \boxed{=}$$

Without parentheses, calculate the power first, and then multiply:

$$4\ .\ 0\ 1\ 3\ \boxed{-}\ 1\ .\ 9\ 6\ 8\ \boxed{=}\ \boxed{y^x}\ 6\ \boxed{\times}\ 5\ .\ 2\ 2 \qquad \underline{381.795123}$$

In algebraic logic, $\boxed{y^x}$ takes priority over binary keys. For example,

$1\ \boxed{+}\ 2\ \boxed{y^x}\ 3\ \boxed{=}$ produces $1 + 2^3 = 9$ not $(1+2)^3 = 27$

$2\ \boxed{y^x}\ 3\ \boxed{\times}\ 5\ \boxed{=}$ produces $2^3 \cdot 5 = 40$ not $2^{15} = 32768$

On our left-to-right model, however,

$1\ \boxed{+}\ 2\ \boxed{y^x}\ 3\ \boxed{=}$ produces $(1+2)^3 = 27$

EXERCISES

Calculate

1. $233.1 + 812.7 - 55.6$
2. $15721 - 6434 - 8399$
3. $14 \cdot 29 \cdot 67 \cdot 73$
4. $\frac{1}{9}(12.8)(15.1)(104.6)$
5. $5375 + (6.1)(388.7)$
6. $(21.9)^2 - (351.5)(0.072)$
7. $\left(57.08 - \frac{33.5}{1.66}\right)^2$
8. $\sqrt{59,740 - 147 \cdot 392}$
9. $\dfrac{79,126}{519 + 625 + 2737}$
10. $\dfrac{1}{(42.7)^2 - (38.4)^2}$
11. $[1 + (8.91 + 4.27 - 3.66)^2]^2$
12. $\left[\left(\dfrac{1}{4.47}\right)^2 + \left(\dfrac{1}{0.83}\right)^2\right]^2$
13. $\dfrac{\sqrt{95} + \sqrt{331}}{12}$
14. $\sqrt{1 + 3\sqrt{2}}$
15. $\dfrac{1}{\dfrac{1}{0.78} + \dfrac{1}{2.09}}$
16. $\dfrac{1}{\dfrac{1}{16} + \dfrac{1}{\dfrac{1}{17} + \dfrac{1}{18}}}$
17. $5 \cdot 3^{13}$
18. $8.3 \times (1.001)^{17}$
19. $(4.2)^6 + (3.8)^7$
20. $(3.5)^{12}(0.49)^8$
21. $\dfrac{(8.59)^2}{5.71 + (6.43)^3}$
22. $\dfrac{32,877}{(9.8)(54.6)^2 - (15.1)(26.3)}$
23. $1 - [1 + \frac{1}{12}(0.05)]^{-120}$
24. $35 \cdot 12^{-4} + 6 \cdot 11^{-2.6}$
25. $\dfrac{656 + 919}{294 + 708}$
26. $\dfrac{491}{588} + \left(\dfrac{857}{136}\right)^2$
27. $\dfrac{6.512}{12.384 - 9.889} + \dfrac{12.384 - 9.889}{6.512}$
28. $\left(\dfrac{37 \cdot 849 + 27 \cdot 651}{10336}\right)^2 + \left(\dfrac{37 \cdot 849 + 27 \cdot 651}{10336}\right)^4$
29. $\left(\dfrac{1}{1 + \sqrt{2}} + \dfrac{1}{1 + \sqrt{3}}\right)^4$
30. $\left(\dfrac{8 - 2\sqrt{5}}{1 + 2\sqrt{7}}\right)^9$
31. $(4.6102 - 8.1522)(9.8914 - 7.6639)^2$
32. $[3(6.47)^2 - (1.98)^2][31.86 + 24.49]$
33. A clerk has to total the prices of 10 items and add on 6% sales tax. First he adds the 10 prices by hand. Then he computes 6% of the total on his calculator and adds this on by hand. How can he make better use of the calculator?

2 FURTHER CALCULATOR TECHNIQUES

Parentheses

If your calculator has parentheses, they can make various computations easier. For example,

$\dfrac{a+b}{c+d}$: $\boxed{(}\,a\,\boxed{+}\,b\,\boxed{)}\ \boxed{\div}\ \boxed{(}\,c\,\boxed{+}\,d\,\boxed{)}\ \boxed{=}$

$(a+b)(c+d)(e+f)$: $\boxed{(}\,a\,\boxed{+}\,b\,\boxed{)}\ \boxed{\times}\ \boxed{(}\,c\,\boxed{+}\,d\,\boxed{)}\ \boxed{\times}\ \boxed{(}\,e\,\boxed{+}\,f\,\boxed{)}\ \boxed{=}$

Closing an open parenthesis automatically completes all operations since the last $\boxed{(}$, without $\boxed{=}$. For example,

$\boxed{(}\,4\,\boxed{\times}\,3\,\boxed{+}\,1\,\boxed{)}$ produces $4 \cdot 3 + 1 = 13$

EXAMPLE 1 Calculate $\left(\dfrac{893 + 2206}{48 \cdot 67 - 1519}\right)^{-4/7}$

SOLUTION

$\boxed{(}\,8\,9\,3\,\boxed{+}\,2\,2\,0\,6\,\boxed{)}\ \boxed{\div}\ \boxed{(}\,4\,8\,\boxed{\times}\,6\,7\,\boxed{-}\,1\,5\,1\,9\,\boxed{)}\ \boxed{=}$
$\boxed{y^x}\,\boxed{(}\,4\,\boxed{\div}\,7\,\boxed{+/-}\,\boxed{)}\ \boxed{=}$ $\underline{0.7088408}$

Remark Avoid $\boxed{=}$ within parentheses. For example, a correct sequence for calculating $4\sqrt{59 + 22}\ (=36)$ is

$4\ \boxed{\times}\ \boxed{(}\,5\,9\,\boxed{+}\,2\,2\,\boxed{)}\ \boxed{\sqrt{\ }}\ \boxed{=}$ 36

An incorrect sequence (on some models) is

$4\ \boxed{\times}\ \boxed{(}\,5\,9\,\boxed{+}\,2\,2\,\boxed{=}\,\boxed{\sqrt{\ }}\,\boxed{)}$ 18

The first $\boxed{=}$ completes the addition $59 + 22 = 81$, but it also completes the pending multiplication by 4, giving $4 \cdot 81 = 324$. The next step gives $\sqrt{324} = 18$, and the final $\boxed{)}\,\boxed{=}$ have no effect. (Check your calculator on this.)

Further Uses of $\boxed{\text{STO}}$ and $\boxed{\text{RCL}}$

A major source of errors is keying in numbers incorrectly. Efficient use of $\boxed{\text{STO}}$ and $\boxed{\text{RCL}}$ helps reduce such errors.

EXAMPLE 2 Let $c = 9.69699$. Calculate $\dfrac{(c+1)(c^2+2)}{c^4+4}$

SOLUTION The obvious calculation requires keying in c three times, but that increases the risk of an error, especially with a number such as c. Instead, we enter c once, carefully. Then whenever we need it, we just hit $\boxed{\text{RCL}}$. (This is like giving c

the new name $\boxed{\text{RCL}}$.) The calculation goes this way:

$c\,\boxed{\text{STO}}\ \boxed{(}\,\boxed{\text{RCL}}\,\boxed{+}\,1\,\boxed{)}\ \boxed{\times}\ \boxed{(}\,\boxed{\text{RCL}}\,\boxed{x^2}\,\boxed{+}\,2\,\boxed{)}\ \boxed{\div}$
$\boxed{(}\,\boxed{\text{RCL}}\,\boxed{x^2}\,\boxed{x^2}\,\boxed{+}\,4\,\boxed{)}\ \boxed{=}$ <u>0.11612657</u>

Memory Plus

The key $\boxed{\text{M+}}$ (also called $\boxed{\text{SUM}}$) adds the displayed number to the memory (without changing the display). Thus,

$5\ \boxed{\text{M+}}\ \boxed{\times}\ 3\ \boxed{=}$ $\begin{cases}\text{adds 5 to the memory}\\ \text{produces 15 in the display}\end{cases}$

Usually, $\boxed{\text{M+}}$ completes a pending operation before adding:

$18\,\boxed{\div}\,9\,\boxed{\text{M+}}$ adds 2 to the memory

(Check your manual; you may need $\boxed{=}$ before $\boxed{\text{M+}}$.)

EXAMPLE 3 Compute $3 \cdot 5 + 3 \cdot 5 \cdot 7 + 3 \cdot 5 \cdot 7 \cdot 9 + 3 \cdot 5 \cdot 7 \cdot 9 \cdot 11$

SOLUTION Using $\boxed{\text{M+}}$ streamlines the computation. Assuming the memory is clear, start with $3\,\boxed{\times}\,5\,\boxed{\text{M+}}$. This computes $3 \cdot 5$ and also adds it to the memory. Next $\boxed{\times}\,7\,\boxed{\text{M+}}$ computes $3 \cdot 5 \cdot 7$ and also adds it to the memory. Continue this way. The terms add up in the memory. At the end, $\boxed{\text{RCL}}$ brings the answer into the display:

$3\,\boxed{\times}\,5\,\boxed{\text{M+}}\ \boxed{\times}\,7\,\boxed{\text{M+}}\ \boxed{\times}\,9\,\boxed{\text{M+}}\ \boxed{\times}\,11\,\boxed{\text{M+}}\ \boxed{\text{RCL}}$ <u>11460</u>

To subtract a number from the memory, use the key $\boxed{\text{M}-}$ if you have one. If not, change its sign and add, using $\boxed{+/-}\,\boxed{\text{M+}}$. When doing a computation with $\boxed{\text{M+}}$ be sure to clear the memory first.

Constant Factors

Some calculators have a "constant factor" feature for multiplying each of a batch of numbers by the same constant. To multiply $a, b, c, \ldots$ by k, start with $k\,\boxed{\times}\,a\,\boxed{=}$. This gives ka, as usual. But then $b\,\boxed{=}$ gives kb, and $c\,\boxed{=}$ gives kc, etc. (Check your manual.)

EXAMPLE 4 A record store offers $27\frac{1}{2}\%$ off all marked prices. Figuring in a 4% sales tax, find the actual cost of records and tapes at $3.19, $3.89, $4.89, $5.57, $7.29, $8.19.

SOLUTION The cost of each item is its marked price times the discount factor $(1 - 0.275) \times 1.04$. So key in

$\boxed{(}\,1\,\boxed{-}\,.275\,\boxed{)}\,\boxed{\times}\,1.04\,\boxed{\times}$

then continue

$3.19\,\boxed{=}\quad 3.89\,\boxed{=}\quad 4.89\,\boxed{=}\quad 5.57\,\boxed{=}\quad 7.29\,\boxed{=}\quad 8.19\,\boxed{=}$

Round each answer to the next higher cent.
  Answer $2.41, $2.94, $3.69, $4.20, $5.50, $6.18

EXERCISES

Calculate

1. $(844 - 296)(18.9 + 25.4)$
2. $[1 + (7.3)(8.5)][29.8 + 14.6]$
3. $(1.297 - 0.845)(4.322 + 9.061)(12.467 - 9.506)$
4. $34(2075 + 3961) - 29(1163 + 2566)$
5. $(1.77 + 6.45)(8.23 - 3.09)^2$
6. $(3.4 + 12.8 + 6.7)^2(25.1 - 18.6)^3$
7. $\dfrac{8611 + 3904}{2787 - 1662}$
8. $\dfrac{13 \cdot 577^2}{(85 + 49)(396 - 187)}$
9. $\dfrac{13 \cdot 59 + 14 \cdot 67}{(106 - 34)(519 - 488)}$
10. $\dfrac{(1.23 + 6.82)(14.07 + 29.54)}{(3.11 + 2.59)(3.66 + 4.17)}$
11. $\dfrac{\sqrt{5} + \sqrt{6}}{\sqrt{11} - \sqrt{10}}$
12. $\dfrac{(25.4)^2 - (19.6)^2}{(3.7)^2 + (11.8)^2}$
13. $\dfrac{7106}{649 + 1283} - \dfrac{1594}{867 + 1192}$
14. $\dfrac{26^2 + 27^2 + 28^2}{16^2 + 17^2 + 18^2}$
15. $\dfrac{(19.5 + 38.6)(52.1 - 13.8)}{\sqrt{47^2 + 63^2}}$
16. $\dfrac{3(1 + \sqrt{2})^4}{(1 + \sqrt{5})(1 + \sqrt{6})}$
17. $\left(\dfrac{433 + 696}{207 + 441}\right)^{5/12}$
18. $[(59 + 7 \cdot 86)(1012 - 9 \cdot 19)]^{-1/7}$
19. $1 + 1 \cdot 2 + 1 \cdot 2 \cdot 3 + 1 \cdot 2 \cdot 3 \cdot 4 + \cdots + 1 \cdot 2 \cdot 3 \cdots 10$
20. $1 - \dfrac{1}{1 \cdot 2} + \dfrac{1}{1 \cdot 2 \cdot 3} - \dfrac{1}{1 \cdot 2 \cdot 3 \cdot 4} + - \cdots - \dfrac{1}{1 \cdot 2 \cdot 3 \cdots 10}$

Calculate, keying in the number a only once

21. $a^3 + a^2 + a \qquad a = 1.7177$
22. $\sqrt{a} + \sqrt{1 + a^2} \qquad a = 0.099494$
23. $a^a \qquad a = 3.838$
24. $\sqrt{a + \sqrt{a + \sqrt{a}}} \qquad a = 411.141$
25. $\dfrac{\sqrt{a}}{3 + \sqrt{a}} \qquad a = 10.011$
26. $(1 + \sqrt{a})(2 + \sqrt{a})(3 + \sqrt{a}) \qquad a = 9.69699$
27. $\dfrac{1 + a}{(2 + a)(3 + a^2)} \qquad a = 0.811188$
28. $\left(\dfrac{a^2}{a^4 - 36}\right)^3 \qquad a = 2.5525$
29. Show that with $\boxed{M+}$ you can do Exercise 21 without using $\boxed{+}$.
30. On the first day of Christmas my true love gave to me, a partridge in a pear tree. On the second day of Christmas my true love gave to me, two calling birds and a partridge in a pear tree. On the third day, she gave me $3 + 2 + 1$ gifts, etc. Calculate the total number of gifts she gave me over all 12 days of Christmas.

3 TABLES OF LOGARITHMS

Recall that common logarithms are logarithms to the base 10. Given a positive number x, we write x in scientific notation:
$$x = m \times 10^c$$
where $1 \leq m < 10$ and c is an integer. By rules of logarithms,
$$\log x = \log m + c$$
The integer c is the **characteristic** of x, and the number $\log m$ is the **mantissa** of x.

EXAMPLE 1 Given $\log 2.7 \approx 0.4314$, find

(a) $\log 27$ (b) $\log 2700$ (c) $\log 0.00027$

SOLUTION Express the numbers in scientific notation:
$$27 = 2.7 \times 10^1 \quad\quad 2700 = 2.7 \times 10^3 \quad\quad 0.00027 = 2.7 \times 10^{-4}$$
They all have the same mantissa, $\log 2.7 \approx 0.4314$. Their characteristics are 1, 3, and -4, respectively.

Answer (a) 1.4314   (b) 3.4314   (c) $0.4314 - 4 = -3.5686$

Given a positive number, it is not obvious how to find its mantissa. Fortunately, there are tables of mantissas (sometimes called log tables). The one in the back of this book gives (approximate) values of $\log m$ for $m = 1.00, 1.01, 1.02, \ldots$, 9.98, 9.99. A portion of that table looks like this:

N	0	1	2	3	4	5	6	7	8	9
3.0	.4771	.4786	.4800	.4814	.4829	.4843	.4857	.4871	.4886	.4900
3.1	.4914	.4928	.4942	.4955	.4969	.4983	.4997	.5011	.5024	.5038
3.2	.5051	.5065	.5079	.5092	.5105	.5119	.5132	.5145	.5159	.5172
3.3	.5185	.5198	.5211	.5224	.5237	.5250	.5263	.5276	.5289	.5302
3.4	.5315	.5328	.5340	.5353	.5366	.5378	.5391	.5403	.5416	.5428
3.5	.5441	.5453	.5465	.5478	.5490	.5502	.5514	.5527	.5539	.5551
3.6	.5563	.5575	.5587	.5599	.5611	.5623	.5635	.5647	.5658	.5670
3.7	.5682	.5694	.5705	.5717	.5729	.5740	.5752	.5763	.5775	.5786
3.8	.5798	.5809	.5821	.5832	.5843	.5855	.5866	.5877	.5888	.5899
3.9	.5911	.5922	.5933	.5944	.5955	.5966	.5977	.5988	.5999	.6010

To find $\log 3.68$, go to the line with 3.6 at the left. Then move across that line to column 8. You find 0.5658, which is $\log 3.68$ rounded off to 4 places.

EXAMPLE 2 From the table find   (a) $\log 337$ (b) $\log 0.394$

SOLUTION (a) Since $337 = 3.37 \times 10^2$, the characteristic is 2 and the mantissa is $\log 3.37$. Go to the line in the table with 3.3 at the left, then across to column 7. You find $\log 3.37 \approx 0.5276$. Hence
$$\log 337 = \log 3.37 + 2 \approx \underline{2.5276}$$

(b) Since $0.394 = 3.94 \times 10^{-1}$, the characteristic is -1 and the mantissa is log 3.94. From the table, log 3.94 ≈ 0.5955. Hence
$$\log 0.394 = \log 3.94 - 1 \approx 0.5955 - 1 = -0.4045$$

Remark For computations, it may be better to write log 0.394 as $0.5955 - 1$ rather than -0.4045. Both are correct, but the first form shows the characteristic -1, while the second form does not.

Linear Interpolation

Suppose we want log 2.153. That's not in the table at the back of the book. The table gives the values
$$\log 2.15 \approx 0.3324 \quad \text{and} \quad \log 2.16 \approx 0.3345$$
Because log x is increasing, log 2.153 is between 0.3324 and 0.3345. To estimate it, we use the method of **linear interpolation.** We pretend that the function log x is *linear* from $x = 2.15$ to $x = 2.16$. Then log 2.153 would be $\frac{3}{10}$ of the way from 0.3324 to 0.3345 because 2.153 is $\frac{3}{10}$ of the way from 2.15 to 2.16. Therefore,
$$\log 2.153 \approx \log 2.15 + \tfrac{3}{10}(0.3345 - 0.3324)$$
$$\approx 0.3324 + \tfrac{3}{10}(0.0021)$$
$$\approx 0.3324 + 0.0006 = 0.3330$$

EXAMPLE 3 From the table, find (a) log 1776 (b) log 0.00006527

SOLUTION (a) Since $1776 = 1.776 \times 10^3$, the characteristic is 3 and the mantissa is log 1.776. Our table shows
$$\log 1.77 \approx 0.2480 \qquad \log 1.78 \approx 0.2504$$
Now, 1.776 is $\frac{6}{10}$ of the way from 1.77 to 1.78. By linear interpolation, we estimate that log 1.776 is $\frac{6}{10}$ of the way from 0.2480 to 0.2504:
$$\log 1.776 \approx \log 1.77 + \tfrac{6}{10}(0.2504 - 0.2480)$$
$$\approx 0.2480 + (0.6)(0.0024)$$
$$\approx 0.2480 + 0.0014 = 0.2494$$
Hence
$$\log 1776 = \log 1.776 + 3 \approx 3.2494$$

(b) Since $0.00006527 = 6.527 \times 10^{-5}$, the characteristic is -5 and the mantissa is log 6.527. Our table shows
$$\log 6.52 \approx 0.8142 \qquad \log 6.53 \approx 0.8149$$
By linear interpolation,
$$\log 6.527 \approx 0.8142 + \tfrac{7}{10}(0.8149 - 0.8142)$$
$$\approx 0.8142 + 0.0005 = 0.8147$$
Hence
$$\log 0.00006257 = \log 6.257 - 5 \approx 0.8147 - 5 = -4.1853$$

Antilogs

Let us reverse the problem: given that log $x = r$, find x. We say that x is the **antilogarithm** of r and write $x =$ antilog r. Actually antilog r is nothing new. The statement log $x = r$ is equivalent to $x = 10^r$. Therefore,

$$\boxed{\text{antilog } r = 10^r}$$

To find antilogs by tables, use a table of antilogs (values of 10^x) if one is available, or try to locate log x in a log table.

EXAMPLE 4 Use the log table in the back of the book to find x, given
(a) log $x = 1.7959$ (b) log $x = 0.3778$ (c) log $x = -8.5203$

SOLUTION The number x has characteristic 1 and mantissa 0.7959. We look for 0.7959 in the log table and find it exactly. It is the mantissa of 6.25. Therefore,

$$x \approx 6.25 \times 10^1 = \underline{62.5}$$

(b) The characteristic is 0. The mantissa 0.3778 does not appear in our table but falls between

$$0.3766 \approx \log 2.38 \quad \text{and} \quad 0.3784 \approx \log 2.39$$

Again we use linear interpolation. We have

```
            0.0012
       ┌──────────┐
  0.3766       0.3778         0.3784
       └─────────────────────────┘
                 0.0018
```

Thus, log x is $\tfrac{12}{18} = \tfrac{2}{3} \approx 0.7$ of the way from log 2.38 to log 2.39. Hence x should be about $\tfrac{7}{10}$ of the way from 2.38 to 2.39, that is, $x \approx \underline{2.387}$.

(c) Be careful; the characteristic is not -8. Add and subtract 9:

$$-8.5203 = 9 - 8.5203 - 9 = 0.4797 - 9$$

Hence the characteristic is -9. The mantissa is log $m = 0.4797$, which falls between

$$0.4786 \approx \log 3.01 \quad \text{and} \quad 4.800 \approx \log 3.02$$

Thus, log m is $\tfrac{11}{14} \approx 0.8$ of the way from log 3.01 to log 3.02; hence $m \approx 3.018$; and

$$x \approx \underline{3.018 \times 10^{-9}}$$

Remark Our table of mantissas is a 4-place table. Its entries are approximations that have been rounded off to 4 decimal places; hence its accuracy is limited. From the table we can find 4-place logarithms for numbers with 3 significant figures and, by interpolation, we can estimate 4-place logarithms for numbers with 4 significant figures. No further accuracy is justified.

Conversely, given log x to 4 places, we can find x to 3 significant figures and estimate the fourth significant figure. For greater accuracy, 5-, 6-, or 7-place tables are required. Such tables exist, but they are big and awkward, and not always available.

EXERCISES

Find the characteristic
1. 94.1
2. 608
3. 1066
4. 37.22
5. 0.085
6. 0.61
7. 0.0009
8. 0.002892
9. 55,306
10. 3,194,077
11. 6.02×10^{23}
12. 1.008×10^{-31}

Find the common logarithm to 4 decimal places. Use linear interpolation if needed
13. 2.08
14. 7.56
15. 1980
16. 132
17. 854.1
18. 6205
19. 0.6549
20. 0.01117
21. 10^{5000}
22. 10^{-844}
23. 1.616×10^{-21}
24. 8.045×10^{15}

Find the antilog to 4 significant figures. Use linear interpolation if needed
25. 0.9562
26. 0.8555
27. 2.6042
28. 3.7466
29. 0.3729 − 4
30. 0.6665 − 2
31. 5.2700
32. 4.5765
33. 0.9001
34. 0.4300
35. 0.8076 − 3
36. 0.1420 − 3

By looking at the log table show that
37. $93.3 < 10^{1.97} < 93.4$
38. $602 < 10^{2.78} < 603$

39. Looking at the log table, we find 6.31 ≈ 0.8000, a nice round number. How does it follow that $(6.31)^5 \approx 10{,}000$?

40. Compute log (log 2).

4 COMPUTATIONS WITH LOGARITHMS

Using the rules for logarithms together with a log table, we can convert various computations into easier computations.

EXAMPLE 1 Compute $x = 2.94 \times 596 \times 0.0738$ to three significant figures.

SOLUTION By the rules for logarithms,
$$\log x = \log 2.94 + \log 596 + \log 0.0738$$
Find the three logarithms in the table and add them:
$$\log x \approx 0.4683 + 2.7752 + (0.8681 - 2) = 2.1116$$
Hence x = antilog 2.1116 = <u>129</u> to 3 significant figures.

EXAMPLE 2 Compute to four significant figures
$$x = \frac{(47.5)^3}{(63.8)(906.2)}$$

SOLUTION By the rules for logarithms,
$$\log x = 3 \log 47.5 - \log 63.8 - \log 906.2$$
We find the logs from the table, using linear interpolation for $\log 906.2$:
$$\log x \approx 3(1.6767) - 1.8048 - 2.9572 = 0.2681$$
Again from the table, using interpolation we find
$$x = \text{antilog } 0.2681 \approx \underline{1.854}$$

EXAMPLE 3 Compute to three significant figures
        (a) $7^{1.56}$ (b) $\sqrt[3]{0.000142}$

SOLUTION Use the rules of logarithms for powers and roots:
  (a) $\log 7^{1.56} = 1.56 \log 7$
$$\approx 1.56 \times 0.8451 \approx 1.3184$$
To three significant figures, antilog $1.3184 = \underline{20.8}$

  (b) $\log \sqrt[3]{0.000142} = \tfrac{1}{3} \log 0.000142$
$$\approx \tfrac{1}{3}(0.1523 - 4) = \tfrac{1}{3}(-3.8477) \approx -1.2826$$
To find the antilog write
$$-1.2826 = 2 - 1.2826 - 2 = 0.7174 - 2$$
To three significant digits, antilog $(0.7174 - 2) \approx \underline{0.0522}$

Remark Logarithms are of no help in addition or subtraction. To compute $\sqrt[3]{65} + \sqrt[4]{71}$, you can obtain each term by logarithms, but then you would have to add them together by hand.

Calculations such as $(1.04)^{50}$ are not very accurate with 4-place logs. In computing $50 \times \log 1.04$, the round-off error in $\log 1.04 \approx 0.0170$ gets multiplied by 50. We obtain
$$\log(1.04)^{50} \approx 0.8500 \qquad (1.04)^{50} \approx 7.079$$
Using 6-place logs gives better accuracy. We have $\log 1.04 \approx 0.017033$; hence
$$\log(1.04)^{50} \approx 0.851650 \qquad (1.04)^{50} \approx 7.1064$$
Even this result is not too precise. A closer value is $(1.04)^{50} \approx 7.10668$.

EXERCISES

Compute to three significant figures using logarithms

1. $4.81 \times 3.99 \times 81.6$
2. $16.9 \times 28.7 \times 55.4$
3. $10.4 \times 73.6 \times 0.035$
4. $9.18 \times 0.212 \times 0.866$
5. $\dfrac{727}{991}$
6. $\dfrac{0.548}{3.62}$
7. $\dfrac{3.88}{9.15 \times 6.05}$
8. $\dfrac{445 \times 1160}{179}$

Exercises 329

9. $\dfrac{2.37 \times 8.54}{93 \times 0.00615}$

10. $\dfrac{236 \times 77 \times 853}{82 \times 991}$

11. $\dfrac{(4.07)(6.91)^2}{(5.26)^3}$

12. $\dfrac{(433)(0.00816)^2}{(76)(0.424)^2}$

13. $\sqrt[3]{1960}$

14. $\sqrt[4]{31.7}$

15. $\sqrt[5]{102}$

16. $\sqrt[6]{0.0229}$

17. $\dfrac{344\sqrt{91}}{1020}$

18. $\dfrac{26.3}{1.57\sqrt{109}}$

19. $\sqrt{\dfrac{(60.7)(29.1)}{215}}$

20. $\sqrt[3]{\dfrac{3.14}{(12.6)(0.037)}}$

21. $\sqrt{1 + \sqrt[3]{5}}$

22. $\sqrt{\sqrt{14} - \sqrt{13}}$

23. $9^{0.47}$

24. $2^{3.28}$

25. $(157)^{-1.27}$

26. $(0.41)^{-6.3}$

27. $(1.08)^{12}$

28. $(1.06)^{15}$

29. $(12.7)^{2.5}(1.76)^{3.3}$

30. $53^{21} \cdot 68^{-19}$

Compute to four significant figures using logarithms

31. $219 \times 4215 \times 0.316$

32. $95.1 \times 63.6 \times 0.01906$

33. $\left(\dfrac{5340}{1128}\right)^3$

34. $\left(\dfrac{0.0004912}{0.0686}\right)^{-5}$

35. $\sqrt[4]{1991}$

36. $\sqrt[3]{\dfrac{814}{255}}$

37. $(3.055)^{1.2}$

38. $(2854)^{0.031}$

39. $10^{0.6881}$

40. $10^{0.5002}$

Answers to Odd-Numbered Exercises

CHAPTER 1

Section 1.2, page 6

1. Distr. Law   3. Assoc. Law for Mult.   5. Add. Identity   7. Rule (5) for Inverses   9. Rule (4) for Inverses   11. 3775   13. 10,000
15. 5   17. $40 \cdot 20 + 40 \cdot 3 + 2 \cdot 20 + 2 \cdot 3 = 966$ 19. Assoc. Law
21. $2a \cdot 2b = 2(a \cdot 2)b = 2(2a)b = (2 \cdot 2)(ab) = 4ab$ 23. $a + b + (-a)$
$= [a + (-a)] + b = 0 + b = b$ 25. $a(b + c + d) = a([b + c] + d)$
$= a[b + c] + ad = ab + ac + ad$ 27. $a(bc + bd) = a(b[c + d])$
$= ab(c + d)$ 29. $(-a)(-b)(-c) = (-a)[(-b)(-c)] = (-a)(bc) = -abc$
31. Add -4 to both sides.   33. Add $-x$ to both sides.   35. Add -1 to both sides, then multiply by $1/a$.   37. If $(abc)d = 0$, then either $abc = 0$ or $d = 0$. If $d \neq 0$, then one of a, b, c is 0 by Example 4 of the text.
39. Either $a + 1 = 0$ or $a + 2 = 0$, hence, either $a = -1$ or $a = -2$.
41. Properties of Zero (3)   43. Yes, 0 (but only 0).   45. $0 \cdot a = 1$ is impossible because $0 \cdot a = 0$ always.

Section 1.3, page 11

1. 8   3. 10   5. -15 7. 7/180   9. 13/72   11. 15/8   13. 18/5
15. 1/15   17. $(ab)^2/(a + b)$ 19. 3168   21. $999\frac{2}{3}$ 23. True
25. False   27. True   29. $\dfrac{a}{1} = a \cdot \dfrac{1}{1} = a$ 31. $\left(\dfrac{a}{d} + \dfrac{b}{d}\right) + \dfrac{c}{d} = \dfrac{a + b}{d} + \dfrac{c}{d}$
$= \dfrac{a + b + c}{d}$ 33. $a \cdot \dfrac{b}{c} = \dfrac{ab}{c} = \dfrac{ba}{c} = b \cdot \dfrac{a}{c} = \dfrac{a}{c} \cdot b$ 35. $b + [a - b]$
$= b + [a + (-b)] = [b + (-b)] + a = 0 + a = a$ 37. $\dfrac{3 + 5}{1 + 1} = 4$, not
$\dfrac{3}{1} + \dfrac{5}{1} = 8$ 39. Add c to both sides.   41. Multiply both sides by c.
43. $\dfrac{a}{b} + \dfrac{c}{d} = \dfrac{ad}{bd} + \dfrac{bc}{bd} = \dfrac{ad + bc}{bd}$ by Rule (1) for quotients   45. Use

the definition of quotients: $a \cdot \dfrac{b}{c} = a \cdot \left(b \cdot \dfrac{1}{c}\right) = (ab) \cdot \dfrac{1}{c} = \dfrac{ab}{c}$
47. Multiply both sides by bd.

Section 1.4, page 15

1. 10   3. -22 5. 28   7. 8   9. 51   11. 15   13. 12
15. 77   17. 71   19. 53   21. $a + 2c - 2$ 23. 17   25. (a) -1,
(b) -2 27. (a) 7, (b) 7   29. (a) -6, (b) 114   31. (a) $a - (bc/d)$,
(b) $(a - b)c/d$ 33. (a) $a\;\boxed{+}\;2\;\boxed{\times}\;b\;\boxed{=}$, (b) $2\;\boxed{\times}\;b\;\boxed{+}\;a\;\boxed{=}$
35. (a) $a\;\boxed{-}\;b\;\boxed{=}\;\boxed{\times}\;5\;\boxed{=}$, (b) $a\;\boxed{-}\;b\;\boxed{\times}\;5\;\boxed{=}$
37. (a) $a\;\boxed{+}\;b\;\boxed{\times}\;c\;\boxed{\div}\;d\;\boxed{=}$, (b) $b\;\boxed{\times}\;c\;\boxed{\div}\;d\;\boxed{+}\;a\;\boxed{=}$
39. (a) $a\;\boxed{\div}\;b\;\boxed{+}\;c\;\boxed{=}\;\boxed{\times}\;6\;\boxed{=}$, (b) $a\;\boxed{\div}\;b\;\boxed{+}\;c\;\boxed{\times}\;6\;\boxed{=}$

Section 1.5, page 21

1. 4   3. 5   5. 5   7. $a^2 + 6$ 9. -8 11. 1   13. 3   15. 9
17. 0.3   19. $a > 0$ 21. $c \leq 7$ 23. $22/7 > \pi$ 25. $9 < (3.019)^2 < 9.12$ 27. $0 < x \leq 5$ 29. $a < b < c$ 31. $a \leq 1 < b < 2$
33. $|a - b| = 3$ 35. $x < y - 2.1$ 37. $|x + 1| < |x - 6|$ 39. $a^2 \geq 0$, $b^2 \geq 0$, hence $a^2 + b^2 \geq 0$ 41. $-a^2 - b^2 \leq 0$. Now add 5 to both sides.
43. If $a = 0$ or $a = 3$, then one term is positive and one is 0. Otherwise, both terms are positive.   45. $abc = (ab)c = (\text{pos})(\text{neg}) = \text{neg}$ 47. $b^2 > 0$ so ab^2 has the same sign as a.   49. $|a| = \pm a$ so $|a|^2 = (\pm a)^2 = a^2$
51. $a^2 + b^2 + c^2 \geq 0$. The sum is positive unless $a = b = c = 0$.

Review Exercises, page 22

1. 13   3. 20   5. 2π 7. $ab + 2a + b + 2$ 9. $a < b < 0$
11. $b - a = 6$ 13. $|a - b| > 4$ 15. If $a^2b^2 = (ab)^2 = 0$, then $ab = 0$. Hence $a = 0$ or $b = 0$ or both.   17. Not necessarily, $a = b$ or $a = -b$.
19. $a = b = c = 0$

CHAPTER 2

Section 2.1, page 28

1. 27   3. $1/36$ 5. $-32/243$ 7. 1   9. $1/72$ 11. $1/2$ 13. 2^4
15. 2^{-7} 17. 2^{14} 19. 2^{12} 21. $6a^5$ 23. $5x^3$ 25. $-80c^9$
27. $3/5u^6$ 29. $-5x^3/4$ 31. a^5b^4 33. b^2/a^6 35. $8c^4/9b^2$
37. $2r^3s^2/9$ 39. $2y^2/x^2$ 41. $4r^3/p^3$ 43. 6.2×10^1
45. 8.1×10^{-3} 47. 2.154×10^5 49. 2.37   51. 4.52×10^{14}
53. 8.13×10^{11} 55. 6.48×10^5 57. 9.6×10^{-9} 59. 6.4×10^{31}
61. 6.4×10^{-29} 63. 6.36×10^6 65. 4.8×10^{-18} 67. 275,854.7354
69. 7.76243×10^{20} 71 1.71756×10^{-65} 73. 5^{50}

Section 2.2, page 32

1. $6x + 1$ 3. $8x^2 + 8x + 2$ 5. $-6x^2 + 6x - 3$
7. $(5/3)x^2 - 2x - 13/2$ 9. $x^3 + 8x^2 - 16x - 4$
11. $-x^5 - 2x^4 - 2x^3 - 2x^2 - x + 7$
13. $5x^8 + 36x^6 - 30x^5 + 20x^4 - 24x^3 + 10x^2$ 15. $x^2 - x - 6$
17. $6x^2 + 17x + 5$ 19. $x^3 + 7x^2 + 5x - 6$ 21. $x^5 + 2x^3 + x$
23. $x^7 - x^6 - \frac{1}{2}x^5 + \frac{1}{2}x^4 - \frac{1}{2}x^3 - \frac{1}{4}x^2$ 25. $x^4 + 2x^3 + 2x^2 + x$
27. $x^5 - 1$ 29. $x^3 + x^2 + x + 1$ 31. $(1 + x)(1 + x^2) \cdots (1 + x^{2^n})$
$= 1 + x + x^2 + \cdots + x^{2^{n+1}-1}$ 33. $1 - x^4$
35. $(1 - x)(1 + x)(1 + x^2) \cdots (1 + x^{2^n}) = 1 - x^{2^{n+1}}$ 37. $x^4 + x^2 + 1$
39. $2x^3 + 4x^2 - x + 5$ 41. $18x^2 + 6x + 25$
43. $x^3 + (a + b + c)x^2 + (ab + ac + bc)x + abc$ 45. $10x^6$ 47. $3x^9$
49. $2x^4$ 51. 2 53. -8 55. 1 57. False, the degree is $m + n$.
59. x^n occurs in the product as many times as n can be expressed as
$p + 5q + 10r$. Each such expression represents one way of making n cents
from p pennies, q nickels, and r dimes.

Section 2.3, page 36

1. $5x + y + 2$ 3. $-5x + 6y + 5z - 2$ 5. $2x^2 + 4y^2 - 3y$
7. $-2x^2 + yz + xz - x - y$ 9. $2x^2 + 5xy + 3y^2$ 11. $25x^2 - 4y^2$
13. $2x^2 - y^2 + xy + 2xz - zy$ 15. $x^8 - y^8$ 17. $a^2 + 4ab + 4b^2 - 9$
19. $x^4 - 2x^2y^2 - 3y^4$ 21. $rst + rs + rt + st + r + s + t + 1$
23. $x^3 - y^3$ 25. $x^3 + 3x^2y + 3xy^2 + y^3$ 27. 16 29. 12
31. $x^2 - 8xy + 16y^2$ 33. $36a^2 + 60ab + 25b^2$ 35. $u^4 + 2u^2v^2 + v^4$
37. $x^2 + y^2 + z^2 + 2xy - 2xz - 2yz$ 39. $11{,}025$ 41. $999{,}991$
43. Yes; $(x + 8)^2$ 45. Yes; $(z^3 + 2)^2$ 47. No 49. Yes; $((x^2 + 1) + 1)^2$
51. Yes; $(ab^4 - 2c^2)^2$ 53. $\dfrac{(n + 1)^2 - (n - 1)^2}{n} = \dfrac{4n}{n} = 4$
55. $3n^2 + 6n + 5 = n^2 + (n + 1)^2 + (n + 2)^2$

Section 2.4, page 42

1. $3x(y - 3z)$ 3. $4a^2c(3b + 2b^2c + c^2)$ 5. $(2x + 1)(3x + 1)(3x)$
7. $(x + 1)(x + 3)$ 9. $(x - 2)(x - 7)$ 11. $(3x + 1)(x + 1)$
13. $(9x - 2)(x - 2)$ 15. $(3x + 2)(2x + 7)$ 17. $(x - 4y)(x + 9y)$
19. $(2x + y)(2x + y)$ 21. $(x^2 + 3)(x^2 + 2)$ 23. $(2y - 9)(2y + 9)$
25. $(a^2 - 4b)(a^2 + 4b)$ 27. $[r - 2(s + t)][r + 2(s + t)]$
29. $-(x + y + z)(x + y + z + 10)$ 31. $u^2(u - 3v)^2(u + 3v)^2$
33. $(3x - 2)(3x^2 + 6x + 4)$ 35. $(x - 3)^2(x + 3)^2$
37. $(a + 2b - c)(a + 2b + c)[(a + 2b)^2 + c^2]$ 39. $(b - 3)(b^2 + 3b + 9)$
41. $(2x - y^2)(4x^2 + 2xy^2 + y^4)$ 43. $(x - 1)(x^4 + x^3 + x^2 + x + 1)$
45. $(x - 2)(x + 2)(x^2 + 2x + 4)(x^2 - 2x + 4)$ 47. $(x - 2)^2(x + 2)^2$
49. $(50 - 1)(50^2 + 50 + 1) = 49(2551)$
51. $(x - 1)(x - 2)(x^2 + x + 1)(x^2 + 2x + 4)$ 53. $(a + c)(a + b)$
55. $v(1 + uv)(u^2 - 2)$ 57. $(x + 1)(x - 1)^2(x^2 + x + 1)$

59. $(x-y)(x+y)(1+x^2+y^2)$ **61.** $(411.3 - 410.3)(411.3 + 410.3) = 821.6$ **65.** $x^6 - y^6 = (x-y)(x^5 + x^4y + x^3y^2 + x^2y^3 + xy^4 + y^5)$

Section 2.5, page 45

1. $\dfrac{3x}{x^2+5}$ **3.** $\dfrac{x}{x+5}$ **5.** $\dfrac{1}{x+4}$ **7.** $\dfrac{x+y}{x+3y}$ **9.** $\dfrac{4x+3}{6x-1}$
11. $\dfrac{x^2}{x^2-x+1}$ **13.** $\dfrac{1}{x+y+2z}$ **15.** $\dfrac{x+y}{x^2y^2}$ **17.** x^n **19.** $\dfrac{1}{x(x-1)}$
21. $\dfrac{2}{(2x+3)(2x+5)}$ **23.** $\dfrac{7x-8}{(3x-4)^2}$ **25.** $\dfrac{5x+4}{(x+2)(x-2)}$
27. $\dfrac{(3x-2)(x-1)}{(x-2)^2(x+2)}$ **29.** $\dfrac{x^2+y^2+xy}{xy}$ **31.** $\dfrac{-2}{x(x+1)(x+2)}$
33. $\dfrac{2(2x^2-4x+3)}{(2x-1)^3}$ **35.** $x+1$ **37.** $\dfrac{x-2}{(x+1)(x-3)}$ **39.** $\dfrac{1}{xy}$
41. $\dfrac{1}{(x-2y)(x-y)}$ **43.** $\dfrac{x+3}{2x}$ **45.** $\dfrac{b+a}{b-a}$ **47.** $\dfrac{z-1}{z}$
49. $\dfrac{4y+2}{9y+13}$ **51.** $\dfrac{4u-3}{2u-4}$ **53.** $\dfrac{x-y}{xy}$ **55.** $\dfrac{x+1}{2x+1}$

Section 2.6, page 52

1. 9   **3.** 3/8   **5.** 2/3   **7.** 0.4   **9.** 3×10^6 **11.** $6a^4$
13. $x^2/5$ **15.** $2\sqrt{6}$ **17.** $2\sqrt[3]{4}$ **19.** $6c^2\sqrt{2c}$
21. $ab^2\sqrt{3ab}/2c^2$ **23.** $ab^2\sqrt[3]{a}/cd^3$ **25.** 9   **27.** $a^2 - 2b^2$ **29.** $\sqrt{6}$
31. $\tfrac{1}{2}(\sqrt{3}-1)$ **33.** $\tfrac{1}{19}(16+6\sqrt{5})$ **35.** $(a - 2\sqrt{ab} + b)/(a-b)$
37. $\sqrt{a^2-x^2} = \sqrt{a^2(1+x^2/a^2)} = a\sqrt{1+x^2/a^2}$ **39.** $\sqrt[n]{abc} = \sqrt[n]{(ab)c} = \sqrt[n]{ab}\sqrt[n]{c} = \sqrt[n]{a}\sqrt[n]{b}\sqrt[n]{c}$ **41.** $(4+5a+7a^2)a\sqrt{a}$ **43.** $\sqrt[3]{2}$
45. a^2 **47.** $2y\sqrt{10xz}$ **49.** $4yz^2\sqrt[3]{x}$ **51.** $v^2\sqrt{\dfrac{2u}{3}}$ **53.** $5/2\sqrt{3x}$
55. Square both sides.   **57.** 441.3785   **59.** 1.961571   **61.** $r^{mn} = s^{nm} = a$

Section 2.7, page 56

1. 9   **3.** 1/5   **5.** 100,000   **7.** 343/8   **9.** $2^{7/3}$ **11.** $2^{7/2}$
13. $2^{7/9}$ **15.** $2^{(2n-k)/4}$ **17.** $1/125x^6$ **19.** $27u^3/v^9$ **21.** x^4y^{-2}
23. $x^{10}y^{15}z^{-20}$ **25.** $32x^{5/2}$ **27.** $x + 2x^2 + x^3$ **29.** $\sqrt[6]{2^5}$ **31.** $\sqrt[4]{b^3}$
33. $\sqrt[4]{b}$ **35.** $\sqrt[6]{x^7} = x\sqrt[6]{x}$ **37.** $\sqrt[8]{a}$ **39.** $6a\sqrt{a}$ **41.** $125/u^6$
43. $4p^2q^8$ **45.** $1/a^{1/5}b^{2/5}$ **47.** $y^{9/20}/x^{3/2}$ **49.** 3.693
51. 2.294×10^{15} **53.** 1.797   **55.** 1.286

Section 2.8, page 59

1. $4x$ **3.** $6a^2$ **5.** $7x - 7$ **7.** x^2 **9.** Leave as $\sqrt{x^2+1}$
11. $(1+\sqrt{2})\sqrt{x}$ **13.** $\sqrt{85}$ **15.** $2^{9/2}$ **17.** a^{-4} **19.** $\sqrt[9]{a}$

21. Leave as $\dfrac{1}{x+3}$ 23. $\dfrac{c}{c+e} + \dfrac{d}{c+e}$ 25. $a^3 + 3a^2b + 3ab^2 + b^3$
27. Leave as $x^{3/2}$ 29. $4x^2 - 7x + 1/x$

Review Exercises, page 60

1. 3 3. $\dfrac{4^2 \cdot 2^3}{2^3 + 4^2} = \dfrac{16}{3}$ 5. $4u^2 - 2uv + \tfrac{1}{4}v^2$
7. $1 - x + x^2 - 2x^3 + x^4 - x^5 + x^6$ 9. $(2y - 3)(2y + 3)(4y^2 + 9)$
11. 2.0×10^{-18} 13. $(5x/y^2z^3)\sqrt[3]{2x/z}$ 15. $3/(x - 3)$
17. $-(x^2y^2 + 1)/y^4$ 19. $-5/14x$

CHAPTER 3

Section 3.1, page 64

1. $.15x$ dollars 3. $36y$ 5. $200/r$ hours 7. $s\sqrt{2}$ 9. $80 + 20(d - 7)$
11. $[5n + (100 - n)]$ cents 13. $5 + x + \sqrt{25 + x^2}$
15. $\left(\dfrac{700}{x} + \dfrac{300}{x - 40}\right)$ hours 17. $x + 1/x$ 19. $\tfrac{1}{2}[n^2 + (n + 1)^2]$
21. $\tfrac{1}{2}(b)(1.5b) = .75b^2$ 23. $(r\sqrt{2})^2 = 2r^2$ 25. $2\left(\dfrac{l}{4}\right)^2 + \left(\dfrac{30 - 2l}{4}\right)^2$ ft^2
27. $\dfrac{150}{n} - \dfrac{150}{n + 10}$ 29. $[(10 + \tfrac{3}{4}n)/(16 + n)] \times 100$
31. $[(50 + .7(x - 50)) - (40 + .6(x - 40))] = 0.1x - 1$ dollars 33. Think!

Section 3.2, page 68

1. $.12x = 62$ 3. $x - .3x = 29.75$ 5. $80 + 20(x - 7) = 15x$
7. $x - .2x + 3500 = x + 500$ 9. $x - (50 - x) = -6$
11. $(x + 1)^2 - x^2 = 97$ 13. $x - 1/x = 4$
15. $150/x - 150/(x + 10) = 1.25$ 17. $15(\tfrac{1}{2}x)/(1000 + 15x) = 1/10$
19. $(\tfrac{1}{4}x)^2 + [\tfrac{1}{4}(40 - x)]^2 = 82$ 21. $x + 5(56 - x) = 100$
23. $x \cdot 2^{15/3} = 128{,}000$ 25. $\dfrac{x + 1}{x + 4} - \dfrac{x}{x + 3} = \dfrac{1}{10}$ 27. $x^2 + 3^2$
$= [\tfrac{1}{2}(12 - 2x)]^2$ 29. $x^2 + (x + 1)^2 = (x + 2)^2$
31. $\dfrac{63{,}360}{x - 2} - \dfrac{63{,}360}{x} = 132$ 33. $\tfrac{1}{8}\left[\left(\dfrac{x^2 - 1}{8}\right)^2 - 1\right] = 55$

Section 3.3, page 73

1. Conditional equation 3. Identity 5. Identity 7. Conditional equation 9. Identity 11. 6 13. -2 15. 7.5 17. 20/13 19. -1
21. 7/6 23. -2 25. 5/6 27. 1 29. No solution 31. -1
33. 5/3 35. $-9/4$ 37. -10.204 39. 1.767 41. $m = E/c^2$

43. $f_1 = ff_2/(f_2 - f)$ 45. $h = S/\pi r(r + 2)$ 47. $x = a(1 - y/b)$
49. $x = (5y + 1)/(y - 1)$ 51. $x = (1 - 5y)/(3y - 2)$
53. $x = (2y + 1)/(y^4 + y^2 + 1)$ 55. $-75/8$ 57. -4

Section 3.4, page 77

1. 800 3. $42.50 5. 12 days 7. $5000.00 9. 7 11. 32°, 37°, 111° 13. -40 15. 80, 77½ 17. 48, 49 19. $\frac{20}{9}$ hr 21. $\frac{18}{19}$ hr 23. 3 hr 25. $16\frac{2}{3}$ gal/min 27. 15 min 29. 8 mph 31. .04 miles = 70.4 yards 33. $21\frac{9}{11}$ min 35. 375 37. $6250 in X, $3750 in B 39. 1/9 41. 892/999 43. 7438/9900 45. $2\frac{2}{3}$ mi

Section 3.5, page 83

1. 0, 4/3 3. 7, -2 5. 3, -4 7. -4 (double root) 9. 1/2, 4
11. 1/2, 1/3 13. 3/8, -2 15. 5/12, 2 17. $-2 \pm \sqrt{2}$
19. $-\frac{5}{2} \pm \frac{1}{2}\sqrt{5}$ 21. No real solutions 23. $-2 \pm \sqrt{11/3}$
25. 9, -1 27. $-1/4$ (double root) 29. $-\frac{5}{6} \pm \frac{1}{6}\sqrt{19}$ 31. No real solutions 33. $2(x - 3/2)^2 + 5/2$ 35. $5(x + 1/5)^2 + 4/5$
37. $\frac{1}{3}(x - 6)^2 - 14$ 39. $(x - k)^2 - k^2$ 41. $-3(x - 2/3)^2 + 4/3$
43. $x^2 - 2x + 6 = (x - 1)^2 + 5 > 0$ 45. $x^2 + 2x + c = (x + 1)^2 + (c - 1)$, which is positive (not zero) if $c > 1$

Section 3.6, page 89

1. two 3. two 5. one (double root) 7. none 9. $-5 \pm \sqrt{23}$
11. $\frac{1}{4}(-1 \pm \sqrt{33})$ 13. 2, 3/5 15. $\frac{1}{10}(4 \pm \sqrt{6})$ 17. $\frac{1}{3}(1 \pm \sqrt{5})$
19. $\frac{1}{5}(1 \pm \sqrt{2})$ 21. 1 (double root) 23. $x = \frac{1}{5}(2 \pm \sqrt{39})$
25. $x = \frac{1}{2}(3 \pm \sqrt{5})$ 27. $x = \frac{1}{2}(5 \pm \sqrt{17})y$ 29. $x = \frac{1}{4}(5 \pm \sqrt{17})y$
31. $x^2 - 3x - 2 = 0$ 33. $-b/a$ 35. 13, 7 37. $x = 2 \pm \sqrt{5}$
39. 17, 18 41. 4, 36 ft 43. 32 inches 45. 3, 4, 5 47. $\frac{1}{2}(1 \pm \sqrt{5})$
49. $x^2 + (70 - x)^2 = 45^2$ has no real roots

Section 3.7, page 95

1. $\pm 1, \pm \sqrt{6}$ 3. $\pm \sqrt{3 - \sqrt{3}}, \pm \sqrt{3 + \sqrt{3}}$ 5. $\pm \frac{1}{2}\sqrt{2}, \pm \frac{1}{5}\sqrt{5}$
7. 4, 9 9. $-2, \sqrt[3]{10}$ 11. $\pm \frac{1}{2}, \pm \frac{1}{2}\sqrt{2}$ 13. $\pm \sqrt{\frac{1}{2}(1 + \sqrt{13})}$
15. $\pm \sqrt{\frac{1}{2}(1 + \sqrt{37})}$ 17. ± 4 19. 2, $\frac{1}{2}(-3 \pm \sqrt{5})$ 21. 0, -2, -3
23. 1, ± 2 25. 0, 1 27. 1 29. 3 31. 11 33. 1 35. 9
37. 6 39. 9 41. $\frac{9}{2}$ 43. ± 13

Review Exercises, page 97

1. No real solutions 3. 2/3 (double root) 5. 6/25 7. $\pm 2, \pm \sqrt{3}$
9. $3 \pm 2\sqrt{2}$ 11. 0, ± 2 13. $(x + 1/2)^2 + 2(y - 2)^2 - 29/4$
15. 56 pennies, 44 nickels 17. 18 wins, 22 losses 19. 7

CHAPTER 4

Section 4.1, page 104

1. $\sqrt{2} + 9 < 1.5 + 9 = 10.5$ 3. $1 + 4\sqrt{3} > 1 + 4(1.7) = 7.8$
5. $4/\sqrt{3} > 4/2 = 2$ 7. $1.4 + 1.7 < \sqrt{2} + \sqrt{3} < 1.5 + 1.8$
9. $2 - \sqrt{2} > 2 - 1.5 = 0.5$; take reciprocals   11. $9615 < 9761$; add 7854 to both sides   13. $-\frac{1}{3} > -\frac{1}{2}$; add 16/47 to both sides   15. $\pi < 4$, hence $1/\pi > 1/4$ and $3/\pi > 3/4$ 17. $99 \cdot 13 < 100 \cdot 13 = 1300$
19. $(2.07)^2(3.026) > 2^2 \cdot 3 = 12$ 21. $\pi\sqrt{2} < (3.2)(1.5) = 4.8 < 5$
23. $10 \cdot \frac{1}{20} < \frac{1}{11} + \frac{1}{12} + \cdots + \frac{1}{20} < 10 \cdot \frac{1}{11}$ 25. 12/13   27. 4/577
29. $\sqrt{2}/1.5$ 31. Best: family, Worst: large economy   33. Averaged 56.1; must have exceeded 55 at some time   35. $\frac{4}{3}\pi(3.0741)^3 > \frac{4}{3} \cdot 3 \cdot 3^3 = 108$
37. $147.015 <$ area < 153.015 cm²   39. Smallest: $11.11, Largest: $15
41. By Rule (6), enough to check that $(a + 1)b > a(b + 1)$; true because $b > a$ 43. $\sqrt{a}\sqrt{a} < \sqrt{a}\sqrt{b} < \sqrt{b}\sqrt{b}$ 45. $a + c - (b + c) = a - b < 0$, since $a < b$ 47. Use Rule (2) twice

Section 4.2, page 109

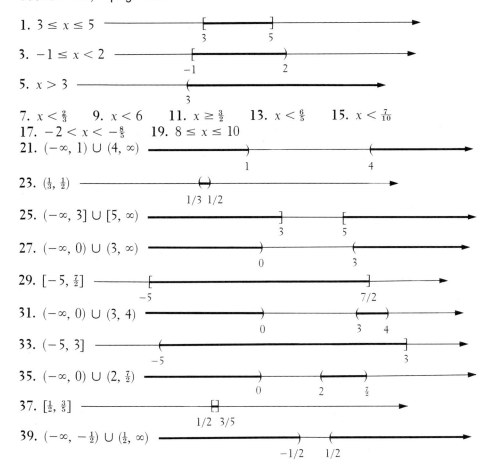

1. $3 \leq x \leq 5$
3. $-1 \leq x < 2$
5. $x > 3$
7. $x < \frac{2}{3}$ 9. $x < 6$ 11. $x \geq \frac{3}{2}$ 13. $x < \frac{6}{5}$ 15. $x < \frac{7}{10}$
17. $-2 < x < -\frac{8}{5}$ 19. $8 \leq x \leq 10$
21. $(-\infty, 1) \cup (4, \infty)$
23. $(\frac{1}{3}, \frac{1}{2})$
25. $(-\infty, 3] \cup [5, \infty)$
27. $(-\infty, 0) \cup (3, \infty)$
29. $[-5, \frac{7}{2}]$
31. $(-\infty, 0) \cup (3, 4)$
33. $(-5, 3]$
35. $(-\infty, 0) \cup (2, \frac{7}{2})$
37. $[\frac{1}{2}, \frac{3}{5}]$
39. $(-\infty, -\frac{1}{2}) \cup (\frac{1}{2}, \infty)$

340 Answers to Odd-Numbered Exercises

37. $\sqrt{61}$, $\sqrt{82}$, 5   39. $y = -\frac{2}{3}x + \frac{13}{3}$ 41. $(2 + \sqrt{51}, 0), (2 - \sqrt{51}, 0)$
43. $(\frac{1}{3}(-5 \pm \sqrt{91}), 1)$ 45. Both distances are 10.
47. The first guess gives $m + n$, the second gives $m + (100 - n)$. This is enough to find m and n.   49. red

Section 5.2, page 130

1. 4/3   3. 0   5. $-3/5$ 7. Yes   9. Yes
11. 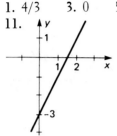   13. 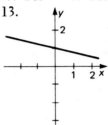   15.

17.   19.   21.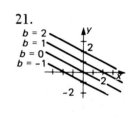

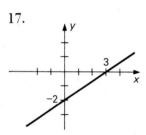

23. 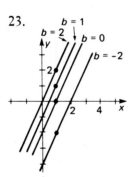   25.

27. Call the points A, B, C, D. Then $\overline{AB} \parallel \overline{CD}$ because both are vertical, and $\overline{AD} \parallel \overline{BC}$ because both have slope 4/3.   29. $y = x - 2$ 31. $x = -7$
33. $y = -\frac{2}{7}x + \frac{32}{7}$ 35. $y = -7x + 3$ 37. $y = 2x - 11$
39. One side has slope $m_1 = 1$. Another side has slope $m_2 = -1$ and $m_1 m_2 = -1$.   41. All sides have length 5.   43. Call the points P, Q. Then $\overline{PQ}$ and $y = -x$ are perpendicular because their slopes are 1 and -1. Also, the midpoint of $\overline{PQ}$ is $(\frac{1}{2}(a - b), -\frac{1}{2}(a - b))$, which lies on $y = -x$.
45. 13,577 ft   47. A lattice point (p, q) is red if $p + q$ is even, black if $p + q$ is odd. If (p, q) is on $y = x + n$, then $p + q = p + (p + n) = 2p + n$, which has the same parity (odd or even) as n. Hence all points are red if n is even, black if n is odd.   49. If (p, q) is on $y = mx + n$, then $p + q$

$= (m + 1)p + n$. If m is odd, then $(m + 1)p$ is even and the parity depends only on n. But if m is even, then $p + q$ can be odd or even; both colors occur. **51.** Call the points P, Q. Then $\overline{PQ}$ and $y = x$ are perpendicular because their slopes are -1 and 1. Also, the midpoint of $\overline{PQ}$ is $(\frac{1}{2}(a + b), \frac{1}{2}(a + b))$, which lies on $y = x$.

Section 5.3, page 135

1. (a) 2 (b) -4 (c) $-.55$ (d) -37 **3.** (a) 1/5 (b) $-1/3$ (c) 4 (d) 1000
5. (a) 0 (b) .34 (c) 5/27 (d) 67   **7.** $4x + 3$ **9.** $\dfrac{1}{x - 5}$ **11.** $x \neq 1$
13. $x \neq 1, 3$ **15.** $|x| \leq 2$ **17.** $x > -5$ **19.** $0 \leq x \leq 6$
21. $0 \leq x \leq 9$ **23.** $f(x) = 3600x$ **25.** $f(x) = x\sqrt{2}$
27. $f(x) = x(1.06) + (10{,}000 - x)(1.09)$ **29.** (a) $h(3)$ (b) $h(2) - h(1)$ (c) $\frac{1}{5}h(5)$
31. The value doubles every eight years.   **33.** The oil level is decreasing $\frac{1}{3}$ ft/min.   **35.** $f(x - 1) = 8x - 3$, $f(-x) = -8x + 5$, $f(2x) = 16x + 5$, $f(x^2) = 8x^2 + 5$ **37.** $f(x - 1) = \dfrac{x - 1}{x}$, $f(-x) = \dfrac{-x}{1 - x} = \dfrac{x}{x - 1}$, $f(2x) = \dfrac{2x}{2x + 1}$, $f(x^2) = \dfrac{x^2}{x^2 + 1}$

39.

x	-5	-4	-3	-2	-1	0	1	2	3	4	5
$f(-x)$	0	0	4	7	7	6	9	8	1	0	0

41. $f(2) = 1$, $f(-5) = -1$, $f(2x) = \begin{cases} 1 \text{ if } x > 0 \\ 0 \text{ if } x = 0 \\ -1 \text{ if } x < 0 \end{cases} = f(x)$ **43.** (a), (b)

45. $f(x_1 x_2) = (x_1 x_2)^n = x_1^n x_2^n = f(x_1)f(x_2)$. A function that does not satisfy the equation is $f(x) = x + 1$.

Section 5.4, page 143

1.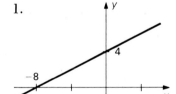

3.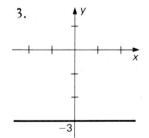

5.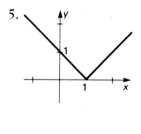

7.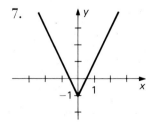

9.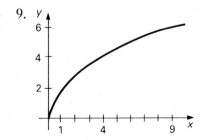

11.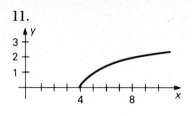
13.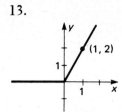
15.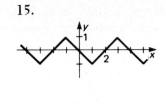

17.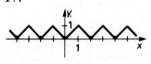
19.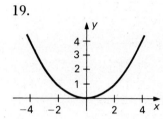
21.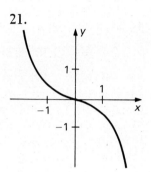

23.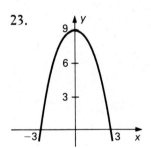
25.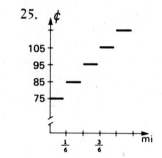
27.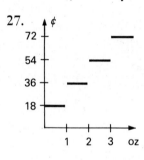

29.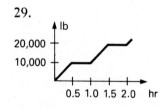
31.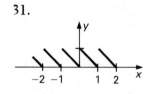

33. Odd   35. Neither   37. Odd   39. Neither   41. Even
43. Even   45. $f(-x) = -x|-x| = -x|x| = -f(x)$. See figure.

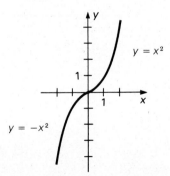

Section 5.5, page 149

1. $4x - 3$, $3x^2 - 11x - 4$, all x; $\dfrac{3x+1}{x-4}$, $x \neq -4$ 3. $2x^2 + 3x - 1$, $x^4 + 3x^3 - x^2 - 3x$, all x; $\dfrac{x^2 + 3x}{x^2 - 1}$, $x \neq \pm 1$ 5. The domains of f and g do not intersect. Hence $f + g$, fg, f/g are not defined. 7. $x^2 + 3 + 1/x$, $x + 3/x$, $x^3 + 3x$, $x \neq 0$ 9. $3x - 5$, $3x - 1$, all x 11. $(2x - 2)^2$, $2(x - 1)^2 - 1$, all x 13. $\sqrt{x^2 - 9}$ for $-3 \leq x \leq 3$; $x - 9$ for $x \geq 0$ 15. x, x, all x 17. $[f \circ g](x) = \dfrac{-11x + 10}{-10x + 9}$, $x \neq \frac{9}{10}, \frac{5}{6}$; $[g \circ f](x) = \dfrac{9x + 10}{-10x - 11}$, $x \neq -\frac{11}{10}, -\frac{5}{4}$ 19. $g(x)$ 21. They give $[f \circ g](x) = x$ and $[g \circ f](x) = x$, where $f(x) = \sqrt{x}$ and $g(x) = x^2$. The domain of $[f \circ g](x)$ is all x, but the domain of $[g \circ f](x)$ is $x \geq 0$. 23. $g(y) = 2y - 12$ 25. $g(y) = \frac{1}{2}y^{1/3}$ 27. $g(y) = (y - 1)^{1/4}$ 29. $g(y) = y^2 - 3$ 31. $g(y) = y^{1/3} + 2$ 33. $g(y) = \begin{cases} \frac{1}{2}y & y \geq 0 \\ y & y < 0 \end{cases}$ 35. $h(x) = 2.5x + 30$, $x(h) = \dfrac{h - 30}{2.5}$ 37. $A(s) = 6s^2$, $s(A) = (\frac{1}{6}A)^{1/2}$ 39. Pressure of y lb/in² is found at $x = g(y)$ ft above sea level. 41. My blood pressure B is the result of grading $n = g(B)$ papers.

43.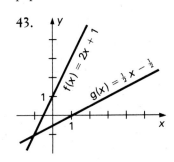

45.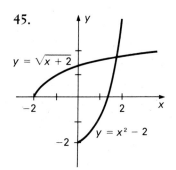

47.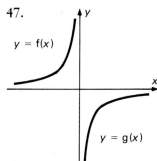

49. If $x_1 < x_2$, then $[f + g](x_1) = f(x_1) + g(x_1) < f(x_2) + g(x_2) = [f + g](x_2)$.

Review Exercises, page 151

1. $\frac{5}{3}$ 3. $y = \frac{2}{3}x + \frac{14}{3}$ 5. $y = -2x + 15$ 7. $y = -\frac{2}{11}x + \frac{19}{22}$ 9. 70, $3x^2 + x$, $(3 - t)/t^2$, $6x + 3h - 1$ 11. (a) Neither (b) odd (c) even

13. 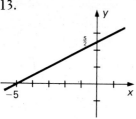   15. 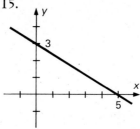   17.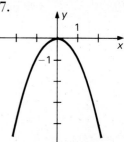

19. $(1/x) - 4$ for $x \neq 0$; $1/(x-4)$ for $x \neq 4$ 21. $g(y) = \frac{4}{3}y + \frac{32}{3}$, all y

CHAPTER 6

Section 6.1, page 158

1. $f(x) = \frac{5}{2}x$ 3. $f(x) = -\frac{1}{2}x + 11$ 5. $f(x) = .06x + 1.38$ 7. 7
9. $r = -\frac{1}{64}t + \frac{3}{8}$ 11. 26,400   13. $p(d) = (14.7/32.6)d + 14.7$
15. $b = (33/40)g + 5/2$

17. 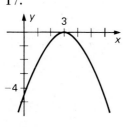   19. 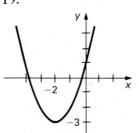   21. $y = (x-2)^2 - 3$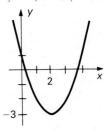

23. $y = -(x - \frac{5}{2})^2 + \frac{33}{4}$ 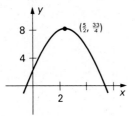   25. $y = 2(x-2)^2 - 13$ 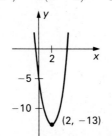   27. $y = -2(x - \frac{1}{4})^2 + \frac{1}{8}$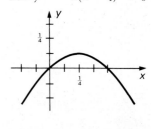

29. $y = (x - \frac{1}{2})^2 - \frac{9}{4}$

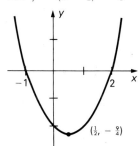

31. $y = (x - \frac{1}{2})^2$

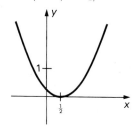

33. 1250 ft²   **35.** 2500/3 ft², length 50, width 50/3 ft
37. $A = (\frac{1}{4}x)^2 + [\frac{1}{4}(40 - x)]^2 = \frac{1}{8}(x - 20)^2 + 50 \geq 50$ **39.** 9

Section 6.2, page 164

1. 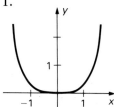   **3.** 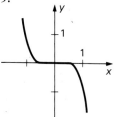   **5.**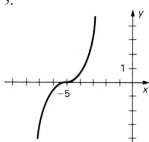

7. 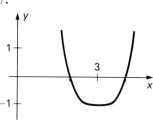   **9.**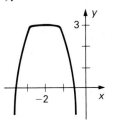

11. $x < 0$, $1 < x < 2$ **13.** $x < 0$, $1 < x < 4$, $x > 9$ **15.** $x \leq 1$, $x \geq 4$
17. $|x| > 3$ **19.** $-3 < x < 0$, $x \neq -2$

21. 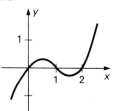   **23.** 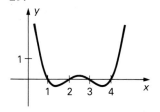   **25.**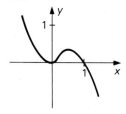

346 Answers to Odd-Numbered Exercises

27.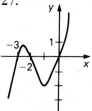
29.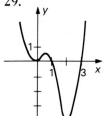
31.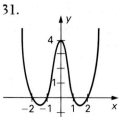

33.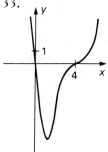
35.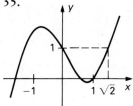

37. If $a > 0$, then $y \to \infty$ as $x \to \infty$, and $y \to -\infty$ as $x \to -\infty$. Hence the graph must cross the x-axis. Similarly if $a < 0$.   39. $y = -\frac{3}{8}(x - 1)(x - 2)(x - 4)$

Section 6.3, page page 170

1.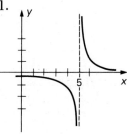
3.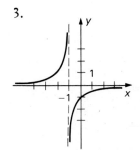
5.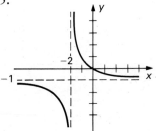

7.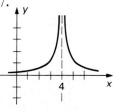
9.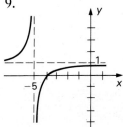
11.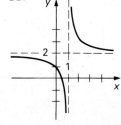

13.

15.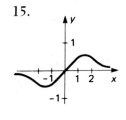

17.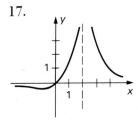

19.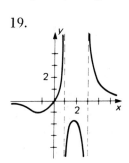

21.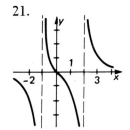

23.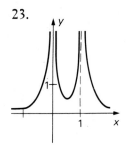

25.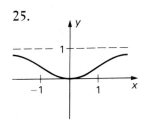

27. $q(x) = \frac{1}{5}(x + 1)(x - 1)$

Section 6.4, page 175

1. $y = \frac{5}{3}x$ **3.** $A = \pi r^2$ **5.** $C = \pi d$ **7.** $y = 3.5/x$ **9.** $z = \dfrac{xy^2}{8}$
11. Decreases by a factor of 4; decreases by a factor of 9. **13.** R is cut to 3/16 its original value. **15.** 20.25 gal **17.** $8\frac{1}{3}$ lb **19.** 53 students
21. 115.2 lb **23.** 5.859 in **25.** \$653.33 **27.** $3\sqrt{2}$ ft **29.** If the circle has radius r and the square has side s, then $s = r\sqrt{2}$ and Area $= s^2 = 2r^2$. **31.** False, y varies directly with x.

Section 6.5, page 184

1. Center: (2, 0); Radius: 2 **3.** Center: (−3, −1); Radius: 2
5. Center: (2, 1); Radius: 2 **7.** Center: $(\frac{1}{4}, -\frac{1}{4})$; Radius: $\frac{1}{4}\sqrt{2}$
9. $(x - 2)^2 + (y - 2)^2 = -1$, impossible **11.** $x^2 + y^2 = 25$
13. $(x - 5)^2 + (y - 5)^2 = 25$

15.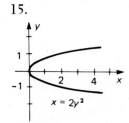

17.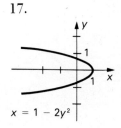

19.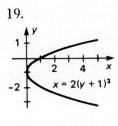

21.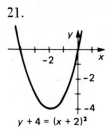

23.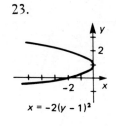

25.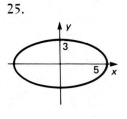

27.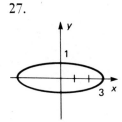

29.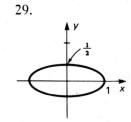

31.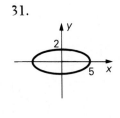

33.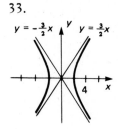

35.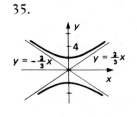

37.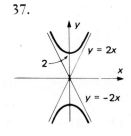

39.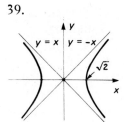

41. $x - 2 = -\frac{1}{18}y^2$ **43.** $\frac{1}{9}x^2 + \frac{1}{25}y^2 = 1$ **45.** Dist^2 from $(1, 0)$ is $(x - 1)^2 + y^2$. Dist^2 from $x = -1$ is $(x + 1)^2$. These are equal because $y^2 = 4x$, so $(x - 1)^2 + y^2 = (x - 1)^2 + 4x = x^2 - 2x + 1 + 4x = (x + 1)^2$.   **47.** The second circle is $(x - 8)^2 + (y - 1)^2 = 6$. The distance between the centers is $\sqrt{65}$, which is greater than the sum of the radii, $3 + \sqrt{6}$.

Answers to Odd-Numbered Exercises

Review Exercises, page 185

1. $y = \frac{3}{2}x + \frac{5}{2}$ 3. $y = \frac{1}{75}x + \frac{400}{3}$

5. 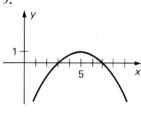 7. $y = (x + 3)^2 - 8$ 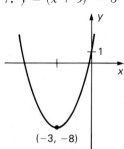 9. $y = x(x - 2)^2$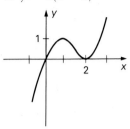

11. $x^2 + (y + 4)^2 = 4^2$ 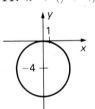 13. 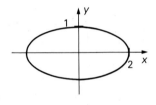 15.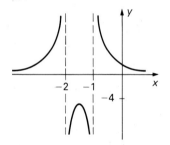

17. $(x - 2)^2 + (y - 5)^2 = 4$ 19. 120 lb

CHAPTER 7

Section 7.1, page 192

1. 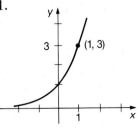 3. 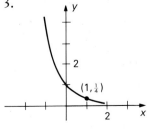 5.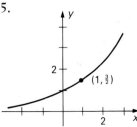

7. 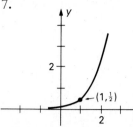 9. 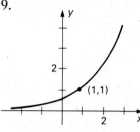 11.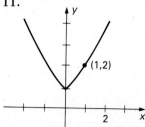

13. 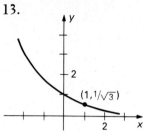 15.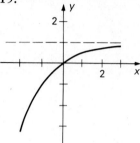

17. $32 = 2^5 < 2^{5.41} < 2^6 = 64$ 19. $27 = 9^{1.5} < 9^{1.66} < 9^2 = 81$
21. $0.04 = 5^{-2} < 5^{-1.3} < 5^{-1} = 0.20$ 23. $M(0) = 10$, $M(t + 5568)$
$= 10 \cdot 2^{-(t+5568)/5568} = 10 \cdot 2^{-t/5568} \cdot 2^{-1} = M(t)/2$ 25. 11,136 years
27. 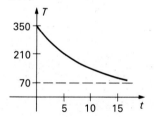   29.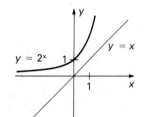

31. $2^{60} = (2^{10})^6 \approx (10^3)^6 = 10^{18}$
33.

x	5	10	20	30	50
2^x	32	1024	1,048,576	1.07×10^9	1.13×10^{15}
3^x	243	59,049	3.49×10^9	2.06×10^{14}	7.18×10^{23}
5^x	3125	9,765,625	9.54×10^{13}	9.31×10^{20}	8.88×10^{34}

35. 5.553×10^8 37. 652   39. 16,096

Section 7.2, page 197

1. $\log_2 1024 = 10$ 3. $\log_{10} 0.01 = -2$ 5. $\log_8 \frac{1}{2} = -\frac{1}{3}$
7. $10^4 = 10,000$ 9. $4^{2.5} = 32$ 11. $3^{-4.5} = 1/81\sqrt{3}$ 13. 8   15. 1

Answers to Odd-Numbered Exercises

17. -3 19. $9/4$ 21. $-1/9$ 23. 12 25. 30 27. 9
29. 125 31. $1/4$ 33. 1.4582 35. 1.5876 37. -1.0390
39. $\log_b 48$ 41. $\log_b(3\sqrt{5}/\sqrt{7})$ 43. $\log_b(r/v)$ 45. 1 47. $1/4$
49. 76 51. $-5/4$ 53. $\pm\sqrt{19}$ 55. 2 57. $x > 125$
59. $0.01 \le x \le 0.1$ 61. 6 63. 3 65. $7/13$ 67. $1/64$

69. 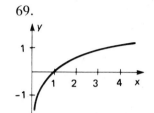   71. 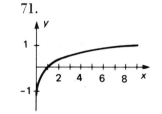   73.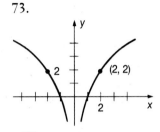

75. $x = 10, 100, 10^{10}$ 77.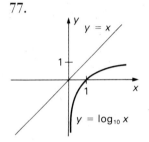

79. By Rule (1), $\log_b(r/s) = \log_b r \cdot \log_b(1/s) = \log_b r - \log_b s$

Section 7.3, page 202

1. 3.32 3. 5.66 5. 1.11 7. 88.7 9. 17.9 11. 9.07
13. 3.32 15. 3.21 17. 0.935 19. -1.46 21. $0, 3.24$ 23. 0
25. 0.802 27. No real solutions   29. 0.565 31. 0.631
33. (a) $1628.89 (b) $1638.62   35. (a) (approx.) $35,235
(b) $4,864,106 (c) $640,534,499   37. $(1.08)^9 = 1.999$; $(1.09)^8 = 1.993$;
$(1.12)^6 = 1.974$ 39. 82 months   41. 13.44% 43. No effect, you
get back the same number.   45. Our calculator will take only numbers less
than 10^{100}. Hence $10^{10^x} < 10^{100} = 10^{10^2}$, so $10^x < 10^2$, $x < 2$. Hence x is the
largest number less than 2 that the calculator can show, e.g., 1.9999999.

Section 7.4, page 207

1. 3.09 3. 1.93 5. 0.237 7. 65 9. 0.434 11. 2.10
13. $-\frac{1}{4}\ln(y/A)$ 15. $\pm\sqrt{-\ln y}$ 17. $\frac{1}{3}(e^y - 5)$ 19. $\ln(y \pm \sqrt{y^2 - 1})$
21. $-\frac{1}{k}\ln\left(\dfrac{a - by}{cy}\right)$ 23. 0.827 25. 8.29 27. -4.76 29. $\ln x$
31. $\log_6 x$ 33. 21.5 35. $(\ln 10)(\log e) = \left(\dfrac{\log 10}{\log e}\right)(\log e) = \log 10 = 1$
37. 1.89 min   39. In the year 2016   41. 8% quarterly   43. 10.99%
45. $12,368$

Answers to Odd-Numbered Exercises

Section 7.5, page 213

1. $P(t) = 10,000(3)^{t/12}$ 3. $5000(1.02)^{-t}$ or $P(t) = 5000(0.98)^t$ 5. 101.92 days 7. 1.16% 9. 10.8 years 11. 2.187 days 13. $M(t) = M_0 e^{(t \ln 2)/5} \approx M_0 e^{(.139)t}$ 15. Let $D(t) = T(t) - B$. Then $D(t) = D_0 a^{-t}$, where $D_0 = D(0) = T_0 - B$. As t increases, $a^{-t} \to 0$, so $T(t) - B \to 0$. Hence $T(t) \to B$, the "ultimate" temperature. 17. 2.51 times stronger 19. 3.01 decibels 21. (a) 0 (b) 5 23. 1.318×10^{10} brighter 25. $5,805.43 27. 139 29. $49.72 31. $88.12 33. $37,978.62 35. 165 months (approx.)

Review Exercises, page 215

1. (a) 16/3 (b) -3 (c) 8 (d) 1

3.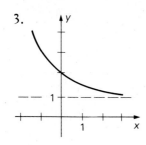

5.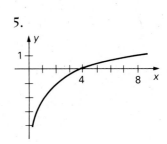

7. $f(x) = k(3/2)^x$ 9. 2/3 11. 0.213 13. $x = \frac{1}{3}(e^{2y} + 2)$
15. (a) $10,007.99, (b) $10,048.31 17. $46,271.37 19. 62.1 days

CHAPTER 8

Section 8.1, page 221

1. (4, 1) 3. (19/5, 7/5) 5. (10/29, 7/29) 7. (9/2, -3)
9. (5.4, 0.8) 11. (0, 0) 13. No solution 15. (2 + 3t/2, t)
17. (-1, 1/2) 19. (25, 4)

21. No points of intersection 23. $(2/\sqrt{37}, 6/\sqrt{37}), (-2/\sqrt{37}, -6/\sqrt{37})$

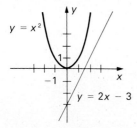

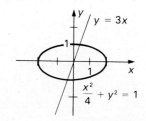

25. $(-2, 0)$, $(1, 3)$ **27.** $(-2, 0)$ **29.** $(3, 0)$, $(-5, -20/3)$

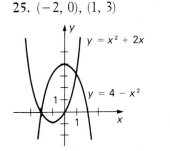
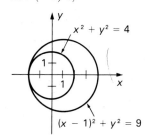

31. $(4, -3)$, $(-3, 4)$

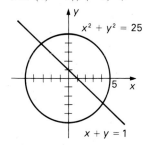

33. $c \geq -13/4$ **35.** Jogging 8 min, walking 18 min **37.** Boat 8 mph, current 1 mph **39.** Burger \$1.28, fries \$0.52 **41.** 7 wins, 9 losses **43.** Copper, 8.9 gm/cc; zinc, 6.9 gm/cc

Section 8.2, page 227

1. $(1, 1, 3)$ **3.** $(1, 0, -4)$ **5.** $(3/14, -1/4, -3/28)$ **7.** $(0, 0, 0)$
9. $(-5/2, -3/2, 9/2)$ **11.** $(-1, 2, 4)$ **13.** $(1, 1, 1)$
15. $(1.0, 2.1, -3.2)$ **17.** $(3, -1, -4, 0)$ **19.** $y = \frac{3}{2}x^2 + \frac{1}{2}x$
21. $y = -\frac{11}{8}x^2 + \frac{5}{4}x + 3$ **23.** $(\frac{13}{3} - \frac{5}{3}t, -\frac{5}{3} + \frac{4}{3}t, t)$
25. $(t, -\frac{1}{2}t - \frac{1}{2}, -\frac{5}{2}t + \frac{1}{2})$ **27.** No solution **29.** 31, 29, 9 **31.** 150 cheap seats, 550 medium-priced seats, 300 expensive seats **33.** Let s, r, c denote minutes of swimming, running, and calisthenics. Choose any s satisfying $10 \leq s \leq 35$, then choose $r = 70 - 2s$ and $c = s - 10$.

Section 8.3, page 231

23. $(-\frac{1}{5}t, \frac{2}{5}t, t)$ **25.** No solution **27.** $(1, 1, -2, -3, 2)$
29. $y = \frac{1}{2}x^3 - \frac{3}{2}x^2 + 1$

Section 8.4, page 236

1. 7 **3.** -13 **5.** 0 **7.** ab **9.** 1
21. $-3 \begin{vmatrix} -2 & 1 \\ 2 & 1 \end{vmatrix} + 0 + 3 \begin{vmatrix} 4 & -2 \\ 5 & 2 \end{vmatrix} = -3(-4) + 3(18) = 66$
$-4 \begin{vmatrix} 0 & 3 \\ 2 & 1 \end{vmatrix} - 2 \begin{vmatrix} -3 & 3 \\ 5 & 1 \end{vmatrix} - 1 \begin{vmatrix} -3 & 0 \\ 5 & 2 \end{vmatrix} = -4(-6) - 2(-18) - (-6) = 66$

354 Answers to Odd-Numbered Exercises

23. 24   **25.** -16 **27.** $(-16/9, -24/9, -20/9)$
29. $(16/7, -5/7, 13/7)$ **31.** $\begin{vmatrix} a & b \\ c & d \end{vmatrix} = ad - bc = -(bc - ad) = -\begin{vmatrix} c & d \\ a & b \end{vmatrix}$
33. $\begin{vmatrix} a & b \\ c + ka & d + kb \end{vmatrix} = (ad + kab) - (bc + kab) = ad - bc = \begin{vmatrix} a & b \\ c & d \end{vmatrix}$
35. Expand by the first column   **37.** Expand by the first column:
$a \begin{vmatrix} b & c \\ y & z \end{vmatrix} - a \begin{vmatrix} b & c \\ y & z \end{vmatrix} + x \begin{vmatrix} b & c \\ b & c \end{vmatrix} = 0 + x(bc - bc) = 0$ **39.** -4

Section 8.5, page 241

1. 12   **3.** -6 **5.** -65 **7.** 0   **9.** -43 **11.** $x^3 + 2x$
13. 116   **15.** -1 **17.** -6 **19.** 240   **31.** False. Doubling one row doubles the determinant. Doubling n rows multiplies it by 2^n.   **33.** 20
35. $\begin{vmatrix} a & b & c \\ d & e & f \\ g & h & i \end{vmatrix} = aei + bfg + cdh - (ceg + bdi + ahf) = -\begin{vmatrix} d & e & f \\ a & b & c \\ g & h & i \end{vmatrix}$ and similarly for any other pair of rows or columns.   **37.** 1, 2   **39.** Subtract Row 2 from Rows 1 and 3, then expand: $(x - x_1)(y_2 - y_1) - (x_2 - x_1)(y - y_1) = 0$, or $\dfrac{y - y_1}{x - x_1} = \dfrac{y_2 - y_1}{x_2 - x_1}$, the point-slope form of the line through (x_1, y_1) and (x_2, y_2).

Section 8.6, page 247

1.
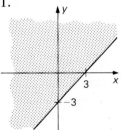

3.
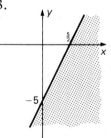

5.
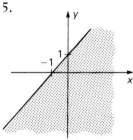

7.
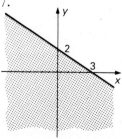

9.
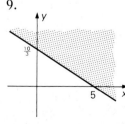

11.
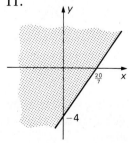

13. 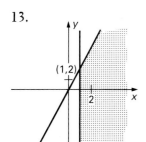   15. 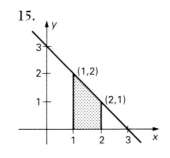   17.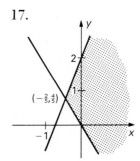

19. 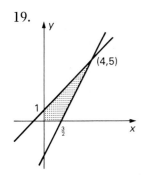   21. 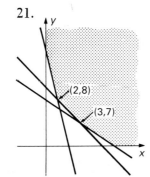   23.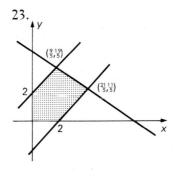

25. Max = 24 at (0, 4), Min = 0 at (0, 0)   27. Max = 80 at (8, 0), Min = 5 at (2, 5)   29. Max = 50 at (10, 0), Min = -30 at (6, 6)   31. Max = 23/7 at (15/7, 8/7), Min = 1 at (1, 0)   33. 300 type A and 300 type B   35. 20,000 barrels of gas and 10,000 barrels of oil   37. 60 units of P and 160 units of Q   39. 20 oz of F and 40 oz of G

Review Exercises, page 249

1. (18/13, 19/13)   3. ($-26/60$, $-79/60$, $35/60$)   5. No solutions
7. (2, 3, -1, -1)   9. 251   11. -149 13. 82   15. (67, 73, 91)
17.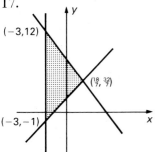

19. If $x = 1$, columns 1 and 3 are proportional; if $x = 2$, columns 2 and 3 are proportional. The equation is quadratic, so these two solutions are the only ones.

CHAPTER 9

Section 9.1, page 256

1. $q(x) = x + 1, r(x) = 8$ 3. $q(x) = 2x^2 - 4x + 14, r(x) = -29$
5. $q(x) = x^3 - 3x^2 + 9x - 35, r(x) = 107$ 7. $q(x) = x^3 + 4x^2 + 4x + 7, r(x) = 13$ 9. $q(x) = 2x^4 - x^3 + 2x^2 - 7x - 1, r(x) = 3$ 11. $q(x) = 2x, r(x) = -7x + 1$ 13. $q(x) = x^2 + 3x + 7, r(x) = x - 17$
15. $q(x) = x^3 - 2x^2 + 4x - 12, r(x) = 24x + 9$
17. $(x - 3)(x^2 - 4x - 10) - 29$ 19. $(x - 3)(x^3 + 3x^2 + 9x + 27) + 81$
21. -47 23. -1248 25. 127 27. 3 29. 64 31. 28
33. 0 35. $f(1) = 0$ 37. $f(a) = 0$ 39. $(x - 3)(x - 2)(x - 1)$
41. $(x + 3)(x + 5)(x - 3)(x - 1)$ 43. $1, \frac{1}{2}(-3 \pm \sqrt{13})$
45. $\pm 1, \frac{1}{2}(-3 \pm \sqrt{13})$

Section 9.2, page 259

1. $[(2r + 4)r + 6]r + 1$ 3. $\{[(r + 2)r - 6]r - 8\}r + 9$
5. $r^2(2r^3 + 1) - 6$ 7. $f(1.4) = 24.344, f(-32) = -27,870$
9. $f(167) = 1,316,982.8, f(5.6) = -160.2488$ 11. $f(7) = 2,117,724, f(-3.8) = 180,676.262$ 13. $f(0.3) = -2.67357, f(26) = 11,905,703$
15. $f(4) = 86,691, f(5) = 494,283$ 17. 3
19. $(\ r\ \boxed{STO}\ \boxed{-}\ a_1\ )\ \boxed{\times}\ (\ \boxed{RCL}\ \boxed{-}\ a_2\ )\ \ldots\ \boxed{\times}\ (\ \boxed{RCL}\ \boxed{-}\ a_n\ )\ \boxed{=}$

Section 9.3, page 263

1. $0, 1, 5$; mult 1 3. $1, -1$; mult 1 5. -2, mult 1; 3, mult 2 7. -1, mult 3; -2, mult 4 9. -2, mult 1; -1, mult 3; 1, mult 4 11. $-2, 2$; mult 2 13. $-2(x + 1)(x - 3)$ 15. $\frac{17}{100}x(x - 5)(x - 8)$ 17. -1
19. $-4, -3, 1$ 21. -1 23. No rational zeros 25. $-1, \frac{1}{2}$
27. -1 29. $x^3 - 2 = 0$ has no rational roots 31. $3x^4 - 2 = 0$ has no rational roots 33. $a_7x^7 + \cdots + a_0 = bx + c$ is a 7-th degree equation having at most 7 real roots. 35. $ax^2 + bx + c = dx + e$ is a quadratic equation with at most 2 real roots.

Section 9.4, page 267

1. $8 + 5i$ 3. $3 + 6i$ 5. $19 + 7i$ 7. $-21 + 20i$ 9. 100
11. -4 13. -20 15. i 17. $\frac{2}{29} - \frac{5}{29}i$ 19. $\frac{4}{25} + \frac{3}{25}i$
21. $-\frac{9}{5} - \frac{13}{5}i$ 23. $\frac{22}{533} - \frac{7}{533}i$ 25. $-\frac{37}{145} - \frac{9}{145}i$ 27. $-\frac{57}{325} + \frac{1}{325}i$
29. $\frac{82}{85} + \frac{39}{85}i$ 31. $f(i) = 0, f(1 + i) = 2 + 9i$ 33. $[i(3 + 2i)](1 - i) = (-2 + 3i)(1 - i) = 1 + 5i; i[(3 + 2i)(1 - i)] = i(5 - i) = 1 + 5i$
35. $(a + bi) + (c + di) = (a + c) + (b + d)i = (c + a) + (d + b)i = (c + di) + (a + bi)$ 37. $(a + bi)[(c + di) + (e + fi)] = (a + bi)[(c + e) + (d + f)i] = [a(c + e) - b(d + f)] + [a(d + f) + b(c + e)]i = [(ac - bd) + (ae - bf)] + [(ad + bc) + (af + be)]i$

$= [(ac - bd) + (ad + bc)i] + [(ae - bf) + (af + be)i]$
$= (a + bi)(c + di) + (a + bi)(e + fi)$ 39. $\overline{(a + bi) + (c + di)}$
$= \overline{(a + c) + (b + d)i} = (a + c) - (b + d)i = (a - bi) + (c - di)$
$= \overline{(a + bi)} + \overline{(c + di)}$ 41. $\overline{3(a + bi) + i} = \overline{3a + (3b + 1)i} = 3a - (3b + 1)i$
$= 3(a - bi) - i = 3\overline{(a + bi)} - i$ 43. $(\sqrt{3} + i)^3$
$= 3\sqrt{3} + 9i + 3\sqrt{3}i^2 + i^3 = 8i$; since $(\bar{z})^3 = \overline{z^3}$, $(\sqrt{3} - i)^3 = -8i$
45. $\frac{32}{53} - \frac{47}{53}i$ 47. $z = \pm(3 - i)$

Section 9.5, page 273

1. $z^2 - 2z + 10$ 3. $(z - 2)(z^2 - 6z + 10) = z^3 - 8z^2 + 22z - 20$
5. $(z^2 - 2z + 2)(z^2 - 4z + 5) = z^4 - 6z^3 + 15z^2 - 18z + 10$
7. $(z^2 - 2z + 10)^2 = z^4 - 4z^3 + 24z^2 - 40z + 100$ 9. $(z + i\sqrt{7})(z - i\sqrt{7})$
11. $(z + \frac{1}{2} - \frac{1}{2}i\sqrt{11})(z + \frac{1}{2} + \frac{1}{2}i\sqrt{11})$
13. $(z - \sqrt{3})(z + \sqrt{3})(z - \sqrt{2}i)(z + \sqrt{2}i)$
15. $zz(z - 2)(z + 2)(z - 2i)(z + 2i)$ 17. $(z - 3)(z^2 + 3z + 9)$
19. $(z^3 - 1)(z^3 - 8) = (z - 1)(z^2 + z + 1)(z - 2)(z^2 + 2z + 4)$ 21. $2 - i$,
$2 + i, -1$ 23. $i, 3$ 25. $i, -i, 2, 3$ 27. $-i, i, 1 + 2i, 1 - 2i, -1$
29. It is divisible by $(z - i)(z + i)(z - 1)(z + 1) = (z^2 + 1)(z^2 - 1) = z^4 - 1$
31. No; the coefficients are not all real.   33. False; it may have three
real zeros.   35. True; it has four zeros and non-real zeros come in pairs.
37. $f(z) = (z - r_1)^{2k_1} \cdots (z - r_n)^{2k_n} = [(z - r_1)^{k_1} \cdots (z - r_n)^{k_n}]^2 = [g(z)]^2$
39. $f(\bar{z}) = a_n \bar{z}^n + \cdots + a_1 \bar{z} + a_0 = \overline{a_n} \overline{z^n} + \cdots + \overline{a_1} \bar{z} + \overline{a_0}$
$= \overline{a_n z^n} + \cdots + \overline{a_1 z} + \overline{a_0} = \overline{a_n z^n + \cdots + a_1 z + a_0} = \overline{f(z)}$

Section 9.6, page 278

1. Yes, since $x = x^3 + 2$ is an equation of odd degree.   3. $f(0) = -1$,
$f(1) = 4$, opposite signs   5. $f(-1) = 4, f(0) = -3$, opposite signs
7. $f(0.6) = -0.184, f(0.7) = 0.043$, opposite signs   9. $(1, 2)$
11. $(-3, -2)$ 13. $(0, 1)$ 15. 1.613 17. 1.433 19. 2.480
21. -5.03938 23. -1.05812 25. -1.36170 27. 3.425
29. 1.709 31. 0.567 33. If $x > 0$, then each term $a_r x^r > 0$; hence
$f(x) > 0$.   35. $f(x) \to \infty$ as $x \to \infty$; hence $f(b) > 0$ for some $b > 0$. Thus $f(x)$
has a real zero between 0 and b. Similarly, $f(x) \to \infty$ as $x \to -\infty$; hence $f(c) > 0$
for some $c < 0$. Thus $f(x)$ has another real zero between c and 0.

Review Exercises, page 279

1. 74 3. $-3/2$ 5. $\frac{2}{85} + \frac{26}{85}i$
7. $(x - 2)(x^4 + 2x^3 + 4x^2 + 10x + 14) + 31$ 9. only 0   11. 3, mult 1;
-1, mult 2; 1, mult 3   13. $-1, \frac{1}{2}(-1 \pm i\sqrt{19})$ 15. between 5 and 6
17. For $x \neq 0$, the equation is equivalent to $x^7 + 6x^4 = x^2 + 1$ (odd degree).   19. $f(0) = -1$. As $x \to +\infty, f(x) \to +\infty$; and as $x \to -\infty, f(x) \to +\infty$.
Therefore there must be at least two roots, one positive and one negative.

CHAPTER 10

Section 10.1, page 283

1. 4, 8, 12, 16, 20 3. $\frac{9}{2}, 4, \frac{7}{2}, 3, \frac{5}{2}$ 5. $\frac{1}{4}, \frac{2}{5}, \frac{3}{6}, \frac{4}{7}, \frac{5}{8}$ 7. 0, 2, 0, 2, 0
9. 3, 6, 12, 24, 48 11. $\sqrt{3}, 4, 3\sqrt{5}, 4\sqrt{6}, 5\sqrt{7}$ 13. 10 15. 72
17. 10 19. 29/12 21. -18 23. 654,321 25. $a_n = 8n$
27. $a_n = n\sqrt{2n-1}$ 29. $a_n = 1/2^{n+2}$ 31. $a_n = (-1)^n \dfrac{n+4}{5n+1}$
33. $a_n = (n+1)(n+2)$ 35. $\sum_{i=4}^{8} 2^i$ 37. $\sum_{i=1}^{15} \dfrac{i}{i+2}$ 39. $\sum_{i=0}^{4} (-1)^i 2^{-i}$
41. $\sum_{i=0}^{6} a_i x^{2i}$ 43. $\sum_{i=101}^{200} (2i-1)$ 45. True 47. False; $\sum_{k=1}^{8} (1+k^3)$
$= 8 + \sum_{k=1}^{8} k^3$ 49. True 51. $\sum_{i=1}^{n} c a_i = c a_1 + c a_2 + \cdots + c a_n$
$= c(a_1 + a_2 + \cdots + a_n) = c \sum_{i=1}^{n} a_i$

Section 10.2, page 288

1. 37, $4 + (n-1)3$ 3. -0.4, $5.1 + (n-1)(-0.5)$ 5. $25 + 11\pi$,
$3 + (n-1)(2+\pi)$ 7. $12a + 45b$, $a + b + (n-1)(a+4b)$ 9. 110
11. 82 13. 13 15. 8, 5.5 17. 62.75 19. 488 21. $9(x+12)$
23. $\frac{1}{2}(3n^2 + n)$ 25. 235 27. 30,000 29. 41,583 31. 117
33. $110,000,000 35. 34 37. There are n^2 dots. Counting them in groups as shown gives $n^2 = 1 + 3 + 5 + \cdots + (2n-1)$. 39. The number of dots in the lower and upper triangles are consecutive triangular numbers. Their sum is a square. 41. $99 \cdot \frac{1}{2}(a_1 + a_{99}) = 99 \cdot \frac{1}{2}(b_1 + b_{99})$ if and only if $\frac{1}{2}(a_1 + a_{99}) = \frac{1}{2}(b_1 + b_{99})$, that is, $a_{50} = b_{50}$. 43. $a_n = a_1 + (n-1)d$, so $a_n = f(n)$, where $f(x) = a_1 + (x-1)d$.

Section 10.3, page 294

1. 729, 3^{n-1} 3. $\frac{1}{2}$, $32(\frac{1}{2})^{n-1}$ 5. 24, $3(\sqrt{2})^{n-1}$ 7. x^{13}, $(-1)^{n-1} x^{2n-1}$
9. $a^{19} b^6$, $a^{3n-2} b^{n-1}$. 11. 10, 50 13. $2\sqrt{2}$, 4 15. -3.5, 35
17. 1008 19. $-85/256$ 21. $15(1 + \sqrt{2})$ 23. 9 25. 1/6
27. $1/(1 - \frac{1}{2}x)$ 29. $29\frac{27}{32}$ ft 31. $\frac{1}{500}(2^{64} - 1)$ in^3 $\approx$ 145.05 mi^3, since $(63,360)^3$ in.^{3} = 1 mi^3 33. 43,690 35. 16/33 37. 301/330
39. 415/999 41. $\sqrt{a_n} = \sqrt{a_1 r^{n-1}} = \sqrt{a_1} (\sqrt{r})^{n-1}$, geometric
43. $a_n = a_{n-1} + d$, so $2^{a_n} = 2^{a_{n-1}} \cdot 2^d$, a geometric sequence with $r = 2^d$.
45. (16/7, 12/7) 47. 100 miles (one hour at 100 mph). A solution using geometric series is possible but much harder.

Section 10.4, page 300

1. True for $n = 1$; $2 = \frac{1}{2}(3 + 1)$. If true for some n, then
$2 + 5 + \cdots + [3(n + 1) - 1] = [2 + 5 + \cdots + (3n - 1)] + (3n + 2)$
$= \frac{1}{2}n(3n + 1) + (3n + 2) = \frac{1}{2}(3n^2 + 7n + 4) = \frac{1}{2}(n + 1)(3n + 4)$
$= \frac{1}{2}(n + 1)[3(n + 1) + 1]$, so true for $n + 1$, hence true for all $n \geq 1$.
3. True for $n = 1$; $9 = 1(4 \cdot 1 + 5)$. If true for some n, then $9 + 17$
$+ \cdots + (8(n + 1) + 1) = 9 + 17 + \cdots + (8n + 1) + (8(n + 1) + 1)$
$= n(4n + 5) + (8n + 9) = 4n^2 + 13n + 9 = (n + 1)(4n + 9)$
$= (n + 1)(4(n + 1) + 5)$, so true for $n + 1$. Hence true for all $n \geq 1$.
5. True for $n = 1$; $1 = \frac{1}{6}(2)(3)$. If true for some n, then
$1^2 + 2^2 + \cdots + n^2 + (n + 1)^2 = \frac{1}{6}n(n + 1)(2n + 1) + (n + 1)^2$
$= \frac{1}{6}(n + 1)[n(2n + 1) + 6(n + 1)] = \frac{1}{6}(n + 1)(2n^2 + 7n + 6)$
$= \frac{1}{6}(n + 1)(n + 2)(2n + 3) = \frac{1}{6}(n + 1)[(n + 1) + 1][2(n + 1) + 1]$, so true for
$n + 1$. Hence true for all $n \geq 1$. **7.** True for $n = 1$; $1/(1 \cdot 2) = \frac{1}{2}$. If true for
some n, then $\frac{1}{1 \cdot 2} + \frac{1}{2 \cdot 3} + \cdots + \frac{1}{n(n + 1)} + \frac{1}{(n + 1)(n + 2)}$
$= \frac{n}{n + 1} + \frac{1}{(n + 1)(n + 2)} = \frac{n^2 + 2n + 1}{(n + 1)(n + 2)} = \frac{(n + 1)(n + 1)}{(n + 1)(n + 2)} = \frac{n + 1}{n + 2}$, so true
for $n + 1$. Hence true for all $n \geq 1$. **9.** True for $n = 1$; $1 \cdot 3 = \frac{1}{6} \cdot 1(2)(9)$.
If true for some n, then $1 \cdot 3 + 2 \cdot 4 + \cdots + n(n + 2) + (n + 1)(n + 3)$
$= \frac{1}{6}n(n + 1)(2n + 7) + (n + 1)(n + 3) = \frac{1}{6}(n + 1)[n(2n + 7) + 6(n + 3)]$
$= \frac{1}{6}(n + 1)(2n^2 + 13n + 18) = \frac{1}{6}(n + 1)(n + 2)(2n + 9) =$ RHS for $n + 1$.
11. True for $n = 1$, $a = \frac{1}{2}[2a + 0d]$. If true for some n, then
$a + (a + d) + \cdots + [a + (n - 1)d] + [a + nd]$
$= \frac{1}{2}n[2a + (n - 1)d] + [a + nd] = (n + 1)a + [\frac{1}{2}n(n - 1) + n]d$
$= \frac{1}{2}(n + 1)2a + \frac{1}{2}n(n + 1)d = \frac{1}{2}(n + 1)(2a + nd) =$ RHS for $n + 1$.
13. True for 2 even integers. Assume true for n. Then the sum of $n + 1$ even
integers is $(a_1 + a_2 + \cdots + a_n) + a_{n+1} =$ even $+$ even $=$ even. **15.** True
for $n = 2$; $\log(a_1 a_2) = \log a_1 + \log a_2$. If true for some n, then
$\log(a_1 a_2 \cdots a_n a_{n+1}) = \log[(a_1 a_2 \cdots a_n) a_{n+1}] = \log(a_1 \cdots a_n) + \log a_{n+1}$
$= \log a_1 + \log a_2 + \cdots + \log a_n + \log a_{n+1} =$ RHS for $n + 1$. Hence true
for all $n \geq 2$. **17.** True for $n = 2$. To tie 2 pieces of string together requires
1 knot. If true for n, then $n + 1$ pieces require $n - 1$ knots for the first n
pieces and one knot to tie these to the $(n + 1)$st piece. So $n + 1$ pieces require
n knots. **19.** $c_n = 2^n - 1$. True for $n = 1$; $2^1 - 1 = 1$. Assume the formula
is true for n. Now add one more person P to a group of n people. There are
three kinds of committees: (1) P alone, (2) an old committee, (3) an old committee plus P. Thus, $c_{n+1} = 1 + c_n + c_n = 2c_n + 1 = 2(2^n - 1) + 1$
$= 2^{n+1} - 1$. So the formula is true for $n + 1$ and hence for all $n \geq 1$.
21. True for $n = 1$ and $n = 2$. If true for some $n \geq 2$, then
$(\frac{3}{2})^{n+1} = \frac{3}{2} \cdot (\frac{3}{2})^n > \frac{3}{2}n \geq n + 1$ (for $n \geq 2$); so true for $n + 1$. Therefore, true for
all $n \geq 1$. **23.** True for $n = 6$; $64 > 42$. If true for some $n \geq 6$, then
$2^{n+1} = 2 \cdot 2^n > 2 \cdot 7n = 14n$. But $14n \geq 7(n + 1)$ for $n \geq 1$, hence
$2^{n+1} \geq 7(n + 1)$; so true for $n + 1$, hence for all $n \geq 6$. **25.** True for $n = 5$;
$120 > 50$. If true for some $n \geq 5$, then $1 \cdot 2 \cdot 3 \cdots n \cdot (n + 1)$

$> (10n)(n + 1) > 10(n + 1)$. Hence true for all $n \geq 5$. 27. $\left(1 - \frac{1}{4}\right)\left(1 - \frac{1}{9}\right)$ $\cdots \left(1 - \frac{1}{n^2}\right) = \frac{n+1}{2n}$ for $n \geq 2$. True for $n = 2$; $\left(1 - \frac{1}{4}\right) = \frac{3}{4}$. If true for some $n \geq 2$, then LHS for $n + 1 = \left(1 - \frac{1}{4}\right)\left(1 - \frac{1}{9}\right)$ $\cdots \left(1 - \frac{1}{n^2}\right)\left(1 - \frac{1}{(n+1)^2}\right) = \frac{n+1}{2n}\left(1 - \frac{1}{(n+1)^2}\right)$ $= \frac{n+1}{2n} \cdot \frac{n^2 + 2n}{(n+1)^2} = \frac{n+2}{2(n+1)} =$ RHS for $n + 1$.   29. At every step, each side is replaced by a length $\frac{4}{3}$ times as long. Hence $P_{n+1} = \frac{4}{3}P_n$ with $P_1 = 3$. Assume $P_n = 3(\frac{4}{3})^{n-1}$, which works for $n = 1$. Then $P_{n+1} = \frac{4}{3}P_n = 3(\frac{4}{3})^n$.
31. True for $n = 1$; suppose true for some $n \geq 1$. Then $a_{n+1} = \sqrt{2 + a_n}$ $< \sqrt{2 + 2} = \sqrt{4} = 2$. So true for $n + 1$.   33. $a_n = n$. True for $n = 0, 1$, and 2. If true up to some n, then $a_{n+1} = 2a_n - a_{n-1} = 2n - (n-1) = n + 1$. So true up to $n + 1$.   35. $a_n = 2^{n-1}$ for $n \geq 1$. True for $n = 1$. If true for some $n \geq 1$, then $a_{n+1} = (a_0 + a_1 + \cdots + a_{n-1}) + a_n = a_n + a_n = 2a_n$ $= 2 \cdot 2^{n-1} = 2^n$. So true for $n + 1$.   37. True for $n = 1$. With three coins, put two on the balance. If one side is heavy, that's the heavy coin. If the sides balance, the third coin is heavy. Suppose true for 3^n. With 3^{n+1} coins, split them into three piles of 3^n and put two piles on the balance. With one weighing, you know which pile the heavy coin is in, and you are back to the preceding case.   39. Let S_n be the number of moves required for n rings. To move the bottom ring to peg 2, the other $n - 1$ rings must be on peg 3 in descending order, which takes S_{n-1} moves. Next, the bottom ring is moved (1 move), and then the remaining $n - 1$ rings are moved on top of it (S_{n-1} moves). Thus $S_n = 2S_{n-1} + 1$. Testing, $S_1 = 1$, $S_2 = 3$, $S_3 = 7$, $S_4 = 15$, $S_n = 2^n - 1$. If true for some n, then $S_{n+1} = 2S_n + 1$ $= 2(2^n - 1) + 1 = 2^{n+1} - 1$, true for $n + 1$.

Section 10.5, page 305

1. $\frac{13!}{8!}$ 3. $\frac{52!}{47!5!}$ 5. $\frac{n!}{3!}$ 7. 72   9. 165   11. 20,020
13. $n(n - 1)$ 15. 24   17. 676,000   19. 4536 (first digit is non-zero)
21. $9 \cdot 10^6$ 23. 8!   25. $\frac{12!}{7!} = 95,040$ 27. 2256
29. $2(5!)^2 = 28,800$ 31. $2(6!) = 1440$ 33. $(26!)^2$
35. $5!3!4! = 17,280$ 37. $\frac{10!}{2!3!3!1!1!} = 50,400$ 39. $\frac{8!}{2^4} = 2520$
41. $\frac{10!}{2^5} = 113,400$

Section 10.6, page 309

1. 105   3. 4495   5. 8568   7. $\frac{52!}{5!47!}$ 9. 35   11. $\frac{9!11!}{4!5!\,4!7!}$

Answers to Odd-Numbered Exercises   **361**

13. $\dfrac{12!}{2!7!3!}$ 15. 252   17. 210   19. $\dfrac{14!}{8!6!}$ 21. $\binom{13}{1}\binom{13}{2}\binom{13}{4}\binom{13}{6}$

23. $\binom{6}{2}\binom{8}{2} = 420$ 25. $\binom{9}{3}\binom{6}{2} = 1260$ 27. $\binom{n}{k} + \binom{n}{k-1}$

$= \dfrac{n!}{k!(n-k)!} + \dfrac{n!}{(k-1)!(n-k+1)!} = \dfrac{n!(n-k+1)}{k!(n-k+1)!} + \dfrac{n!k}{k!(n-k+1)!}$

$= \dfrac{n!(n-k+1+k)}{k!(n-k+1)!} = \dfrac{n!(n+1)}{k!(n-k+1)!} = \binom{n+1}{k}$ 29. In a group of n

people, $\binom{n}{2}$ handshakes are possible. Let $p_1, p_2, \cdots, p_n$ denote the n people. Then p_1 can shake $n-1$ hands, p_2 can shake $n-2$ hands (without repetition), etc. So $\binom{n}{2} = 1 + 2 + \cdots + (n-1)$.

Section 10.7,   page 313

1. $x^7 + 7x^6 + 21x^5 + 35x^4 + 35x^3 + 21x^2 + 7x + 1$
3. $x^5 + 10x^4 + 40x^3 + 80x^2 + 80x + 32$ 5. $16x^4 - 32x^3y + 24x^2y^2 - 8xy^3 + y^4$
7. $a^{12} - 6a^{10}b + 15a^8b^2 - 20a^6b^3 + 15a^4b^4 - 6a^2b^5 + b^6$
9. $x^{10} - 10x^8 + 45x^6 - 120x^4 + 210x^2 - 252 + \dfrac{210}{x^2} - \dfrac{120}{x^4} + \dfrac{45}{x^6} - \dfrac{10}{x^8} + \dfrac{1}{x^{10}}$
11. $32x^5 - 40x^4 + 20x^3 - 5x^2 + \tfrac{5}{8}x - \tfrac{1}{32}$ 13. $-35/27$ 15. 40   17. 448
19. 2016   21. 160   23. 220   25. $84x^3$ 27. $(100 + 2)^3$
$= (100)^3 + 3(100)^2 \cdot 2 + 3(100) \cdot 2^2 + 2^3 = 1{,}061{,}208$
29. $(100 - 1)^4 = 96{,}059{,}601$ 31. 1.082857   33. 0.904382
35. $(1 + .001)^{50} > 1 + \binom{50}{1} 1^{49}(.001)^1 = 1.05$ 37. $(10 + 1)^n = 11^n$

Review Exercises,   page 314

1. $a_n = 5 + 3n$ 3. 3220   5. $6[1 - (\tfrac{2}{3})^{n+1}]$ 7. 3/8   9. True for $n = 1$; $1^2 = \tfrac{1}{3} \cdot 1(2-1)(2+1)$. Suppose true for n. Then for $n + 1$, LHS
$= 1^2 + 3^2 + 5^2 + \cdots + (2n-1)^2 + (2(n+1) - 1)^2$
$= \tfrac{1}{3}n(2n-1)(2n+1) + (2n+1)^2 = \tfrac{1}{3}(2n+1)[n(2n-1) + 3(2n+1)]$
$= \tfrac{1}{3}(2n+1)(2n^2 + 5n + 3) = \tfrac{1}{3}(2n+1)(2n+3)(n+1) = $ RHS for $n + 1$.
11. True for a_1, a_2. Assume true up to n. Then $a_{n+1} = a_n + a_{n-1}$, the sum of two even integers, hence even.
13. $a^7 - 14a^6b + 84a^5b^2 - 280a^4b^3 + 560a^3b^4 - 672a^2b^5 + 448ab^6 - 128b^7$
15. $(4!)(5!) = 2880$ 17. $10 \cdot 9 \cdot 8 \cdot 7 \cdot 6 \cdot 5 = 151{,}200$ 19. 252

APPENDIX

Section 1,   page 320

(Some answers are rounded off.)
1. 990.2   3. 1,985,746   5. 7,746.07   7. 1,361.557   9. 20.388

11. 8,396.13   13. 2.32835   15. 0.56801   17. 7,971,615
19. 16,930.59   21. 2.71722   23. 0.39284   25. 1.57186
27. 2.99316   29. 0.37060   31. −17.57454   33. Add the numbers on the calculator, and then multiply by 1.06.

Section 2, page 323

1. 24,276.4   3. 17.911   5. 217.169   7. 11.124   9. 0.76389
11. 30.357   13. 2.9039   15. 28.311   17. 1.2603   19. 4,037,913
21. 9.7363   23. 174.50   25. 0.513304   27. 0.176127
29. a [M+] [x^2] [M+] [y^x] 1.5 [M+] [RCL]

Section 3, page 327

1. 1   3. 3   5. −2   7. −4   9. 4   11. 23   13. 0.3181
15. 3.2967   17. 2.9315   19. .8162 − 1   21. 5000   23. .2084 − 21
25. 9.041   27. 402.0   29. 2.360×10^{-4} 31. 1.862×10^5
33. 7.945   35. 6.421×10^{-3} 37. log 93.3 < 1.97 < log 93.4, so $93.3 < 10^{1.97} < 93.4$ 39. $6.31 \approx 10^{.8}$, so $(6.31)^5 \approx (10^{.8})^5 = 10^4 = 10,000$

Section 4, page 328

1. 1.57×10^3 3. 2.68×10^1 5. 7.34×10^{-1} 7. 7.01×10^{-2}
9. 3.54×10^1 11. 1.34   13. 1.25×10^1 15. 2.52   17. 3.22
19. 2.87   21. 1.65   23. 2.81   25. 1.63×10^{-3} 27. 2.52
29. 3.71×10^3 31. 2.917×10^5 33. 1.061×10^2 35. 6.680
37. 3.820   39. 4.876

TABLE OF COMMON LOGARITHMS

x	0	1	2	3	4	5	6	7	8	9
1.0	.0000	.0043	.0086	.0128	.0170	.0212	.0253	.0294	.0334	.0374
1.1	.0414	.0453	.0492	.0531	.0569	.0607	.0645	.0682	.0719	.0755
1.2	.0792	.0828	.0864	.0899	.0934	.0969	.1004	.1038	.1072	.1106
1.3	.1139	.1173	.1206	.1239	.1271	.1303	.1335	.1367	.1399	.1430
1.4	.1461	.1492	.1523	.1553	.1584	.1614	.1644	.1673	.1703	.1732
1.5	.1761	.1790	.1818	.1847	.1875	.1903	.1931	.1959	.1987	.2014
1.6	.2041	.2068	.2095	.2122	.2148	.2175	.2201	.2227	.2253	.2279
1.7	.2304	.2330	.2355	.2380	.2405	.2430	.2455	.2480	.2504	.2529
1.8	.2553	.2577	.2601	.2625	.2648	.2672	.2695	.2718	.2742	.2765
1.9	.2788	.2810	.2833	.2856	.2878	.2900	.2923	.2945	.2967	.2989
2.0	.3010	.3032	.3054	.3075	.3096	.3118	.3139	.3160	.3181	.3201
2.1	.3222	.3243	.3263	.3284	.3304	.3324	.3345	.3365	.3385	.3404
2.2	.3424	.3444	.3464	.3483	.3502	.3522	.3541	.3560	.3579	.3598
2.3	.3617	.3636	.3655	.3674	.3692	.3711	.3729	.3747	.3766	.3784
2.4	.3802	.3820	.3838	.3856	.3874	.3892	.3909	.3927	.3945	.3962
2.5	.3979	.3997	.4014	.4031	.4048	.4065	.4082	.4099	.4116	.4133
2.6	.4150	.4166	.4183	.4200	.4216	.4232	.4249	.4265	.4281	.4298
2.7	.4314	.4330	.4346	.4362	.4378	.4393	.4409	.4425	.4440	.4456
2.8	.4472	.4487	.4502	.4518	.4533	.4548	.4564	.4579	.4594	.4609
2.9	.4624	.4639	.4654	.4669	.4683	.4698	.4713	.4728	.4742	.4757
3.0	.4771	.4786	.4800	.4814	.4829	.4843	.4857	.4871	.4886	.4900
3.1	.4914	.4928	.4942	.4955	.4969	.4983	.4997	.5011	.5024	.5038
3.2	.5051	.5065	.5079	.5092	.5105	.5119	.5132	.5145	.5159	.5172
3.3	.5185	.5198	.5211	.5224	.5237	.5250	.5263	.5276	.5289	.5302
3.4	.5315	.5328	.5340	.5353	.5366	.5378	.5391	.5403	.5416	.5428
3.5	.5441	.5453	.5465	.5478	.5490	.5502	.5514	.5527	.5539	.5551
3.6	.5563	.5575	.5587	.5599	.5611	.5623	.5635	.5647	.5658	.5670
3.7	.5682	.5694	.5705	.5717	.5729	.5740	.5752	.5763	.5775	.5786
3.8	.5798	.5809	.5821	.5832	.5843	.5855	.5866	.5877	.5888	.5899
3.9	.5911	.5922	.5933	.5944	.5955	.5966	.5977	.5988	.5999	.6010
4.0	.6021	.6031	.6042	.6053	.6064	.6075	.6085	.6096	.6107	.6117
4.1	.6128	.6138	.6149	.6159	.6170	.6180	.6191	.6201	.6212	.6222
4.2	.6232	.6243	.6253	.6263	.6274	.6284	.6294	.6304	.6314	.6325
4.3	.6335	.6345	.6355	.6365	.6375	.6385	.6395	.6405	.6415	.6425
4.4	.6435	.6444	.6454	.6464	.6474	.6484	.6493	.6503	.6513	.6522
4.5	.6532	.6542	.6551	.6561	.6571	.6580	.6590	.6599	.6609	.6618
4.6	.6628	.6637	.6646	.6656	.6665	.6675	.6684	.6693	.6702	.6712
4.7	.6721	.6730	.6739	.6749	.6758	.6767	.6776	.6785	.6794	.6803
4.8	.6812	.6821	.6830	.6839	.6848	.6857	.6866	.6875	.6884	.6893
4.9	.6902	.6911	.6920	.6928	.6937	.6946	.6955	.6964	.6972	.6981
5.0	.6990	.6998	.7007	.7016	.7024	.7033	.7042	.7050	.7059	.7067
5.1	.7076	.7084	.7093	.7101	.7110	.7118	.7126	.7135	.7143	.7152
5.2	.7160	.7168	.7177	.7185	.7193	.7202	.7210	.7218	.7226	.7235
5.3	.7243	.7251	.7259	.7267	.7275	.7284	.7292	.7300	.7308	.7316
5.4	.7324	.7332	.7340	.7348	.7356	.7364	.7372	.7380	.7388	.7396
x	0	1	2	3	4	5	6	7	8	9

TABLE OF COMMON LOGARITHMS

x	0	1	2	3	4	5	6	7	8	9
5.5	.7404	.7412	.7419	.7427	.7435	.7443	.7451	.7459	.7466	.7474
5.6	.7482	.7490	.7497	.7505	.7513	.7520	.7528	.7536	.7543	.7551
5.7	.7559	.7566	.7574	.7582	.7589	.7597	.7604	.7612	.7619	.7627
5.8	.7634	.7642	.7649	.7657	.7664	.7672	.7679	.7686	.7694	.7701
5.9	.7709	.7716	.7723	.7731	.7738	.7745	.7752	.7760	.7767	.7774
6.0	.7782	.7789	.7796	.7803	.7810	.7818	.7825	.7832	.7839	.7846
6.1	.7853	.7860	.7868	.7875	.7882	.7889	.7896	.7903	.7910	.7917
6.2	.7924	.7931	.7938	.7945	.7952	.7959	.7966	.7973	.7980	.7987
6.3	.7993	.8000	.8007	.8014	.8021	.8028	.8035	.8041	.8048	.8055
6.4	.8062	.8069	.8075	.8082	.8089	.8096	.8102	.8109	.8116	.8112
6.5	.8129	.8136	.8142	.8149	.8156	.8162	.8169	.8176	.8182	.8189
6.6	.8195	.8202	.8209	.8215	.8222	.8228	.8235	.8241	.8248	.8254
6.7	.8261	.8267	.8274	.8280	.8287	.8293	.8299	.8306	.8312	.8319
6.8	.8325	.8331	.8338	.8344	.8351	.8357	.8363	.8370	.8376	.8382
6.9	.8388	.8395	.8401	.8407	.8414	.8420	.8426	.8432	.8439	.8445
7.0	.8451	.8457	.8463	.8470	.8476	.8482	.8488	.8494	.8500	.8506
7.1	.8513	.8519	.8525	.8531	.8537	.8543	.8549	.8555	.8561	.8567
7.2	.8573	.8579	.8585	.8591	.8597	.8603	.8609	.8615	.8621	.8627
7.3	.8633	.8639	.8645	.8651	.8657	.8663	.8669	.8675	.8681	.8686
7.4	.8692	.8698	.8704	.8710	.8716	.8722	.8727	.8733	.8739	.8745
7.5	.8751	.8756	.8762	.8768	.8774	.8779	.8785	.8791	.8797	.8802
7.6	.8808	.8814	.8820	.8825	.8831	.8837	.8842	.8848	.8854	.8859
7.7	.8865	.8871	.8876	.8882	.8887	.8893	.8899	.8904	.8910	.8915
7.8	.8921	.8927	.8932	.8938	.8943	.8949	.8954	.8960	.8965	.8971
7.9	.8976	.8982	.8987	.8993	.8998	.9004	.9009	.9015	.9020	.9025
8.0	.9031	.9036	.9042	.9047	.9053	.9058	.9063	.9069	.9074	.9079
8.1	.9085	.9090	.9096	.9101	.9106	.9112	.9117	.9122	.9128	.9133
8.2	.9138	.9143	.9149	.9154	.9159	.9165	.9170	.9175	.9180	.9186
8.3	.9191	.9196	.9201	.9206	.9212	.9217	.9222	.9227	.9232	.9238
8.4	.9243	.9248	.9253	.9258	.9263	.9269	.9274	.9279	.9284	.9289
8.5	.9294	.9299	.9304	.9309	.9315	.9320	.9325	.9330	.9335	.9340
8.6	.9345	.9350	.9355	.9360	.9365	.9370	.9375	.9380	.9385	.9390
8.7	.9395	.9400	.9405	.9410	.9415	.9420	.9425	.9430	.9435	.9440
8.8	.9445	.9450	.9455	.9460	.9465	.9469	.9474	.9479	.9484	.9489
8.9	.9494	.9499	.9504	.9509	.9513	.9518	.9523	.9528	.9533	.9538
9.0	.9542	.9547	.9552	.9557	.9562	.9566	.9571	.9576	.9581	.9586
9.1	.9590	.9595	.9600	.9605	.9609	.9614	.9619	.9624	.9628	.9633
9.2	.9638	.9643	.9647	.9652	.9657	.9661	.9666	.9671	.9675	.9680
9.3	.9685	.9689	.9694	.9699	.9703	.9708	.9713	.9717	.9722	.9727
9.4	.9731	.9736	.9741	.9745	.9750	.9754	.9759	.9763	.9768	.9773
9.5	.9777	.9782	.9786	.9791	.9795	.9800	.9805	.9809	.9814	.9818
9.6	.9823	.9827	.9832	.9836	.9841	.9845	.9850	.9854	.9859	.9863
9.7	.9868	.9872	.9877	.9881	.9886	.9890	.9894	.9899	.9903	.9908
9.8	.9912	.9917	.9921	.9926	.9930	.9934	.9939	.9943	.9948	.9952
9.9	.9956	.9961	.9965	.9969	.9974	.9978	.9983	.9987	.9991	.9996
x	0	1	2	3	4	5	6	7	8	9

Index

A

Abscissa 117
Absolute
  inequality 106
  value 19
    properties of 112
Addition
  geometric interpretation 17
  complex numbers 265
  functions 144
Additive
  identity 4
  inverse 5
Algebraic
  logic 15, 316
  errors 58
Annuity 212
Antilogarithm 326
Approximating
  solutions of equations 277
  zeros of polynomials 275
Arithmetic progression 285
  sum of 286
Associative laws 3
Asymptote 167
  hyperbola 182
Axis
  coordinate 117
  hyperbola 182
  parabola 179

B

Battleship game 124 (Exs. 47–8)
Bel 211
Bell, Alexander Graham 211
Binary key 316
Black box 132, 145
Binomial 30
  coefficient 310
  theorem 311
Braking distance 173
British Guiana 202

C

Calculator 22
  logic 312
  techniques 321
  use of 316
Carbon-14 dating 211
Cartesian coordinate system 118
Characteristic 324
Closed
  interval 107
  half-plane 243
Coefficient 30
  binomial 310
Cofactor 234
Column (of matrix) 228
Combination 307
Common
  difference 285
  factor 38
  logarithm 199
  ratio 290
Commutative laws 3
Complete factorization
  complex polynomial 272
  real polynomial 270
Completing the square 82
Complex
  arithmetic 265
  conjugate 267
  number system 264
  zero 268
Composite function 145
Composition 145
Compound interest 200
  continuous compounding 205
Conditional
  equation 70
  inequality 106
Conic section 177, 183
Conjugate, complex 267
  rules for 267
  zeros of polynomial 271
Constant
  factor (calculator) 322
  of proportionality 172

i

Constraint 244
  set 245
Cooling 214 (Exs. 15–16)
Coordinate
  axis 17, 117
  of a point 117
Counting principle 302
Cramer's rule
  2×2 system 233
  3×3 system 235
  limitations of 241

D

Decay law 210
Decibel 212
Degree 30, 34
Dependent
  system 219, 226
  variable 132
Determinant
  2×2 232
  3×3 234
  expansion by minors 235
  general 238
Difference
  common 285
  of squares 40
  of cubes 40
Discriminant 85
Distance 20
  formula 120
Distinguishable permutations 304
Distributive law 4
Division 9
  algorithm 251
  of complex numbers 266
  of polynomials 251
Domain 132, 137
Double zero 261
Dummy variable 282

E

e 204
Earp, Wyatt 97
Earthquake 211
Echelon form 233
Elementary row operation 229
Ellipse 180
Equation(s) 61, 70
  conditional 70
  involving radicals 92
  of lines 127
  polynomial 92
  systems of 217
Equivalent
  equations 70
  inequalities 106
  linear systems 223

Estimation 102
Even function 141
Exponent(s) 10, 24
  rules for 25, 54
Exponential
  decay 210
  equation 199
  function 189
  growth 191, 209

F

Factor theorem 255, 269
Factorial 302
Factoring 37
  by grouping 41
  formulas 40
  second degree polynomials 38
Feasible solution 245
Finite sequence 281
Franklin, Ben 203 (Ex. 35)
Function 132
  not given by formula 134
Functional notation 133
Fundamental theorem of algebra 269

G

Galileo 287
Gauss, K. F. 269
Gaussian elimination 223
  vs. Cramer's rule 241
Geometric progression 290
  sum of 291
Geometric (infinite) series 292
  sum of 293
Graph 137
Graphing principles 139
Graph of
  circle 177
  ellipse 181
  equation 124
  even function 141
  factored polynomial 162
  hyperbola 182
  inverse function 148
  odd function 141
  parabola 179
  rational function 165
Growth law 229

H

Half
  -life 211
  -open interval 107
  -plane 243
Hammurabi 211
Hedge against inflation 202

Horizontal
  asymptote   167
  shift   156

I

Identity   71
  additive   4
  multiplicative   4
Inconsistent system   219, 226
Independent variable   132
Inductive definition   298
Inequalities   17
  involving absolute values   111
  properties of   99
  quadratic   113
  rules for   100
  solving   106
  systems of   243
Infinite
  geometric series   292
  interval   107
  sequence   281
Integer   2
  zeros of polynomial   262
Interval   107
Inverse
  additive   5
  function   146
    existence of   147
    graph of   148
  multiplicative   5
  rules for   5
  variation   173
Inversely proportional   171
Irrational number   2
Irreducible quadratic   272

L

Lascaux caves   214 (Ex. 12)
Lattice point   131
Leading
  coefficient   30
  term   30, 161
Left-to-right logic   15, 316
Linear
  equation   71
    applications   75
  function   153
  inequality   243
  interpolation   276, 325
  programming   244
  relation   153
  system   217
Logarithm(s)
  base b   193
  common   199
  computations with   327
  function   196

Logarithm(s) (*Continued*)
  natural   204
  rules for   194
  tables   324
  various bases   205
Logic (calculator)   15, 316
Lowest terms   43, 165

M

Magnitude of stars   214 (Exs. 21–24)
Malthus, Thomas   192
Mantissa   324
Mathematical induction   297
Matrix   228
  notation   232
Memory (calculator)   318, 322
Memory plus key   322
Midpoint formula   122
Minor   234
Monomial   30
Multiplication
  of complex numbers   265
  of functions   144
Multiplicative
  identity   4
  inverse   5
Multiplicity   261, 270
My dear Aunt Sally   13

N

n-th root   50
  rules for   51
Natural
  number   2
  logarithm   204
Negative
  integer   2
  number   17
Non-negative number   17
Non-linear system   220
Number line   17

O

Objective function   244
Odd function   141
Open
  interval   107
  half-plane   243
Optimization   157
Order   17
Ordered pair   117
Ordinate   117

P

Parabola   155, 179
Parentheses   13
  in calculators   321
  nested   14
Parity   161
Partial sum   292
Pascal's triangle   312
Payment of debts   213
Perfect square   35
Permutation   303
Point-slope form   127
Polynomial   30
  algebra   30
  division   251
  evaluation   257
    calculator algorithm   258
  function   159
  in two variables   34
Positive
  integer   2
  number   17
Powers   24
  calculating   28, 56
Priority
  calculator   15
  of operations   13
Problem solving   96
Proportional   171

Q

Quadrant   118
Quadratic
  equations   80
    applications   87
    solution of   80
  formula   85
  function   155
  inequality   111
  type equation   91
Quotient(s)
  equality of   9
  in polynomial division   251
  rules for   10

R

Radical(s) 47
  equations involving   92
Range   132, 137
Rational
  exponents   53
    rules for   54
  expression   43
  function   165
  number   2
  zero   261
Rationalizing the denominator   49

Real
  number   2
  zeros of polynomials   274
Rectangular coordinate system   118
Remainder
  in polynomial division   251
  theorem   255
Repeating decimal   2, 77
Richter scale   187, 212
Root
  cube   50
  n-th   50
  of equation   70, 260
  square   47
Row (of matrix)   228

S

San Francisco earthquake   187, 212
Scientific notation   27
  in calculator display   27
Sections of a cone   184
Semi-major axis   181
Semi-minor axis   181
Sequence   281
Sign of a number   18
  rules for   18
Significant figures   202
Simple zero   270
Slope 126
  -intercept form   129
  of parallel lines   129
  of perpendicular lines   129
Solution
  of an equation   70
    two variables   124
  set of inequality   106
    two variables   243
Square root(s)   47
  by calculator   52
  rules for   48
Straight line depreciation   154
Subtraction   7
  complex numbers   265
  functions   144
Summation
  index   282
  notation   282
Synthetic division   252
System
  linear equations   217
  linear inequalities   243
  matrix notation   228
  non-linear   220

T

Tower of Hanoi puzzle   301 (Ex. 39)
Translating
  words into math   61

Translating (*Continued*)
   problems into algebra   65
Triangular
   form (system)   223
   number   289 (Ex. 38)
Twelve days of Christmas   287, 302 (Ex. 40)

U

Unary key   317
Union of sets   108

V

Variable   132
Variation
   direct   171
   inverse   173
   joint   174

Vertex
   ellipse   180
   hyperbola   182
   parabola   179
Vertical
   asymptote   167
   shift   140

Z

Zero   5
   properties of   6
   of polynomial   260
      complex   268
      conjugate   271
      existence of   274
      integer   262
      rational   261